Entomopathogenic Nematodes in Biological Control

Editors
Randy Gaugler, Ph.D.
Rutgers University
New Brunswick, New Jersey

and

Harry K. Kaya, Ph.D.
University of California
Davis, California

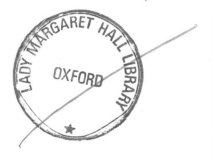

CRC Press
Boca Raton   Ann Arbor   Boston

Library of Congress Cataloging-in-Publication Data

Entomopathogenic nematodes in biological control / editors, Randy Gaugler, Harry K. Kaya.
p.   cm.
Includes bibliographical references and index.
ISBN 0-8493-4541-3
1. Insect pests--Biological control. 2. Insect nematodes. 3. Nematoda as carriers of
disease. 4. Insects--Diseases. 5. Bacterial diseases.
I. Gaugler, Randy. II. Kaya, Harry K.
SB933.334.E57 1990   632'.7—dc20   90-2126

Direct all inquires to CRC Press, Inc., 2000 Corporate Blvd., N.W., Boca Raton, Florida,
33431.

© 1990 by CRC Press, Inc.

International Standard Book Number 0-8493-4541-3

Library of Congress Card Number 90-2126
Printed in the United States

Cover illustration: Scanning electron micrograph of the oral cavity of
Steinernema carpocapsae showing its associated symbiont, *Xenorhabdus
nematophilus*. Courtesy of L. M. LeBeck.

Acquiring Editor: Russ Hall
Production Director: Sandy Pearlman
Coordinating Editor: Anita Hetzler
Cover Design: Chris Pearl
Indexing: Sharon Smith, Professional Indexing
Composition: CRC Press, Inc.

FOREWORD

Among the reasons for increased interest in all facets of biological control has been wide public awareness of the environmental damage resulting from chemical pesticides. Persistent chemicals, intended to destroy pests, often entered the food chain and left their long-lasting impact on nontarget organisms, including humans. The development of insect resistance to chemical pesticides also stimulated interest into biological means of controlling insect pests. Insect pathogens, especially bacterial and viral agents, have received considerable attention as biological insecticides since the 1940s. During the 1980s, however, biological control employing entomopathogenic nematodes has rapidly developed into a subdiscipline of insect pathology, equivalent in rank to the viruses, bacteria, protozoa, and fungi. The exceptional potential of these nematodes has now been recognized, particularly with the recent breakthroughs in mass production technology which have permitted large-scale applications against insect pests. As a reading of this book will demonstrate, significant achievements have been accomplished in a short period of time in using steinernematid and heterorhabditid nematodes as biological control agents, notably against weevils and insects in cryptic habitats. Instances of insufficient control against other insects have served as an impetus for research beyond the applied aspects, including ecological relationships, organismal and molecular genetics, and behavior. Accordingly, readers will welcome this first in-depth treatise concerning fundamental and mission-oriented aspects of entomopathogenic nematodes and their associated bacteria.

The editors of this book, who are world renown for their creativity with entomopathogenic nematodes, have assembled the foremost authorities from four continents to contribute on basic and applied concepts. The authors have taken advantage of this opportunity to express their views to a wide scientific audience. They have combined their international experience so that the latest developments in this fascinating and rapidly expanding field are presented in a comprehensive manner with diverse topics ranging from biological control theory to organismal and molecular biology. Thus, the large body of information brings into sharp focus anticipated new directions in pest control and basic nematode and bacterial biology. Special effort has been made by the editors to minimize the taxonomic confusion of scientific names and to provide a foundation of knowledge, allowing this book to set the bench mark for future research. Therefore, this volume will serve for years to come as a standard source of information for investigators and students of biological control, microbiology, nematology, and molecular biology.

Readers of this book will not fail to realize that the scope of entomopathogenic nematology is largely due to the early efforts of Dr. Rudolf W. Glaser, to whom this treatise is appropriately dedicated. His visionary research has inspired many investigators to continue the search for new entomopathogenic nematodes, and to draw on his published observations and experiments to reveal new and important discoveries.

Karl Maramorosch
Robert L. Starkey Professor
of Microbiology

PREFACE

Although nearly 40 nematode families are associated with insects, very few of these nematodes cause host mortality, and only two families, Steinernematidae and Heterorhabditidae, are widely available for use in biological control. Current research heavily emphasizes steinernematids and heterorhabditids, nematodes characterized by their mutualistic relationship with *Xenorhabdus* bacteria. Because this book is limited to nematodes that serve as vectors of bacteria, we have adopted the term "entomopathogenic" to describe these nematodes rather than more conventional but less restrictive terms such as "entomophilic", "entomogenous", "entomophagous", or "insect-parasitic". Use of entomopathogenic also serves to reinforce the link between this field and its parent discipline, insect pathology.

The research emphasis on entomopathogenic nematodes reflects the unusual middle ground these biological control agents occupy between predators/parasitoids and microbial pathogens; a position that endows them with a unique combination of biological control attributes. Thus, they share a capacity for host-seeking and an exemption from government registration requirements with predators/parasitoids, yet can be mass reared and stored on a scale only imaginable with some microbial pathogens. They also possess extreme virulence for insects, mammalian safety, and a broad host range. These attributes have encouraged several companies over the past ten years to attempt to develop steinernematid and heterorhabditid nematodes as biological insecticides. The resulting availability of large numbers of nematodes attracted scientific investigators and stimulated a tremendous international surge in research efforts.

The explosive growth in research has not been achieved without cost. Long dominated by a small core of homogeneously trained entomologists, entomopathogenic nematology has become as diverse as biology itself as nematologists, bacteriologists, ecologists, physiologists, biochemists, molecular biologists, etc., have entered the field. This sudden diversity is highly desirable, but communication has unavoidably suffered. Similarly, industry has been a powerful positive force in taking the field so far so fast; but industry's influence has also been divisive. As key researchers have become allied, or have been perceived to become allied, with competing companies, the early spirit of cooperation and free exchange of ideas among researchers has begun to erode. A further difficulty is that the field has become so strongly international, that keeping up with the deluge of new information being published, often in obscure journals, is all but impossible.

Entomopathogenic nematology has made enormous strides since Glaser's discovery of nematodes infecting white grubs more than sixty years ago. The 1980s was a decade of unparalleled progress in understanding these nematodes. The field has grown almost exponentially during this period and is presently poised for further growth as recent research breakthroughs and new technologies are exploited; but growth has been so rapid that the present momentum is threatened. Entomopathogenic nematology is still in its infancy and is in need of further organization to avoid fragmentation.

This book is intended to enhance communication among the diverse areas of the discipline by providing an international multi-disciplinary forum for the exchange of information. Contributors were encouraged to review the current status of the discipline, to identify, discuss, and attempt to resolve important obstacles in the development of entomopathogenic nematodes for biological control, and to assess research needs and priorities. The book is also intended to serve as a comprehensive resource for those entering the field.

Although this work is the product of many contributors, every effort was made to unify the book into a single cohesive viewpoint by minimizing differences (e.g., taxonomic). It is our hope that this book will serve as a bench mark, as well as to give direction and focus to the young discipline of entomopathogenic nematology.

We take this opportunity to thank the contributors for their time and effort in making it all possible. Special thanks are due to Leslie Campbell, James Campbell, and Jennifer Woodring for their assistance in bringing this work to fruition swiftly. Finally, we express our deepest appreciation to our families for their support and patience from the inception to the completion of this project.

Randy Gaugler
Harry K. Kaya

CONTRIBUTORS

R. J. AKHURST, CSIRO Division of Entomology, Canberra ACT 2601, Australia

ROBIN BEDDING, CSIRO Division of Entomology, Canberra ACT 2601, Australia

JOE W. BEGLEY, Yoder Brothers, Alva, Florida 33920 U.S.A.

BRUCE BLEAKLEY, Department of Bacteriology, University of Idaho, Moscow, Idaho 83843 U.S.A.

N. E. BOEMARE, INRA-CNRS-USTL, Laboratoire de Pathologie Comparee, 34060 Montpellier Cedex, France

JOHN CURRAN, CSIRO Division of Entomology, Canberra ACT 2601, Australia

GARY B. DUNPHY, Department of Entomology, MacDonald College of McGill University, Ste-Anne-De-Bellevue, Quebec, H9X 1CO Canada

L. E. EHLER, Department of Entomology, University of California, Davis, California 95616 U.S.A.

TIBOR FARKAS, Department of Genetics, Eotvos Lorand University, Budapest, Hungary

ANDRAS FODOR, Biological Research Center of the Hungarian Academy of Sciences, H-6701 Szeged, Hungary

SUSAN FRACKMAN, Center for Great Lakes Research, University of Wisconsin, Milwaukee, Wisconsin 53201 U.S.A.

MILTON J. FRIEDMAN, Biosys, Palo Alto, California 94303 U.S.A.

RAMON GEORGIS, Biosys, Palo Alto, California 94303 U.S.A.

W. M. HOMINICK, Department of Biology, Imperial College of Science, Technology & Medicine, Silwood Park Field Station, Ascot, Berks, SL5 7PY England

N. ISHIBASHI, Department of Applied Biological Sciences, Saga University, Saga 840, Japan

HARRY K. KAYA, Department of Nematology, University of California, Davis, California 96516 U.S.A.

MICHAEL G. KLEIN, Horticultural Insects Research Lab, ARS, USDA, Ohio Agricultural Research & Development Center, Wooster, Ohio 44691 U.S.A.

E. KONDO, Department of Applied Biological Sciences, Saga University, Saga 840, Japan

KENNETH H. NEALSON, Center for Great Lakes Studies, University of Wisconsin, Milwaukee, Wisconsin 53201 U.S.A.

GEORGE O. POINAR, JR., Division of Entomology and Parasitology, University of California, Berkeley, California 94720 U.S.A.

A. P. REID, Department of Biology, Imperial College of Science, Technology & Medicine, London, SW7 2BB England

THOMAS N. SCHMIDT, Department of Biology, University of Indiana, Bloomington, Indiana 47405 U.S.A.

GRAHAM S. THURSTON, Department of Entomology, MacDonald College of McGill University, Ste-Anne-De-Bellevue, Quebec, H9X 1C0

GABRIELLA VECSERI, Biological Research Center of the Hungarian Academy of Sciences, H-6701 Szeged, Hungary

CHRISTOPHER Z. WOMERSLEY, Department of Zoology, University of Hawaii, Honolulu, Hawaii 96822 U.S.A.

Dedicated to the memory of
Rudolf William Glaser (1888-1947)
in recognition of his
pioneering research in insect pathology, and in
particular, for his contributions to our
understanding of nematode diseases of insects

TABLE OF CONTENTS

ECOLOGY

COMMERCIALIZATION AND APPLICATION TECHNOLOGY

8. Commercial Production and Development
   Milton J. Friedman

9. Formulation and Application Technology
   Ramon Georgis

EFFICACY

10. Efficacy Against Soil-Inhabiting Insect Pests
    Michael G. Klein

BIOTECHNOLOGY AND GENETICS

CONCLUSIONS

17. Perspectives on Entomopathogenic Nematology
   W. M. Hominick and A. P. Reid

1. Some Contemporary Issues in Biological Control of Insects and Their Relevance to the Use of Entomopathogenic Nematodes

L. E. Ehler

I. INTRODUCTION

California citrus growers witnessed the first major success in biological control of insect pests 100 years ago when the introduced cottony-cushion scale (*Icerya purchasi*) was brought under complete control by introduced natural enemies from Australia. Although several natural enemies were actually imported, the project's success can be attributed to two species: the vedalia beetle, *Rodolia cardinalis*, and a parasitic fly, *Cryptochaetum iceryae*. As we mark the centennial observance of the cottony-cushion scale project, it is apparent that biological control is entering a new era. Predaceous and parasitic arthropods, which have been the dominant tools of applied biological control of insects, will continue to play a major role. However, other kinds of biological-control agents, such as entomopathogenic nematodes and microbial pathogens modified through biotechnology, can be expected to play a greater role.

Nematodes which serve as vectors of pathogenic bacteria show considerable potential in biological control of insect pests.[1-5] These biological-control agents are similar to insect parasitoids in that the immature form develops at the expense of one host individual; they are ecologically similar to both parasitoids and arthropod predators because the host individual is eventually killed. However, these nematodes differ from predators and parasitoids in at least two ways: (1) the nematode is mutualistically associated with a bacterial pathogen which actually kills the host insect, and (2) the pathogenicity of the bacterium has a great influence on the efficacy of the system. Despite their unique standing among biological-control agents, the use of these nematodes in biological control should not be considered as a science unto itself. Many of the major concepts or issues in modern biological control are relevant to entomopathogenic nematodes and should not be ignored. The purpose of this chapter is to provide an overview of some contemporary issues in biological control, with particular emphasis on those which appear most relevant to the use of entomopathogenic nematodes. In addition, areas where nematologists can make major contributions to the theory of biological control are noted.

Biological control is operationally defined as the action of natural enemies which maintains a host population at levels lower than would occur in the absence of the enemies. Natural enemies would include insect parasitoids, predaceous arthropods, nematodes, and microbial pathogens. With recent advances in molecular biology and genetic engineering, some have proposed that the definition of biological control include not only intact enemies, but their gene products as well.[6] This has already generated considerable controversy, and as substantially more funds become available for research in "biological control", the definition of the discipline will become more than just an academic issue.

Entomologists generally divide biological control into two broad categories: (1) natural biological control, or that which is effected by native (or coevolved) natural enemies in the native home (origin) of a given insect; and (2) applied biological control, or that which exists due to human intervention. Applied biological control is further divided into classical biological control (introduction of exotic natural enemies) and augmentative biological control (enhancement of natural enemies already in place). Although these terms have traditionally been used to describe the use of insects to control other insects, the use of entomopathogenic nematodes can be embraced as well.

II. NATURAL BIOLOGICAL CONTROL

It is important to study and document naturally occurring biological control, regardless of whether the target insect is a pest species. The knowledge gained from such studies can be relevant to both applied biological control and pest management. The examples which follow relate largely to insect predators and parasites, and those who study entomopathogenic nematodes are encouraged to develop comparable ones.

Insects which are nonpests (or minor ones) and their natural enemies may make good model systems for addressing fundamental questions relevant to practical biological control. One example is *Rhopalomyia californica*, a native cecidomyiid midge which develops in terminal galls on *Baccharis pilularis* in northern California. The midge could be considered a minor pest when one subspecies of its host plant is grown as a ground cover in urban areas and on freeway margins; otherwise, it is of no economic significance in the state. Midge larvae are exploited by at least seven species of hymenopterous parasites, and the structure and impact of this parasite "guild" have been the subject of major investigations by Doutt,[7] Ehler,[8-11] Force,[12,13] and Hopper.[14] These studies have contributed significantly to our knowledge and understanding of parasite guild structure and have generated a number of predictions for applied biological control. Insect nematologists should investigate comparable systems which contain one or more entomopathogenic nematodes.

For those insects which are major pests, a study of their natural biological control can provide important insights into potential control strategies.[15] If an insect is a pest in its native home, it would be worthwhile to assess its natural-enemy complex so as to determine why the enemies are unable to keep the pest population at noneconomic levels. With the proper understanding of the system, it should be possible to devise an augmentation program to enhance the efficiency of one or more of the enemies. It is also possible to introduce exotic natural enemies for control of native pests; in this case, candidate enemies would ordinarily be obtained from a closely related host in another region. If this strategy is to be implemented, an analysis of the natural biological control (including structure of the natural-enemy guilds) would be helpful in predicting the kind of species that might be most suitable. For exotic pests, it is often possible to study the ecology and natural control of a given species in its native home, and then utilize this information in devising introduction strategies in classical biological control. Although this is not standard operating procedure in classical biological control, there are at least some circumstances when it would seem to be a sound approach.[15]

In a given agroecosystem, we can expect to find native natural enemies essentially preadapted for survival in such habitats. Often, these natural enemies play an important role in maintaining pest populations at relatively low levels, and thus form a cornerstone for the pest-management programs which are eventually developed for these crop systems. Some California examples include predaceous Hemiptera (e.g., *Orius tristicolor*, *Geocoris pallens*) in cotton[16] and alfalfa,[17] and western predatory mite (*Metaseiulus occidentalis*) in grape vineyards[18] and almond orchards.[19] In this context, the role of entomopathogenic nematodes in natural biological control of insect pests clearly deserves further investigation. The native home of a given nematode should be determined so that ecological investigations can be carried out in the habitat in which the nematode species presumably evolved. Information derived from such studies should enhance our ability to predict the kind of target habitat in which a given species might be most effective. At the generic level, the centers of origin for *Steinernema* and *Heterorhabditis* should be determined, because evolutionary theory suggests the presence of closely related species in the center of origin of a given genus. Such species could hold considerable potential in applied biological control of insects.

III. CLASSICAL BIOLOGICAL CONTROL

The importation of exotic natural enemies for biological control of both native and exotic insect pests has been practiced on an organized basis for about 100 years. As a result, a standard set of practical guidelines has been developed and these are generally adhered to throughout the world. However, the ecological theory attendant to classical biological control remains underdeveloped (see Section V) and warrants more attention in future introductions of both arthropods and entomopathogenic nematodes.

The practice of classical biological control is well developed. It generally consists of the following sequence of procedures: foreign exploration (usually in the native home of the pest), quarantine of imported material, mass production of candidate agents, field colonization, and evaluation. Some protocols for importation of arthropod natural enemies in the U.S. were reviewed by Coulson and Soper,[20] whereas those for insect-parasitic nematodes were summarized by Nickle et al.[21] In both cases, host specificity is of prime concern because of the potential effects of an introduced species on nontarget organisms. As a general rule, the amount of preintroductory investigation required for an arthropod or nematode should be directly proportional to the perceived environmental risk. Biological-control practitioners have often resisted the notion of conducting detailed preintroductory investigations. Although this attitude is improving, the ecological framework for conducting such studies is inadequate and requires more attention. Those involved in the importation of entomopathogenic nematodes are in a good position to exploit this situation and, thereby, assume a position of leadership in the movement to make classical biological control a more predictive science.

In recent years, a number of detailed summaries of projects in classical biological control have been published (see Ehler[22] for a recent list). Although these projects have produced a variety of results, at least two important generalizations emerge from an analysis of the data. First, the number of introductions which fail (i.e., no establishment) is much greater than those which result in establishment. A recent estimate of the rate of establishment for predators and parasites from ca. 1890 to ca. 1968 was 34%.[23] Second, the proportion of projects resulting in complete control of the target pest is relatively low. For the same timeframe, the global average was 16%.[24] Presumably, the rates of establishment and of complete success will increase with theoretical and technological advances. According to Poinar,[5] a few nematodes have been employed in classical biological control, and with the current interest in steinernematid and heterorhabditid species, it is likely that the number of introductions will increase considerably. It will be important to document each of these, not just for historical reasons, but also because each introduction is an experiment in colonization. Analysis of arthropod introductions since 1890 has been extremely valuable in testing hypotheses concerning colonization of exotic species.[23,25] The same should be true for introductions of entomopathogenic nematodes.

IV. AUGMENTATIVE BIOLOGICAL CONTROL

Human intervention designed to enhance or augment the effectiveness of those natural enemies already in place is applicable to both native and exotic species. Such augmentative procedures generally involve manipulation of either the environment or the natural enemy itself.

Coppel[26] recognized nine areas of study fitting the category of environmental management or manipulation: (1) land use, including crop rotation, strip harvesting of hay alfalfa, etc.; (2) habitat provision, such as increased plant diversity, placement of nest boxes, etc.; (3) food provision, such as spraying artificial honeydew, use of nectar/pollen bearing plants, hedgerows, etc.; (4) tillage methods for annual crops, and dust reduction in other situations; (5) modified release strategies, such as caging introduced species with hosts to increase the probability of establishment; (6) host or prey provision, as in the "pest-in-first" technique; (7) reducing enemies of beneficial species, as for example in control of Argentine ant; (8) improved pesticide utilization, particularly more selective usage; and (9) semiochemical and behavioral strategies, such as those which exploit kairomones. The potential for these various strategies in applied biological control is generally recognized, but as Coppel[26] correctly notes, implementation has been neglected. There are numerous reasons for this neglect, and probably foremost is that many proposed environmental manipulations are not compatible with agronomic practices in the target system. Economics may also be a limiting factor in many cases. Environmental modification to enhance the efficacy of entomopathogenic nematodes may be limited in similar fashion; nevertheless, this approach should certainly be explored.

The direct manipulation of a natural enemy consists of mass production and field release of individuals of a given species. Entomologists recognize two kinds of releases, and these are best thought of as two points on a continuum. Inoculative release is the release of a relatively small number of individuals; this is designed so that the progeny of the individuals released will provide season-long pest suppression. In contrast, inundative release is the release of a massive number of individuals with the aim of obtaining immediate pest suppression. Both techniques have been implemented in numerous settings, and there are now many commercially available, biological-control agents for such use in agriculture. Recent reviews on inoculative/inundative release of natural enemies include Ridgway and Vinson,[27] Stinner,[28] Ables and Ridgway,[29] Hussey and Scopes,[30] van Lenteren,[31,32] and van Lenteren and Woets.[33]

Those contemplating augmentative release of entomopathogenic nematodes are encouraged to address conceptual problems in at least two areas. First, the "theory of inundative release", as developed by entomologists, is relatively primitive and very much in need of sophistication. As the number of commercially available nematode species increases, this problem can be expected to become even more critical. (A more detailed discussion of this matter is given in Section VI.) The second area of concern is genetic improvement of natural enemies, either through artificial selection, hybridization, or genetic engineering. Hoy has pioneered genetic improvement of arthropods (chiefly phytoseiid mites), and although considerable progress has been made, much remains to be done.[34,35] The investigations by Gaugler[36,37] on improvement of entomopathogenic nematodes are timely and should lead to important advances in genetic improvement of traits other than just pesticide resistance. Improved tolerance to the soil environment, or increased retention of *Xenorhabdus* bacteria, would seem to be desirable goals.

V. SOME CONTEMPORARY ISSUES

A list of some contemporary issues relevant to the introduction of natural enemies for biological control of insect pests is given in Table 1. Although many of these originated in the context of classical biological control, many are equally applicable to augmentative biological control through inoculative or

TABLE 1. Some Contemporary Issues in the Introduction of Natural Enemies for Biological Control of Insects

Phase of project	Major theme	Variations about major theme	Refs.
Preintroductory	Target pest	Native pests less suitable than exotic ones?	15,38-41
		Direct pests more suitable than indirect ones?	40, 42, 43
		r-selected pests less suitable than more K-selected ones?	16, 41, 44, 45
	Habitat of pest	Biological control better suited for stable environments?	39, 41
		Biological control better suited to islands rather than continents?	39, 40
		Biological control best suited for tropical agroecosystems?	40, 46, 47
		Relative suitability of monocultures and polycultures	41, 48
	Kinds of natural enemies	Attributes of effective natural enemies	15, 31, 32, 49

			References
	Introduction strategy	Old (coevolved) vs. new exploiter-victim systems	41, 50, 51
		Relative value of predators and parasites (=parasitoids)	23, 24, 41
		Relative value of specialist vs. generalist natural enemies	15, 17, 41
		Utility of facultative secondary parasites	15
		Evolutionary strategies (especially reproductive potential and competitive ability)	13, 52-54
		Predictive vs. empirical approaches	15
		Holistic vs. reductionist approaches	51
Introductory	Genetics	Relative importance of founder effect, drift, selection, and inbreeding	31, 55, 56
		Within-species variation	56
	Factors affecting establishment	Relative importance of habitat, no. released, competition, and other nongenetical processes	39, 40, 57
Postintroductory	Evaluation of impact	Ecological impact on pest population	58-60
		Economic impact	61, 62
		Environmental impact	20, 22, 63-66

TABLE 1 (continued). Some Contemporary Issues in the Introduction
of Natural Enemies for Biological Control of Insects

Phase of project	Major theme	Variations about major theme	Refs.
	Explanation of effectiveness	Various hypotheses to account for the operation of successful natural enemies	42, 67-72
	Factors affecting postcolonization performance	Relative importance of climate, competitors, habitat, genetic variation, etc.	15, 39, 41

inundative release. In assembling the references for the various issues, I have chosen recent reviews whenever possible, rather than the many original research papers which may exist on the subject.

A. Preintroductory Phase

There are many issues which might emerge prior to the field release of a natural enemy. These relate to the suitability of the target pest and the habitat in which it occurs, the kinds of natural enemies available, and the choice of introduction strategy to be implemented. With respect to target pest and habitat, most of the issues relate primarily to classical biological control using predaceous and parasitic insects and may be of limited relevance to the use of entomopathogenic nematodes. In contrast, at least four issues listed under "kinds of natural enemies" are relevant (i.e., attributes of effective natural enemies, new vs. old exploiter-victim systems, specialists vs. generalists, and evolutionary strategies). Questions involving introduction strategy are also pertinent. In the limited space available, it is not possible to assess each issue, so the discussion will be restricted to some general aspects of "introduction strategy", an important matter which continues to evoke controversy in biological control of insects, and one which is germane to the use of entomopathogenic nematodes.

Introduction strategy refers to the choice of a species or combination of species to release for control of a given pest in a given situation. This holds for classical, inoculative, and inundative biological control. As noted by van Lenteren[31,32] and van Lenteren and Woets,[33] the attributes required of a successful natural enemy may well vary according to the type of release program. Thus, for a given target pest, there should be a pool of biological-control agents, and it might include predators (insects, spiders, or mites), insect parasitoids, microbial pathogens (viruses, bacteria, fungi, or protozoans), and nematodes. From this pool of enemies, we can devise introduction strategies to maximize either long-term control (i.e., through classical biological control), season-long control (i.e., through inoculative release), or immediate control (i.e., through inundative release). The optimal strategies for each of the approaches could easily be different for the same pool of biological-control agents. In addition, there may be more than one optimal strategy in each category for bringing about the desired outcome. Unfortunately, the theoretical framework necessary for conducting the required preintroductory investigations is not adequately developed, and this remains one of the most pressing conceptual problems in biological control of insects. The development of "predictive" introduction strategies,[15] based on holistic or reductionist approaches,[51] is a worthy scientific problem and those utilizing entomopathogenic nematodes are urged to contribute to its resolution whenever possible. The use of nematodes in inundative-release programs may represent a special opportunity to address the pertinent issues (see Section VI).

B. Introductory Phase

The introductory phase of a classical biological-control program begins with foreign exploration and ends with field colonization of the imported agent. During this phase, genetical issues and the various factors which might affect establishment are of major concern and it would appear that all of the issues listed in Table 1 are relevant to the use of entomopathogenic nematodes. The genetical issues relate to both classical biological control and inundative/inoculative release. The influence of prolonged culture on fitness and field efficacy is of particular concern. Unfortunately, our current understanding of the population genetics of introduction and of within-species variation in natural enemies of insect pests is so limited that realistic predictions for applied biological control are generally not available. Thus, theoretical and empirical contributions resulting from the use of entomopathogenic nematodes would be most welcome. The same holds for factors affecting establishment of natural enemies. In this case, the empirical evidence for or against a given hypothesis tends to be circumstantial and, not surprisingly, this has led to some controversy. Perhaps the most controversial issue is the competitive-exclusion hypothesis: establishment of introduced natural enemies can be precluded by interspecific competition with incumbent species and (or) species released simultaneously. Although the empirical evidence is consistent with the hypothesis,[25] more definitive "proof" would be helpful. Entomopathogenic nematodes, because of their relatively low motility, may prove to be ideal experimental organisms for testing this and related hypotheses.

C. Postintroductory Phase

Once a natural enemy is established there are at least three areas of concern: evaluation of its impact, explanation of its effectiveness (or ineffectiveness), and factors influencing its effectiveness (Table 1). All of the attendant issues relate to both classical biological control and inundative/inoculative release, and should apply to the use of entomopathogenic nematodes as well. Evaluation of ecological impact of an enemy on the host population is relatively straightforward; a number of standard techniques are available. Economic impact is best left to professional economists. Environmental impact of introduced biological-control agents is a recent issue, and because of its relevance to entomopathogenic nematodes, this matter is treated in a later section.

Once a natural enemy is judged to be effective (or successful), the next step is to explain how this success is brought about. In recent years, different hypotheses have been put forward to account for the operation of successful natural enemies in nature, including those emphasizing phenomena such as temporal density dependence, spatial density dependence, or local pest extinction. There is no general agreement on the applicability of a given hypothesis in the real world, and we currently lack a robust theory to account for the operation of successful natural enemies in nature. More empirical testing of the various hypotheses is thus in order, including those situations in which entomopathogenic nematodes are released and eventually provide successful control. We may eventually find that the underlying mechanism varies with the kind of pest-enemy system under investigation. Because of their uniqueness among biological-control agents, nematodes associated with mutualistic bacteria may prove to be a case in point.

The factors which affect the post-colonization performance of an introduced natural enemy have been extensively reviewed, with particular emphasis on classical biological control. In contrast to the issues related to establishment, these issues are less controversial, presumably because the natural enemy in question is permanently established and can be subjected to controlled experimentation, etc. However, experimentation involving insect predators and parasites is often hampered by their mobility. Natural enemies with restricted motility (at least in the short term), such as entomopathogenic nematodes, should be a considerable improvement. It is especially critical to determine why a particular natural enemy is ineffective in a given situation so that future introduction strategies can be adjusted accordingly.

VI. THEORY OF INUNDATIVE RELEASE

As the number of commercially available biological-control agents increases, the need for a coherent theory of inundative release becomes more apparent. This seems especially true in the case of entomopathogenic nematodes in the families Steinernematidae and Heterorhabditidae. As inundative release is likely to be the major approach in the utilization of these nematodes in the immediate future, the theory of inundative release is one of the most critical issues to be considered. A complete account of the theoretical issues would include the preintroductory, introductory, and postintroductory phases of a project; however, the present discussion will deal largely with "introduction strategy". In the case of inundative release, introduction strategy refers to (1) choosing a species or combination of species for release, and (2) determining the number of individuals (per chosen species) to release to maximize short-term biological control. The choice of a species or combination of species is not synonymous with the old controversy over single- vs. multiple-species introductions; in fact, this traditional dichotomy should probably be discarded. The modern issue relates to subsets or combinations of species, and poses such questions as: (1) is one single-species introduction (SSI) better than another, (2) is one multiple-species introduction (MSI) better than another, and (3) is a given SSI better than a given MSI (and vice versa)?

A. Single-Species Release

Recent reviews of practical results obtained from inundative releases with both arthropods and entomopathogenic nematodes reveal that most projects utilized single-species releases. From a theoretical standpoint, such one-host/one-enemy systems are more tractable, so it is not surprising that most theoretical work relates to single-species releases. Most of this work deals with release rates rather than choice of species to release. For example, Knipling[73] has constructed a number of theoretical "dose-response" type models, in which the impact on the target pest population is a simple function of the number of individuals released per unit of space. In contrast, Barclay et al.[74] compared various mathematical models of host/parasitoid population dynamics and examined the theoretical inundation rates required for eradication of the target host population. Unfortunately, empirical verification of these models has been neglected. This should receive priority in future research, equal to (if not greater than) that given to developing still more models. The ecological basis for choosing a given species for mass production and release has received limited attention. Whereas there is a relatively large body of literature on the various attributes of an effective natural enemy, this is of limited value because these attributes relate primarily to long-term, self-sustaining control (i.e., classical biological control).[75] The attributes of an effective natural enemy for use in an inundative-release program should be explored. In this regard, van Lenteren and co-workers have made considerable advances with respect to inundative/inoculative releases in greenhouses,[33] and their efforts should serve as a useful guide for future investigations. In the end, we may find that the attributes of an effective enemy for inundative release will vary with the nature of the habitat, the kind of target pest, etc. The same is probably true for classical biological control.

Those concerned with inundative release of entomopathogenic nematodes can make important theoretical contributions to introduction strategy. Because of their relatively low motility, nematodes may be ideal organisms for testing theoretical predictions concerning inundation rates. However, new models may be required for a single species which serves as a vector of one or more pathogens. Similarly, a new approach to deriving attributes of an effective natural enemy may be required, one which considers the unique features of entomopathogenic nematodes. The relationship between the choice of a species to release and the number of individuals required for a given level of control should also be explored. As cost of a release program is likely to be a limiting factor, selection of the most appropriate species could reduce the number of individuals required. Within-species variation in a candidate natural enemy should also be assessed. Tauber and Tauber[76] have argued that different geographic races of a predaceous insect can have different potentials in inundative releases, and the same should apply to entomopathogenic nematodes. Such variation adds another layer of complexity to the hypothetical questions of introduction strategy posed earlier, and raises new questions (e.g., can a multiracial SSI be the equivalent of a MSI utilizing one race/species?). Clearly, numerous scenarios are possible and these should be experimentally evaluated whenever possible.

B. Multispecies Release

In the case of single-species releases, practical application preceded the attendant theory. For multispecies release, the opposite seems to be true: theoretical explorations have preceded practical application. Perhaps the main reason for this latter predicament is that arthropod natural enemies, because of their dispersal ability, are poor organisms for testing introduction strategies involving different subsets of species. Even where inundative or inoculative releases of more than one species were made,[77] the main concern was suppression of the target pest, rather than critical testing of specified introduction strategies. Whereas the ecological impact of different combinations of enemy species on a pest population is amenable to laboratory experimentation, the pertinent studies involving arthropod enemies, such as those by Flanders and Badgley,[78] Flanders,[79] Force,[12] White and Huffaker,[80,81] Hassell and Huffaker,[82] and Laing and Huffaker,[83] are of marginal value because they were designed to assess long-term host/enemy dynamics. A notable exception is the experimental work of Press et al.[84] Finally, it is encouraging that mathematical ecologists have begun to recognize that a suitable theoretical basis for biological control cannot be developed solely around single-species releases. However, current models incorporating two or more enemy species tend to be unrealistic, relatively simple, and of little or no immediate value in practical biological control. Whether this situation will improve in the foreseeable future is questionable.

With respect to the use of entomopathogenic nematodes, the potential for meaningful theoretical contributions appears to be good. Because of their relatively low mobility, these enemies appear to be good organisms for testing the efficacy of various introduction strategies, especially those involving multispecies combinations. This could be an excellent opportunity to gather long-needed empirical evidence, and those investigators who are in the position to conduct the appropriate experiments are encouraged to do so.

VII. ENVIRONMENTAL IMPACT

The biology of invasions has been of major interest to ecologists in recent years, and the subject of environmental impact of introduced species has emerged as one of the major concerns. Although much of the debate at present concerns environmental risks of genetically engineered organisms,[85,86] biological-control agents have not escaped scrutiny. When viewed from a broad ecological perspective (see below), introduced predators and parasites of insect pests and those phytophagous insects imported for weed control can be expected to have an environmental impact. The same assumption should hold for entomopathogenic nematodes.

A. Ecological Perspective

In the present context, environmental impact can be defined as any effect on a nontarget organism which results from the intentional introduction of a natural enemy.[22,63] Thus, an introduction resulting in permanent establishment of a biological-control agent will probably have an environmental impact; however, the impact will not necessarily be an adverse one. Impacts can be classified according to a number of factors, including: (1) duration (i.e., short-term as in inundative release vs. long-term as in classical biological control), (2) predictability (i.e., totally unexpected vs. predictable), (3) outcome (i.e., negative or positive), (4) magnitude (i.e., minor or subtle effects of little or no practical concern vs. major effects of considerable practical consequence), (5) interaction (i.e., direct or indirect [e.g., involving a third species]), and (6) timing (i.e., immediate vs. delayed). There are good examples from the empirical record in classical biological control for most of these categories; the case histories in question have been discussed elsewhere.[20,22,63,64,66]

B. Entomopathogenic Nematodes

It would be premature to claim that releases of entomopathogenic nematodes will have no environmental impact. The environmental impact of a biological-control agent is a function of (1) the attributes of the introduced species, (2) the nature of the target zone, and (3) the introduction strategy employed.[63] Thus, a nematode capable of parasitizing many different host species which is introduced into a habitat containing susceptible nontarget hosts will presumably have an environmental impact. However, environmental impact and environmental risk are different matters. Risk applies primarily to negative impacts of sufficient magnitude to warrant the concern of regulatory officials. Entomopathogenic nematodes may pose little environmental risk, although considerably more evidence may be required before a generalization can be made. It is recommended that all types of environmental impact be investigated, because in the process we can gain a better understanding of the structure of both natural and managed communities or ecosystems. This holds for inundative/inoculative release as well as classical biological control. Each release can be viewed as a perturbation experiment, and if attendant changes in community structure are detected and measured, these will enhance our understanding of the target habitat. As steinernematid and heterorhabditid nematodes are especially suited for control of soil-dwelling stages of pests, the use of these nematodes in inundative release programs should provide a good opportunity to gather needed information on the structure of the soil ecosystem.

VIII. PREDICTIVE BIOLOGICAL CONTROL

Apart from its obvious practical value, predictability in an applied science is significant because it is a measure of the development or intellectual maturity of the discipline in question. Biological control is no exception to this, and as should be evident from recent reviews of both arthropod natural enemies[15] and entomopathogenic nematodes,[1] considerable research remains to be done. Although predictability applies to many aspects of biological control, the present discussion is restricted to predictions prior to introduction. This includes both classical biological control and augmentative control through inundative/inoculative release.

The need for predictive capability is critical in at least two areas. First, it would be helpful in classical biological control to be able to predict the attributes of species which are likely to (1) permanently establish in a new region, and (2) subsequently provide the kind of ecological impact desired. In inundative release, permanent establishment is not an issue, so it would be helpful to just be able to predict the attributes of an effective species and the release rate providing the level of immediate control desired. Second, it is important to be able to predict the environmental impact of the introduction of either a new species (as in classical biological control) or large numbers of an incumbent species (as in inundative release). Unfortunately, modern biological control is limited in these matters, and does not fully satisfy the requirements for being a predictive science. The improvement of this condition should be one of the top priorities for future research, and this will likely prove to be one of the most intellectually challenging issues to be faced.

In the past 20 years, mathematical models of host/enemy dynamics have played a major role in the development of biological-control theory. This trend is likely to continue. From a practical standpoint, these models have been of value in at least two areas: explanation of successful biological control and justification of current practices, particularly in classical biological control. Unfortunately, there has been little effort devoted to predicting realistic introduction strategies. It is perhaps time for mathematical modelers to make major shift in emphasis, from explaining what we have already observed to predicting what we will observe if we pursue a particular introduction strategy. Biological-control practitioners should encourage mathematical modelers to tackle this problem, for it should be of benefit to both parties.

IX. CONCLUSIONS

Modern insect control is shifting away from reliance on synthetic organic insecticides in favor of a more integrated approach to pest management. Biological control is a major tactic in integrated pest management (IPM), and with the current movement toward "sustainable" agriculture, biological control can be expected to play an even more substantial role in IPM. Entomopathogenic nematodes are a welcome addition to the natural-enemy pool and should further enhance our ability to truly integrate the various control measures for management of those target pests where individual tactics (such as chemical control, host-plant resistance) alone are inadequate. However, the IPM movement has in some ways overshadowed biological control,[26] and may well have adversely affected the intellectual development of this critical discipline. Thus, biological-control workers must seek an optimal allocation of effort towards both biological control and IPM so that ecologically sound pest control can be maximized. The IPM movement and the practical contributions it demands from biological control must not be allowed to take precedence over the intellectual or theoretical needs of the science of biological control. Otherwise, we may be forced to set aside some of the most intellectually challenging problems in modern biological control. This would be especially unfortunate because there is no guarantee that the next generation of scientists will be in any better position to address these problems.

REFERENCES

1. **Gaugler, R.,** Ecological considerations in the biological control of soil-inhabiting insects with entomopathogenic nematodes, *Agric. Ecosystems Environ.*, 24, 351, 1988.
2. **Kaya, H. K.,** Entomogenous nematodes for insect control in IPM systems, in *Biological Control in Agricultural IPM Systems*, Hoy, M. A., and Herzog, D. C., Eds., Academic Press, Orlando, 1985, 283.
3. **Kaya, H. K.,** Diseases caused by nematodes, in *Epizootiology of Insect Diseases*, Fuxa, J. R., and Tanada, Y., Eds., John Wiley & Sons, New York, 1987, 453.
4. **Nickle, W. R.,** Nematodes with potential for biological control of insects and weeds, in *Biological Control in Crop Production*, Papavizas, G. C., Ed., Allanheld/Osmun, Totowa, 1981, 181.
5. **Poinar, G. O., Jr.,** Entomophagous nematodes, in *Biological Plant and Health Protection*, Franz, J. M., Ed., G. Fischer Verlag, Stuttgart, 1986, 95.
6. **National Academy of Sciences,** *Report of the Research Briefing Panel on Biological Control in Managed Ecosystems*, National Academy Press, Washington, 1987.
7. **Doutt, R. L.,** The dimensions of endemism, *Ann. Entomol. Soc. Am.*, 54, 46, 1961.
8. **Ehler, L. E.,** Ecology of *Rhopalomyia californica* Felt (Diptera: Cecidomyiidae) and its parasites in an urban environment, *Hilgardia*, 50, 1, 1982.
9. **Ehler, L. E.,** Foreign exploration in California, *Environ. Entomol.*, 11, 525, 1982.
10. **Ehler, L. E.,** Species-dependent mortality in a parasite guild and its relevance to biological control, *Environ. Entomol.*, 14, 1, 1985.
11. **Ehler, L. E.,** Effect of malathion-bait sprays on the spatial structure of a parasite guild, *Entomol. Exp. Appl.*, 52, 15, 1989.
12. **Force, D. C.,** Competition among four hymenopterous parasites of an endemic insect host, *Ann. Entomol. Soc. Am.*, 63, 1675, 1970.
13. **Force, D. C.,** Ecology of insect host-parasitoid communities, *Science*, 184, 624, 1974.
14. **Hopper, K. R.,** The effects of host-finding and colonization rates on abundances of parasitoids of a gall midge, *Ecology*, 65, 20, 1984.
15. **Ehler, L. E.,** Introduction strategies in biological control of insects, in *Critical Issues in Biological Control*, Mackauer, M., Ehler, L. E., and Roland, J., Eds., Intercept, Andover, 1990, chap. 6.
16. **Ehler, L. E., and Miller, J. C.,** Biological control in temporary agroecosystems, *Entomophaga*, 23, 207, 1978.
17. **Bisabri-Ershadi, B., and Ehler, L. E.,** Natural biological control of western yellow-striped armyworm, *Spodoptera praefica* (Grote), in hay alfalfa in northern California, *Hilgardia*, 49(5), 1, 1981.
18. **Flaherty, D. L., Wilson, L. T., Stern, V. M., and Kido, H.,** Biological control in San Joaquin Valley vineyards, in *Biological Control in Agricultural IPM Systems*, Hoy, M. A., and Herzog, D. C., Eds., Academic Press, Orlando, 1985, 501.
19. **Hoy, M. A.,** Integrated mite management for California almond orchards, in *Spider Mites, Their Biology, Natural Enemies and Control*, Vol. 1B, Helle, W., and Sabelis, M. W., Eds., Elsevier, Amsterdam, 1985, 299.
20. **Coulson, J. R., and Soper, R. S.,** Protocols for the introduction of biological control agents in the U. S., in *Plant Protection and Quarantine*, Vol. 3, Kahn, R. P., Ed., CRC Press, Boca Raton, FL, 1989, 1.

21. **Nickle, W. R., Drea, J. J., and Coulson, J. R.,** Guidelines for introducing beneficial insect-parasitic nematodes into the United States, *Ann. Appl. Nematol.,* 2, 50, 1988.
22. **Ehler, L. E.,** Planned introductions in biological control, in *Assessing Ecological Risks of Biotechnology,* Ginzburg, L. R., Ed., Butterworths, Boston, 1990, in press.
23. **Hall, R. W., and Ehler, L. E.,** Rate of establishment of natural enemies in classical biological control, *Bull. Entomol. Soc. Am.,* 25, 280, 1979.
24. **Hall, R. W., Ehler, L. E., and Bisabri-Ershadi, B.,** Rate of success in classical biological control of arthropods, *Bull. Entomol. Soc. Am.,* 26, 111, 1980.
25. **Ehler, L. E., and Hall, R. W.,** Evidence for competitive exclusion of introduced natural enemies in biological control, *Environ. Entomol.,* 11, 1, 1982.
26. **Coppel, H. C.,** Environmental management for furthering entomophagous arthropods, in *Biological Plant and Health Protection,* Franz, J. M., Ed., G. Fischer Verlag, Stuttgart, 1986, 57.
27. **Ridgway, R. L., and Vinson, S. B., Eds.,** *Biological Control by Augmentation of Natural Enemies,* Plenum, New York, 1977.
28. **Stinner, R. E.,** Efficacy of inundative releases, *Annu. Rev. Entomol.,* 22, 515, 1977.
29. **Ables, J. R., and Ridgway, R. L.,** Augmentation of entomophagous arthropods to control pest insects and mites, in *Biological Control in Crop Production,* Papavizas, G. C., Ed., Allanheld/Osmun, Totowa, 1981, 273.
30. **Hussey, N. W., and Scopes, N. E. A., Eds.,** *Biological Pest Control: The Glasshouse Experience,* Cornell University Press, Ithaca, 1985.
31. **van Lenteren, J. C.,** Evaluation, mass production, quality control and release of entomophagous insects, in *Biological Plant and Health Protection,* Franz, J. M., Ed., G. Fischer Verlag, Stuttgart, 1986, 31.
32. **van Lenteren, J. C.,** Parasitoids in the greenhouse: successes with seasonal inoculative release systems, in *Insect Parasitoids,* Waage, J. K., and Greathead, D. J., Eds., Academic Press, London, 1986, 341.
33. **van Lenteren, J. C., and Woets, J.,** Biological and integrated pest control in greenhouses, *Annu. Rev. Entomol.,* 33, 239, 1988.
34. **Hoy, M. A.,** Biological control of arthropod pests: traditional and emerging technologies, *Am. J. Altern. Agric.,* 3, 63, 1988.
35. **Hoy, M. A.,** Recent advances in genetics and genetic improvement of the Phytoseiidae, *Annu. Rev. Entomol.,* 30, 345, 1985.
36. **Gaugler, R.,** Entomogenous nematodes and their prospects for genetic improvement, in *Biotechnological Advances in Invertebrate Pathology and Cell Culture,* Maramorosch, K., Ed., Academic Press, New York, 1987, 457.
37. **Gaugler, R., Campbell, J. F., and McGuire, T. R.,** Selection for host-finding in *Steinernema feltiae, J. Invertebr. Pathol.,* 54, 363, 1989.
38. **Carl, K. P.,** Biological control of native pests by introduced natural enemies, *Biocont. News Info.,* 3, 191, 1982.
39. **Ehler, L. E., and Andres, L. A.,** Biological control: exotic natural enemies to control exotic pests, in *Exotic Plant Pests and North American Agriculture,* Wilson, C. L., and Graham, C. L., Eds., Academic Press, New York, 1983, 395.
40. **Greathead, D. J.,** Parasitoids in classical biological control, in *Insect Parasitoids,* Waage, J. K., and Greathead, D. J., Eds., Academic Press, London, 1986, 289.
41. **Hokkanen, H. M. T.,** Success in classical biological control, *CRC Crit. Rev. Pl. Sci.,* 3, 35, 1985.
42. **Huffaker, C. B., Messenger, P. S., and DeBach, P.,** The natural enemy component in natural control and the theory of biological control, in *Biological Control,* Huffaker, C. B., Ed., Plenum, New York, 1971, 16.
43. **Turnbull, A. L., and Chant, D. A.,** The practice and theory of biological control of insects in Canada, *Can. J. Zool.,* 39, 697, 1961.
44. **Conway, G.,** Man versus pests, in *Theoretical Ecology,* May, R. M., Ed., Saunders, Philadelphia, 1976, 257.

45. **Southwood, T. R. E.,** Entomology and mankind, *Am. Sci.,* 65, 30, 1977.
46. **Huffaker, C. B.,** Some implications of plant-arthropod and higher-level, arthropod-arthropod food links, *Environ. Entomol.,* 3, 1, 1974.
47. **Janzen, D. H.,** The role of the seed predator guild in a tropical deciduous forest, with some reflections on tropical biological control, in *Biology in Pest and Disease Control,* Price Jones, D., and Solomon, M. E., Eds., Blackwell, Oxford, 1974, 3.
48. **van Emden, H. F.,** Plant diversity and natural enemy efficiency in agroecosystems, in *Critical Issues in Biological Control,* Mackauer, M., Ehler, L. E., and Roland, J., Eds., Intercept, Andover, 1990, chap. 4.
49. **Huffaker, C. B., Luck, R. F., and Messenger, P. S.,** The ecological basis of biological control, in *Proc. XV Int. Cong. Entomol.,* Vancouver, 1977, 560.
50. **Hokkanen, H., and Pimentel, D.,** New approach for selecting biological control agents, *Can. Entomol.,* 116, 1109, 1984.
51. **Waage, J. K.,** Ecological theory and the selection of biological control agents, in *Critical Issues in Biological Control,* Mackauer, M., Ehler, L. E., and Roland, J., Eds., Intercept, Andover, 1990, chap. 7.
52. **Horn, D. J., and Dowell, R. V.,** Parasitoid ecology and biological control in ephemeral crops, in *Analysis of Ecological Systems,* Horn, D. J., Mitchell, R. D., and Stairs, G. R., Eds., Ohio State University, Columbus, 1979, 281.
53. **Pschorn-Walcher, H.,** Biological control of forest insects, *Annu. Rev. Entomol.,* 22, 1, 1977.
54. **Zwolfer, H.,** The structure and effect of parasite complexes attacking phytophagous host insects, in *Dynamics of Populations,* den Boer, P. J., and Gradwell, G. R., Eds., Center Agric. Public. Doc., Wageningen, 1971, 405.
55. **Mackauer, M.,** Genetic problems in the production of biological control agents, *Annu. Rev. Entomol.,* 21, 369, 1976.
56. **Roush, R. T.,** Genetic variation in natural enemies: critical issues for colonization practices in biological control, in *Critical Issues in Biological Control,* Mackauer, M., Ehler, L. E., and Roland, J., Eds., Intercept, Andover, 1990, chap. 13.
57. **Hoy, M. A.,** Improving establishment of arthropod natural enemies, in *Biological Control in Agricultural IPM Systems,* Hoy, M. A., and Herzog, D. C., Eds., Academic Press, Orlando, 1985, 151.
58. **DeBach, P., Huffaker, C. B., and MacPhee, A. W.,** Evaluation of the impact of natural enemies, in *Theory and Practice of Biological Control,* Huffaker, C. B., and Messenger, P. S., Eds., Academic Press, New York, 1976, 255.
59. **Gutierrez, A. P., Hagen, K. S., and Ellis, C. K.,** Evaluating the impact of natural enemies: a multitrophic perspective, in *Critical Issues in Biological Control,* Mackauer, M., Ehler, L. E., and Roland, J., Eds., Intercept, Andover, 1990, chap. 5.
60. **Luck, R. F., Shepard, B. M., and Kenmore, P. E.,** Experimental methods for evaluating arthropod natural enemies, *Annu. Rev. Entomol.,* 33, 367, 1988.
61. **Reichelderfer, K. H.,** Economic feasibility of biological control of crop pests, in *Biological Control in Crop Production,* Papavizas, G. C., Ed., Allanheld/Osmun, Totowa, 1981, 403.
62. **Tisdell, C. A.,** Economic impact of biological control of weeds and insects, in *Critical Issues in Biological Control,* Mackauer, M., Ehler, L. E., and Roland, J., Eds., Intercept, Andover, 1990, chap. 15.
63. **Ehler, L. E.,** Environmental impact of introduced biological-control agents: implications for agricultural biotechnology, in *Risk Assessment in Agricultural Biotechnology,* Marois, J. J., and Bruening, G., Eds., University of California Division of Agriculture & Natural Resources, Oakland, 1990, in press.
64. **Harris, P.,** Environmental impact of introduced biological control agents, in *Critical Issues in Biological Control,* Mackauer, M., Ehler, L. E., and Roland, J., Eds., Intercept, Andover, 1990, chap. 14.

65. **Howarth, F. G.,** Classical biological control: panacea or Pandora's Box, *Proc. Hawaiian Entomol. Soc.*, 24, 239, 1983.
66. **Legner, E. F.,** Importation of exotic natural enemies, in *Biological Plant and Health Protection*, Franz, J. M., Ed., G. Fischer Verlag, Stuttgart, 1986, 19.
67. **Beddington, J. R., Free, C. A., and Lawton, J. H.,** Characteristics of successful natural enemies in models of biological control of insect pests, *Nature*, 273, 513, 1978.
68. **Hassell, M. P.,** *The Dynamics of Arthropod Predator-Prey Systems*, Princeton University Press, Princeton, 1978.
69. **May, R. M., and Hassell, M. P.,** Population dynamics and biological control, *Phil. Trans. R. Soc. Lond.* B., 318, 129, 1988.
70. **Murdoch, W. W.,** The relevance of pest-enemy models to biological control, in *Critical Issues in Biological Control*, Mackauer, M., Ehler, L. E., and Roland, J., Eds., Intercept, Andover, 1990, chap. 1.
71. **Murdoch, W. W., Chesson, J., and Chesson, P. L.,** Biological control in theory and practice, *Am. Nat.*, 125, 344, 1985.
72. **Waage, J. K., and Greathead, D. J.,** Biological control: challenges and opportunities, *Phil. Trans. R. Soc. Lond.* B, 318, 111, 1988.
73. **Knipling, E. F.,** The theoretical basis for augmentation of natural enemies, in *Biological Control by Augmentation of Natural Enemies*, Ridgway, R. L., and Vinson, S. B., Eds., Plenum, New York, 1977, 79.
74. **Barclay, H. J., Otvos, I. S., and Thomson, A. J.,** Models of periodic inundation of parasitoids for pest control, *Can. Entomol.*, 117, 705, 1985.
75. **Huffaker, C. B., Rabb, R. L., and Logan, J. A.,** Some aspects of population dynamics relative to augmentation of natural enemy action, in *Biological Control by Augmentation of Natural Enemies*, Ridgway, R. L., and Vinson, S. B., Eds., Plenum, New York, 1977, 3.
76. **Tauber, M. J., and Tauber, C. A.,** Criteria for selecting *Chrysopa carnea* biotypes for biological control: adult dietary requirements, *Can. Entomol.*, 107, 589, 1975.
77. **Parker, F. D., Lawson, F. R., and Pinnell, R. E.,** Suppression of *Pieris rapae* using a new control system: mass releases of both the pest and its parasites, *J. Econ. Entomol.*, 64, 721, 1971.
78. **Flanders, S. E., and Badgley, M. E.,** Prey-predator interactions in self-balanced laboratory populations, *Hilgardia*, 35, 145, 1963.
79. **Flanders, S. E.,** Mechanisms of population homeostasis in *Anagasta* ecosystems, *Hilgardia*, 39, 367, 1968.
80. **White, E. G., and Huffaker, C. B.,** Regulatory processes and population cyclicity in laboratory populations of *Anagasta kuhniella* (Zeller) (Lepidoptera: Phycitidae). I. Competition for food and predation, *Res. Popul. Ecol.*, 11, 57, 1969.
81. **White, E. G., and Huffaker, C. B.,** Regulatory processes and population cyclicity in laboratory populations of *Anagasta kuhniella* (Zeller) (Lepidoptera: Phycitidae). II. Parasitism, predation, competition, and protective cover, *Res. Popul. Ecol.*, 11, 150, 1969.
82. **Hassell, M. P., and Huffaker, C. B.,** Regulatory processes and population cyclicity in laboratory populations of *Anagasta kuhniella* (Zeller) (Lepidoptera: Phycitidae). III. The development of population models, *Res. Popul. Ecol.*, 11, 186, 1969.
83. **Laing, J. E., and Huffaker, C. B.,** Comparative studies of predation by *Phytoseiulus persimilis* Athias-Henriot and *Metaseiulus occidentalis* (Nesbitt) (Acarina: Phytoseiidae) on populations of *Tetranychus urticae* Koch (Acarina: Tetranychidae), *Res. Popul. Ecol.*, 11, 105, 1969.
84. **Press, J. W., Flaherty, B. R., and Arbogast, R. T.,** Interactions among *Plodia interpunctella*, *Bracon hebetor*, and *Xylocoris flavipes*, *Environ. Entomol.*, 3, 183, 1974.
85. **Marois, J. J., and Bruening, G., Eds.,** *Risk Assessment in Agricultural Biotechnology*, University of California Division of Agriculture & Natural Resources, Oakland, 1990, in press.
86. **Ginzburg, L. R., Ed.,** *Assessing Ecological Risks of Biotechnology*, Butterworths, Boston, 1990, in press.

Biology and
Taxonomy

2.   Taxonomy and Biology of Steinernematidae and Heterorhabditidae

George O. Poinar, Jr.

I. INTRODUCTION

Many types of associations exist between nematodes and insects, ranging from phoresis to parasitism and pathogenesis.[1] The families Steinernematidae and Heterorhabditidae, which occur in the latter category, are unique because: (1) they are the only nematodes which have evolved the ability to carry and introduce symbiotic bacteria into the body cavity of insects, (2) they are the only insect pathogens with a host range which includes the majority of insect orders and families, and (3) they can be cultured on a large scale on, or in, artificial solid or liquid media. Steinernematid and heterorhabditid nematodes have other significant attributes. They can kill insects within 48 hr, can form a durable, infective stage which can be stored for long periods and applied by conventional methods, and persist in the natural environment. Because the nematodes are adaptable biological organisms, natural populations of insects would not be expected to acquire immunity against them. In addition, plants and mammals are not adversely affected.[2] Therefore, these pathogens are attractive from a biological and commercial perspective. There are currently nine recognized species of *Steinernema* and three of *Heterorhabditis*; these figures will certainly increase as new populations of these nematodes are discovered.

Steinernematid and heterorhabditid nematodes are becoming accepted as biological control agents, especially against insects in the soil environment. They have survived the tests of production, application, field efficacy, and safety standards. A number of commercial enterprises worldwide are now producing them.

The present chapter deals with the biology and taxonomy of representatives of these two families. A listing of the presently recognized species together with their synonyms and strains is followed by a diagnostic description of each species. Keys based on characters found in the infective juveniles and males, respectively, are presented.

II. HISTORICAL ASPECTS

The Swiss-born scientist, Gotthold Steiner (1886-1961), reported the first steinernematid, *Steinernema kraussei*.[3] Steiner came to the U.S. in 1921 to study at Yale University for 1 year and remained in the U.S. to work under N. A. Cobb. In 1932, he became head of the Nematology Section of the U.S. Department of Agriculture (USDA) and retired from the USDA in 1956 with a Distinguished Service Medal. Steiner could well be considered the "Father of Entomogenous Nematodes" in America because of his contributions to the systematics of many insect-parasitic nematode groups.

Rudolf William Glaser (1888-1947) pioneered the culture and field application of entomopathogenic nematodes, especially of the steinernematid, *Steinernema glaseri*. Glaser, who worked for the Rockefeller Institute for Medical Research in Princeton, New Jersey, discovered a nematode parasitizing the Japanese beetle (*Popillia japonica*) in New Jersey in May 1929. He sent it to Steiner, who described the nematode as a new genus and species, *Neoaplectana glaseri (S. glaseri)*.[4] Glaser became the first to cultivate an entomopathogenic species on solid media[5] and axenically,[6] and the first to conduct field experiments with cultured nematodes against an insect pest, the Japanese beetle.[7-9]

Significant developments have continued over the last 60 years. The relationship between nematodes and their symbiotic bacteria has been revealed and explored. Additional nematode species have been discovered and studied. Finally, field experiments have shown that these nematodes have potential as biological insecticides against a range of soil and other insects. Unfortunately, much of the information concerning production and application of these nematodes is currently found in patents and confidential reports of commercial companies, and is unavailable to the public.

III. TAXONOMY AND SYSTEMATICS
A. Steinernematidae

Steiner described the first steinernematid as *Aplectana kraussei*, isolated from Germany in 1923.[3] In 1927, Travassos erected a new genus, *Steinernema*, for the species. However, it is apparent that two different species are included in the original description of *S. kraussei*. Thus, the species *kraussei* should presently stand as a species inquirenda and the name *kraussei* as a nomen dubium. In 1929, Steiner erected the genus *Neoaplectana*[4] but did not give clear differences separating this genus from *Steinernema*. In 1934, Filipjev noted the resemblance between the two genera and placed them in the subfamily Steinernematidae. This subfamily was erected to the family level by Chitwood and Chitwood in 1937. Both genera, *Neoaplectana* and *Steinernema*, were considered valid, but all recognized species were placed in the former genus because it was more completely defined, and the type species (*glaseri*) was intensively studied in the 1930s and 1940s.

An examination of the original type species (*S. kraussei*) revealed no difference with *Neoaplectana* in regards to the number and arrangement of the head papillae and *Neoaplectana* was synonymized under *Steinernema*.[10] This action resulted in some confusion and it was hoped that a form fitting Steiner's original generic description of *Steinernema* might be found, but up until the present, it has not. In order to avoid further confusion and establish consistency, the use of *Steinernema* should be employed until a nematode with Steiner's original characters is recovered. *S. glaseri* will serve as the designated type species.

The double use of the specific epithet *"feltiae"* in the literature has also been confusing. The nematode *S. feltiae* was first described by Filipjev in 1934,[11] and additional populations of this species were maintained in the USSR.[12] Stanuszek[13] isolated what he determined to be *S. feltiae* from caterpillars, hybridized his *feltiae* with the DD-136 strain of *S. carpocapsae,* and synonymized the latter under the former taxon.[14,15] Poinar[16] showed that Stanuszek had actually isolated a population of *S. carpocapsae* so this synonymy is not valid. This invalid synonymy was also confirmed by esterase patterns.[17] Moreover, Kozodoi et al.[12] and Poinar[16] showed that the *feltiae* originally described by Filipjev is conspecific with *bibionis* Bovien. Thus, *feltiae* has priority and replaces *bibionis*. Ordinarily this would not cause a great problem; however, because the epithet *feltiae* has also been confused with *carpocapsae* in the literature, there have been two nematodes recognized as *feltiae*, one that is now *carpocapsae* and the other that was *bibionis*.

In previous publications, it may be difficult to determine which nematode was meant when *feltiae* was used, unless the nematode strain was given. This demonstrates the importance of providing information about the strain being used in all studies. The proposed nomenclature (and synonyms) for *Steinernema* species is presented in Table 1. An emended description of the family Steinernematidae follows the table.

TABLE 1. Recognized Species of *Steinernema* and Synonyms

Recognized species (all originally described in the genus *Neoaplectana* except *kushidai* and *scapterisci*)	Synonyms
affinis (Bovien, 1937)	—
anomali (Kozodoi, 1984)	?*N. arenaria* Artyukhovsky, 1967
carpocapsae (Weiser, 1955)	*N. belorussica* Veremchuk, 1966
	N. chresima Steiner in Glaser, McCoy and Girth, 1942
	N. dutkyi Turco et al., 1971
	N. dutkyi Jackson, 1965
	N. dutkyi Welch, 1963
	N. elateridicola Veremchuk, 1970
	N. semiothisae Veremchuk and Litvinchuk, 1971
	N. agriotos Veremchuk, 1969
	N. feltiae Filipjev sensu Stanuszek, 1974

TABLE 1 (continued). Recognized Species of *Steinernema* and Synonyms

Recognized species (all originally described in the genus *Neoaplectana* except *kushidai* and *scapterisci*	Synonyms
feltiae (Filipjev, 1934) sensu Filipjev	*N. bibionis* Bovien, 1937 *N. bothynoderi* Kirjanova and Puchkova, 1955 *N. georgica* Kakulia and Veremchuk, 1965 *N. kirjanovae* Veremchuk, 1969 *N. leucaniae* Hoy, 1954 *N. menozzii* Travassos, 1932
glaseri (Steiner, 1929)	—
intermedia (Poinar, 1985)	—
kushidai Mamiya, 1988	—
rara (Doucet, 1986)	—
scapterisci Nguyen and Smart, 1990	*N. carpocapsae*, Uruguay strain Nguyen and Smart, 1988

Steinernematidae Chitwood and Chitwood 1937, 1950; Rhabditoidea (Oerley), Rhabditida (Oerley) (syn. Neoaplectanidae Sobolev)

Obligate entomopathogenic nematodes capable of infecting a wide variety of insects. Life cycle with a third-stage infective juvenile (often enclosed in a second-stage cuticle) (Figure 1); containing cells of a symbiotic bacterium (*Xenorhabdus* spp.) in its alimentary tract. Infective juvenile capable of surviving in the environment, entering the body cavity of a host, and then developing into a male or female. One, two, or more generations possible in host.

Adults: Amphimictic: In nature found only inside the infected insect cadaver. Stylet absent. Six lips partially or completely fused; each lip with a labial papilla at its tip. Four cephalic papillae. Two, often inconspicuous, lateral amphids present. Cuticle smooth. Head rounded, not offset. Stoma partially collapsed with the posterior portion surrounded by pharyngeal tissue. Cheilorhabdions usually pronounced and represented by a thick ring of sclerotized material lining the fused area of the lips. Prorhabdions and mesorhabdions usually present, but metarhabdions usually vestigial. Pharynx muscular with a cylindrical procorpal area (sometimes with a swollen nonvalvated metacorpal area) followed by a narrow isthmus which expands into a nearly spherical basal bulb containing a reduced valve lined with refractive ridges. Nerve ring conspicuous and normally surrounding the isthmus portion of the pharynx. Ventral excretory pore distinct. Lateral fields and phasmids inconspicuous.

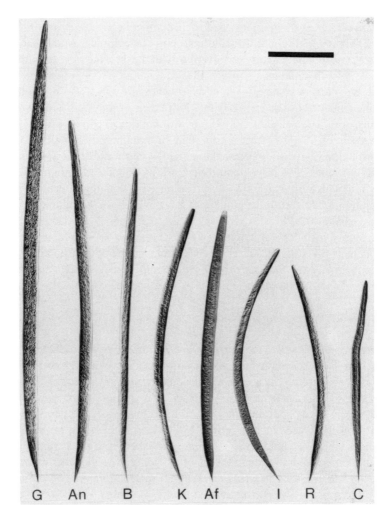

Figure 1. Infective-stage juveniles of *Steinernema* spp. G = *S. glaseri*; An = *S. anomali*; B = *S. feltiae*; K − *S. kraussei* (here a population of *S. feltiae*); Af = *S. affinis*; I = *S. intermedia*; R = *S. rara*; C = *S. carpocapsae* (all presented at the same scale; bar = 177 μm).

Females: Amphidelphic with opposed reflexed ovaries. Vulva with lips, located in midbody region, functional during mating and initial stages of oviposition. Vagina deteriorating in older females which become ovoviviparous, the young juveniles consuming the body contents and eventually killing the females. Size variable depending on the amount of nourishment available.

Males: Testis single, reflexed at tip. Spicules paired, separate. Gubernaculum present (Figure 2). Bursa absent. Tail tip with or without mucron. Spermatozoa beginning maturation in testis of fourth-stage male. Spermatozoa amoeboid in shape. Male tail usually with a complement of 23 genital papillae (rarely 21).

Infective-stage juvenile: The "dauer" or "infective" stage is a third-stage juvenile (Figure 1), often still inside a second-stage cuticle. Much narrower than the corresponding parasitic juvenile. Cuticle bearing 4-8 longitudinal striations. Mouth and anus closed. Pharynx and intestine collapsed. Tail pointed. Excretory pore anterior to nerve ring (Figure 3D). Cells of symbiotic bacteria maintained in a specialized anterior portion of intestine. Mouth region not armed with a dorsal hook but with or without dorsal cuticular thickening. Development of infective stage into a male or amphimictic female. Kozodoi and Spiridonov[18] have shown that the cuticular ridges differ on *Steinernema* infective juveniles. The ridges could be a useful character for distinguishing species although care and high magnification with scanning electron microscope are required because the number of ridges varies according to the part of the body examined.

The family includes the single genus *Steinernema* Travassos, 1927 (syn. *Neoaplectana* Steiner, 1929). The type species is hereby designated as *S. glaseri*. Normally *S. kraussei* would be the type species, but "species inquirenda" cannot be validly designated as type species.

Species of *Steinernema* are listed below in chronological order:

S. kraussei (Steiner, 1923) syn. *Aplectana kraussei* Steiner, 1923.[3]

The original description of *S. kraussei* by Steiner[3] includes two separate nematodes, one resembling *S. feltiae*, the other resembling *S. carpocapsae*. For this reason, the species *kraussei* is assigned to a species inquirenda, and the name *kraussei* is a nomen dubium. On the basis of the original description by Steiner[3] and the emended description by Mrácek,[19] there are no consistent differences which separate *kraussei* from other species in the genus. In their key to the *Steinernema* species, Wouts et al.[10] only state the size of the male spicules (45-55 µm long in *kraussei* vs. 60-65 µm long in *feltiae* [=*bibionis*]) for distinguishing between *kraussei* and *feltiae*. However, the 60-65 µm spicule length value of *feltiae* was based on the New Zealand strain of *feltiae*[20] and not the Danish strain which Bovien stated had a spicule length of 53 µm, clearly within the range for *kraussei*.

In studies on a strain labelled *S. kraussei* from Czechoslovakia, the present author and Sha[17] could find no clear differences from *feltiae*. Breeding experiments showed that *S. kraussei* could cross (viable F_1) with the SN strain of *S. feltiae*. Thus, at this point, the status of *kraussei* is unclear, especially since in the original description, Steiner[3] figured the male tails of what appear to be two separate species of *Steinernema*.

Geographical range. This species has been collected only from the sawfly, *Cephalcia abietis*, in Czechoslovakia.[3,19]

Steinernema glaseri (Steiner, 1929) (Figures 1G, 2A)

This species is characterized by its large infective juveniles. Aside from the original description of Steiner,[4] the species was re-described from a natural population infecting a scarabaeid larva in North Carolina.[21] Although females have no diagnostic characters, the males can be distinguished by the lack of a terminal tail mucron and spicules which have a terminal ventral notch, hook, or scar. The infective stage can be distinguished by its length (average = 1130 µm; range = 864-1448), the distance from the head to the excretory pore (87-110 µm), and the ratio E (1.22-1.38).

Geographical range. This species occurs in North and South America. It is uncertain whether the Soviet *arenaria* Artyukhovsky[22] is more closely related to *anomali* Kozodoi or *glaseri*. Strains of *S. glaseri* are listed in Table 2.

Steinernema feltiae (Filipjev, 1934) (Figures 1B, 1K, 2G)

Syn. *N. bibionis* Bovien, 1937; *N. bothynoderi* Kirjanova and Puchkova, 1955; *N. georgica* Kakulia and Veremchuk; *N. kirjanovae* Veremchuk, 1969; *N. leucaniae* Hoy, 1954; *N. menozzii* Travassos, 1932.

It is now clear that this species is conspecific with *S. bibionis* (Bovien)[23] (see Kozodoi et al.[12] and Poinar[16]). In addition to the descriptions by Filipjev[11] and Bovien,[23] Wouts[20] provided a description of a New Zealand strain of *S. feltiae*. This species is characterized by the shape of the spicules and male tail. There is a terminal male tail mucron ranging from 4-13 µm in length (rarely shorter). The spicules are yellow orange and lack a distinct capitulum and rostrum. Adults can also be distinguished from *S. carpocapsae* by the location of the excretory pore (further posterior in *feltiae*). The infective juvenile can be distinguished by its length (average = 849 µm; range = 736-950) and their ratio B (length of pharynx divided by total length).

Geographical range. This species has been recovered from Europe, Australia, and New Zealand. Strains of *S. feltiae* (=*bibionis*) are presented in Table 2.

Steinernema affinis (Bovien, 1937) (Figures 1Af, 2E)

Descriptions of this species include the original by Bovien[23] and a re-description by Poinar.[24]

Diagnostic characters in the male include gray or colorless spicules, the presence of a minute tail mucron, and the shape of the spicules (short capitulum, curved calamus and sometimes a small, anteriorly located rostrum). The infective stage is intermediate in length (average = 693 µm; range = 608-800) and contains a refractile spine in the tip of the tail.

Geographical range. This species has only been reported from northern Europe (Denmark and West Germany). Strains of *S. affinis* are listed in Table 2.

Steinernema carpocapsae (Weiser, 1955) (see Table 1 for synonyms) (Figures 1C, 2F)

This species is widely distributed and represented by many strains (Table 2).

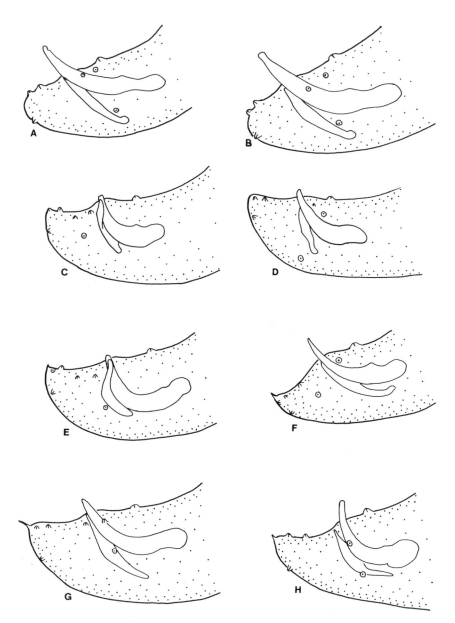

Figure 2 (above). Schematic drawings of the male tails of *Steinernema* spp. A = *S. glaseri*, B = *S. anomali*, C = *S. intermedia*, D = *S. kushidai*, E = *S. affinis*, F = *S. carpocapsae*, G = *S. feltiae*, H = *S. rara*.

Figure 3. Morphological structures of *Heterorhabditis* and *Steinernema*. A. Anterior tip of an infective stage *H. bacteriophora* (HP88 strain) showing large dorsal tooth (arrow) and smaller subventral tooth. B. Ventral view of male tail of *H. bacteriophora* (HP88 strain) showing the bursa and bursal papillae. C. Excretory pore opening (arrow) occurs posterior to the nerve ring (N) in *Heterorhabditis* spp. D. Excretory pore opening (arrow) occurs anterior to the nerve ring (N) in *Steinernema* spp.

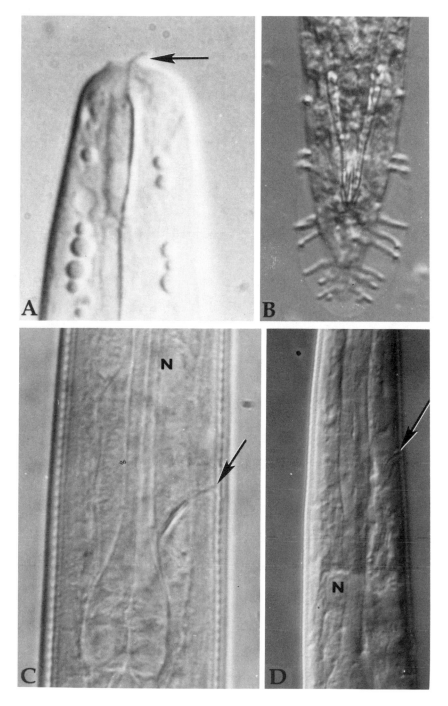

Figure 3

TABLE 2. Strains of *Steinernema* Species (combined with those cited by Poinar[60])

Species	Strain	Source	Geographic locality	Ref.
affinis (Bovien)	DK	larvae of Bibionidae	Denmark	23
	SK	larvae of Bibionidae	Skive, Denmark	24
	KL	soil	Kiel, Germany	Present study
anomali (Kozodoi)	Riazan	*Anomala dubia*	Riazon Province, USSR	29
	Voronez	*A. dubia*	Voronez Province, USSR	29
carpocapsae (Weiser)	Czechoslovakian	*Cydia pomonella*	Czechoslovakia	25
	DD-136	*C. pomonella*	Virginia, U.S.	see 60
	Mexican	*C. pomonella*	Allende, Chihuahua, Mexico	see 60
	Sierra	*C. pomonella*	California, U.S.	see 60
	Agriotos (Leningrad)	*Agriotes lineatus*	Leningrad, USSR	28
	All	*Vitacea polistiformis*	Georgia, U.S.	2
	XI	*C. pomonella*	Poland	14
	X-III	*Agrotis segetum*	Poland	14
	X-IV (Pieridarum)	*Pieris brassicae*	Poland	14, see 60

TABLE 2 (continued). Strains of *Steinernema* Species (combined with those cited by Poinar[60])

Species	Strain	Source	Geographic locality	Ref.
	Breton	*Otiorhynchus sulcatus*	France	see 60
	Umea	Soil	Sweden	see 60
	42	Cross between Breton and DD-136	—	see 60
	Italian	Soil	Sweden	see 60
	Hopland	Soil	California	see 60
	Quebec	*Listronotus oregonensis*	Quebec, Canada	see 60
	N.C.	Soil	North Carolina, U.S.	see 60
	Nelson	*Vespula* sp.	Tasmania, Australia	49
	Powranna	Soil	Tasmania	see 60
	Murrumbateman	Soil	New South Wales, Australia	see 60
	P7	Soil	Tasmania	49
	N55	Soil	Tasmania	49
	Argentinian	*Graphognathus leucoloma*	Argentina	see 60
	Rhagolites	*Rhagolites pomonella*	Massachusetts, U.S.	see 60
	Stanuszek	Agrotine caterpillars	Poland	14

Species	Strain	Host	Location	Ref.
feltiae (Filipjev) (=*bibionis* (Bovien))	KL	*Bibio* spp.	Denmark	23, see 60
	NZ	*Graphognathus leucoloma*	New Zealand	20
	N60	Soil	Canberra, Australia	see 60
	T335	*Otiorhynchus sulcatus*	Tasmania	see 60
	Dutch	Soil	Holland	see 60
	Murrumbateman	Soil	New South Wales	see 60
	Dover	Soil	Tasmania	see 60
	T231 (Risdon)	Soil	Tasmania	see 60
	Nive	Soil	Tasmania	see 60
	T298 (Plenty)	Soil	Tasmania	see 60
	Bruny	Soil	Tasmania	see 60
	T319 (Wellington)	Soil	Tasmania	see 60
	VI	Soil	Victoria, Australia	49
	CA	*Agrotis ipsilon*	Christchurch, NZ	55
	DU	Soil	Dunedin, NZ	Present study
	CZ	Soil	Czechoslovakia	61
	FN	Soil	Kuminki, Finland	Present study
	SN	Soil	France	Collected by C. Scotto la Massesse
feltiae (Filipjev)	Pinnock	*Heliothis armigera*	Getton, Queensland, Australia	Present study

TABLE 2 (continued). Strains of *Steinernema* Species (combined with those cited by Poinar[60])

Species	Strain	Source	Geographic locality	Ref.
glaseri (Steiner)	N.J.	*Popillia japonica*	New Jersey, U.S.	7
	N.C.	*Strigoderma arboricola*	North Carolina, U.S.	see 60
	FL	Soil	Florida, U.S.	see 60
	Araros	*Migdolus fryanus*	Santa Rosa, Brazil	62
	Klein	*Anomala flavipennis*	Wilmington, NC, U.S.	Present study
	KG, NC 32, NC 34, NC 40, NC 50, NC 52	Soil	North Carolina, U.S.	63
intermedia (Poinar)	SC	Soil	South Carolina	31
kushidai Mamiya	Hamakita	*Anomala cupre*	Hamakita (Shizuoka), Japan	35
rara (Doucet)	Cordoba	*Heliothis* sp.	Rio Cuarto, Cordoba, Argentina	33
scapterisci Nguyen and Smart	Uruguay	*Scapteriscus* sp.	Uruguay	36

Descriptions of the following strains exist: Czechoslovakian,[25,26] DD-136,[2,26] Agriotos,[2,27,28] and Stanuszek.[2,14,15] Diagnostic characters in males include the presence of a small mucron on the tail tip (usually from 1-4 μm in length) and gray-yellow spicules with a distinct capitulum and rostrum. The infective juvenile can be recognized by its length (average = 558 μm; range = 438-650), anteriorly positioned excretory pore, and small ratio E (0.54-0.64) (distance from head to excretory pore divided by tail length).

Geographical range. This species occurs in Europe, North America, South America, Australia, and New Zealand. Strains of *S. carpocapsae* are presented in Table 2.

Steinernema anomali (Kozodoi, 1984) (Figures 1An, 2B)

This species was found parasitizing the chafer, *Anomala dubia*, in the Riazan and Voronez provinces of the USSR.[29] Descriptions have been made by Kozodoi[29] and Poinar and Kozodoi.[30] This species is morphologically and biologically so similar to *S. glaseri* that they could be considered the same species except that they do not interbreed. A distinguishing character is the slightly swollen tip of the spicule in *S. anomali.* In *S. glaseri*, the spicule tip appears to be notched or even hooked. The infective juvenile can be separated from that of all *Steinernema* species except *glaseri* by its length (average = 1034 μm; range = 724-1408). *S. anomali* can be separated from the infective stage of *S. glaseri* by the shorter distance from the head to the excretory pore and ratio D (distance from head to excretory pore divided by distance from head to base of pharynx).

Geographical range. S. anomali has been reported only from the USSR. It is similar to *S. arenaria* and may, in fact, be a junior synonym. However, insufficient information to make this determination is available at this time.[22] Strains of *S. anomali* are listed in Table 2.

Steinernema intermedia (Poinar, 1985) (Figures 1I, 2C)

S. intermedia was described from a strain collected in South Carolina (U.S.). The original description provides the only characterization of this species[31] aside from a separate study on sperm development and morphology.[32] Males can be separated from other *Steinernema* species by the absence of a terminal tail mucron, bluntly rounded spicule tips, and the strongly curved spicules. The infective juvenile is longer (average = 671 μm; range = 608-800) than those of *S. carpocapsae, S. kushidai,* and *S. rara,* yet shorter than those of all the other *Steinernema* species. The absence of a spine in the tail tip of the infective juvenile separates this species from *S. affinis.*

Geographical range. This species has been reported only from the original collection site in South Carolina.[31]

Steinernema rara (Doucet, 1986) emd. Poinar, Mrácek and Doucet, 1988 (Figures 1R, 2H)

This species was isolated from *Heliothis* larvae in Córdoba, Argentina.[33] In addition to the original description,[33] further characterization was provided by Poinar et al.[34] This unusual species is the only known steinernematid which lacks one pair of male genital papillae (21 instead of the normal 23 papillae). Other diagnostic characters include the presence of a male tail mucron, the lemon yellow spicules, and the degree of spicule curvature. The infective stage can be separated from that of other steinernematids by its shorter overall length (average = 511 μm; range = 443-573) and length of the pharynx.

Geographical range. S. rara has been reported only from the original collection site in Argentina.[33]

Steinernema kushidai Mamiya 1988 (Figure 2D)

This species was obtained from cadavers of *Anomala cuprea* larvae which had been reared in soil collected from Hamakita, Shizuoka Prefecture, Japan.[35] The original description was presented by Mamiya and provides the only characterization of this species.[35] Diagnostic characters provided by Mamiya include a constriction of the gubernaculum, a rounded manubrium on the colorless spicules, the location of the excretory pore, and the length of the infective stage (average = 589 μm; range = 524-662).

Geographical range. S. kushidai has been reported only from the original collection site in Japan.[35]

Steinernema scapterisci Nguyen and Smart 1990[36]

Steinernema scapterisci, referred to earlier as the Uruguay strain of *S. carpocapsae*,[37] was described recently as a new species.[36] It is similar to *S. carpocapsae*, differing from the latter species mainly in physiological characters related to its adaptations for parasitizing mole crickets of the genus *Scapteriscus* (Gryllotalpidae, Orthoptera). The infective juvenile can be distinguished from *S. carpocapsae* by its greater ratio E.

Geographical range. S. scapterisci has been reported only from its original site of discovery in Uruguay.[36]

On the basis of the published descriptions of the above-reported species of *Steinernema* and the author's own observations, measurements of the infective stages of the nine recognized species are listed in Table 3. Keys to the infective juveniles and males are presented below.

Key to the Infective Juveniles of *Steinernema* spp.

Included here are all available measurements of different strains of the species to present normal variability within the species. Average refers to the arithmetic mean of 10 individuals. All measurements are in microns. Steinernematid infective juveniles are illustrated in Figure 1.

1. Average length of infective stages greater than 1000 (range = 724-1500) — 2

1'.   Average length of infective stages less than 1000 (range = 438-1200) — 3

2.   Average distance from head to excretory pore from 87-110; ratio D varies from 0.58-0.71 — *S. glaseri* (Steiner)

2'.   Average distance from head to excretory pore from 76-86; ratio D varies from 0.52-0.59 — *S. anomali* (Kozodoi)

3.   Average length of infective stage from 800-900 (range = 736-950) - *S. feltiae* (Filipjev)

3'.   Average length of infective stages less than 800 (range = 438-800) — 4

4.   Average length of infective stages from 660-700 (range = 608-800) — 5

4'.   Average length of infective stages from 500-600 (range = 438-662) — 6

5.   Minute refractile spine in tail tip present — *S. affinis* (Bovien)

5'.   Minute refractile spine in tail tip absent — *S. intermedia* (Poinar)

6.   Average distance from head to pharynx base from 115-127 (range = 103-190) — 7

6'.   Average distance from head to pharynx base from 102-111 (range = 89-120) — 8

7.   Ratio E is 0.73 (range = 0.60-0.80) — *S. scapterisci* Nguyen and Smart

7'.   Ratio E is 0.60 (range = 0.54-0.66) — *S. carpocapsae* (Weiser)

8.   Average distance from head to excretory pore is 38, range is 32-40; average length of infective stage is 511, range is 443-573; ratio D is 0.35 (0.30-0.39) — *S. rara* (Doucet)

8'.   Average distance from head to excretory pore is 46, range is 42-50; average length of infective stage is 589, range is 524-662; ratio D is 0.41 (0.38-0.44) — *S. kushidai* Mamiya

Key to Males of *Steinernema* spp.

Illustrations of male tails appear in Figure 2.

1.   Tip of tail containing a cuticular mucron (=spine) — 2

1'.   Tip of tail lacking a cuticular mucron (=spine) — 5

2.   Average length of tail mucron (N = 10) is 1-4; capitulum on spicule mostly distinct — 3

2'.   Average length of tail mucron (N = 10) is 4-13; capitulum on spicule mostly indistinct — *S. feltiae*

3.   Length of tail mucron equal to or shorter than that of surrounding genital papillae; spicule colorless, tips blunt; distal tip of gubernaculum overlaps spicule tip — *S. affinis*

3'.   Length of tail mucron longer than surrounding genital papillae; spicules yellow, tips blunt or pointed; distal tip of gubernaculum does not overlap spicule tips — 4

TABLE 3. Collective Measurements of the Infective-Stage Juveniles of *Steinernema* Species[a]

Character	*carpocapsae* (Weiser) (after Poinar[60])	*feltiae* (Filipjev) (after Poinar[60])	*glaseri* (Steiner) (after Poinar[60])	*intermedia* (Poinar) (after Poinar[60])	*affinis* (Bovien) (after Poinar[24])	*anomali* (Kozodoi) (after Poinar and Kozodoi[30])	*scapterisci* Nguyen and Smart (after Nguyen and Smart[36])	*rara* (Doucet) (after Doucet and Poinar,[33] present study)	*kushidai* Mamiya (after Mamiya[35])
Total length	558(438-650)	849(736-950)	1130(864-1448)	671(608-800)	693(608-800)	1034(724-1408)	572(517-609)	511(443-573)	589(524-662)
Greatest width	25(20-30)	26(22-29)	43(31-50)	29(25-32)	30(28-34)	46(28-77)	24(18-30)	23(18-26)	26(22-31)
Distance: head to excretory pore	38(30-56)	62(53-67)	102(87-110)	65(59-69)	62(51-69)	83(76-86)	39(36-48)	38(32-40)	46(42-50)
Distance: head to nerve ring	85(75-99)	99(88-112)	120(112-126)	93(85-99)	95(88-104)	109(100-120)	97(83-106)	70(60-88)	76(70-84)
Distance: head to pharynx base	120(103-190)	136(115-150)	162(158-163)	123(110-133)	126(115-134)	138(123-160)	127(113-134)	102(89-120)	111(106-120)
Length tail	53(46-61)	81(70-92)	78(62-87)	66(53-74)	66(64-74)	75(64-84)	54(48-60)	51(44-56)	50(44-59)
Ratio A[b]	21(19-24)	31(29-33)	29(26-35)	23(20-26)	23(21-28)	26(17-34)	—	23(20-26)	22.5(19-25)
Ratio B[c]	4.4(4.0-4.8)	6.0(5.3-6.4)	7.3(6.3-7.8)	5.3(5.0-6.0)	5.5(5.1-6.0)	7.6(5.9-10.8)	—	4.7(4.1-5.6)	5.3(4.9-5.9)
Ratio C[d]	10.0(9.1-11.2)	10.4(9.2-12.6)	14.7(13.6-15.7)	10.0(9.3-10.8)	10.5(9.5-11.5)	13.8(9.4-16.9)	—	9.8(8.7-11.0)	11.7(9.9-12.9)
Ratio D[e]	0.26(0.23-0.28)	0.45(0.42-0.51)	0.65(0.58-0.71)	0.51(0.48-0.58)	0.49(0.43-0.53)	0.55(0.52-0.59)	0.31(0.27-0.40)	0.35(0.30-0.39)	0.41(0.38-0.44)
Ratio E[f]	0.60(0.54-0.66)	0.78(0.69-0.86)	1.31(1.22-1.38)	0.96(0.89-1.08)	0.94(0.74-1.08)	1.19(1.06-1.30)	0.73(0.60-0.80)	0.72(0.63-0.80)	0.92(0.84-0.95)

a All measurements in microns, and range is given in parentheses following the average value (N = 25).
b Length divided by width.
c Length divided by distance from head to pharynx base.
d Length divided by tail length.
e Distance from head to excretory pore divided by distance from head to pharynx base.
f Distance from head to excretory pore divided by tail length.

4.   Spicule tips pointed; six pairs of tail genital papillae; spicule with prominent ventral arch — *S. carpocapsae* and *S. scapterisci* (characters separating the males of these species have yet to be described)

4′.   Spicule tips blunt; five pairs of tail genital papillae; spicules with inconspicuous ventral arch (rostrum) — *S. rara*

5.   Tail tip rounded; spicules notched or swollen at tip, moderately curved (angle between calomus and lamina usually between 50 and 70°) — 6

5′.   Tail tip pointed; spicules not notched or swollen at tip, strongly curved (angle between calomus and lamina usually between 70 and 90°) — 7

6.   Spicule tips swollen — *S. anomali*

6′.   Spicule tips notched or scarred — *S. glaseri*

7.   Gubernaculum with a median constriction, distal portion not surrounding spicule tips — *S. kushidai*

7′.   Gubernaculum lacking a median constriction; distal portion surrounding spicule tips — *S. intermedia*

B. Heterorhabditidae

The family Heterorhabditidae Poinar was erected in 1976 with *H. bacteriophora* Poinar as the type species.[38] These are obligately parasitic rhabditoids with a biology similar to the steinernematids except that they have a heterogenic life cycle with hermaphroditic and amphimictic females. Their morphology is similar to that of the microbotrophic rhabditids. The nematode *H. heliothidis* (=*Chromonema heliothidis* Khan, Brooks, and Hirschmann[39]) originally described in the family Steinernematidae, is now synonymized with *Heterorhabditis bacteriophora* Poinar. Only the single genus *Heterorhabditis* is presently included in the family.

The ability of two nematode populations to mate and produce fertile F$_1$ progeny determines a true biological species and has been used to identify new species of Steinernematidae. However, because of technical difficulties in handling the fragile males of *Heterorhabditis*, it has not been practical to utilize this tool for this genus. As a result, new species are determined by morphology,[40] electrophoretic analysis of enzymes,[41] and DNA fingerprinting.[40] Table 4 lists the present species and strains of *Heterorhabditis*. An emended description of the family Heterorhabditidae is presented below.

Heterorhabditidae Poinar 1976; Rhabditoidea (Oerley), Rhabditida (Oerley).

Obligate entomopathogenic nematodes capable of infecting a wide variety of insects. Life cycle with a third-stage infective juvenile (Figures 3A, 3C, 4) often enclosed in a second-stage cuticle. Cells of a symbiotic bacterium (*Xenorhabdus* sp.) in infective juvenile's alimentary tract. Infective juvenile capable of entering the body cavity of a host and developing into a hermaphroditic female.

Adults: Hermaphroditic and amphimictic populations found only inside

TABLE 4. Species (with Synonyms) and Strains of *Heterorhabditis*

Species	Strain	Original source	Geographical locality	Ref.
bacteriophora Poinar, 1976 (=*heliothidis* (Khan, Brooks, and Hirschmann, 1976)	HB1	*Heliothis punctigera*	Brecon, South Australia	38,41
	NC1	*Heliothis zea*	Clayton, NC, U.S.	39,41
	C8406-	—	Hailing Is., Guangdong, China	41
	It145	—	Forli, Italy	41
	V15	—	Geelong, Victoria, Australia	41
	NC14	—	Clayton, NC, U.S.	41
	NC69	—	Raleigh, NC, U.S.	41
	NC127	—	Lewiston, NC, U.S.	41
	NC162	—	Rocky Mount, NC, U.S.	41
	Arg1	soil	Rio Cuarto, Argentina	64
	Hamm	*Cuculio caryae*	Bryon, GA, U.S.	Present study
	Forschler	Scarabaeidae	Lexington, KY, U.S.	Present study
	Ali	*Cyclocephala hirta*	Apple Valley, CA, U.S.	Present study
	Overholt	*Diatrea grandiosella*	Dimmit, TX, U.S.	Present study
	McCoy	*Diaprepes abbreviatus*	Plymouth, FL, U.S.	Present study
	Schroeder	*Diaprepes abbreviatus*	Orlando, FL, U.S.	Present study

TABLE 4 (continued). Species (with Synonyms) and Strains of *Heterorhabditis*

Species	Strain	Original source	Geographical locality	Ref.
	Brazil	soil	Pernambrico, Brazil	Present study
	Creighton	*Diabrotica balteata*	Charleston, SC, U.S.	65
	NC 200	—	Reidsville, NC, U.S.	41
	NC 323	—	Salisburg, NC, U.S.	41
	NC 405	—	Clinton, NC, U.S.	41
	NC 476	—	Whiteville, NC, U.S.	41
	HP 88	*Phyllophaga* sp.	Logan, UT, U.S.	42
megidis Poinar, Jackson, and Klein, 1987	HO 1	*Popillia japonica*	Ohio, U.S.	Present study
Heterorhabditis zealandica Poinar (=New Zealand population of *H. heliothidis* sensu Wouts (1979))	HLit	—	Lithuania	41
	HNach	—	Nachodka, USSR	41
	HQ614	—	Bundaberg, Queensland, Australia	41
	T310	—	Sandy Bay, Tasmania, Australia	41
	T327	—	Nicholl's Rivulet, Tasmania, Australia	41
	NZH, NZ	*Heteronychus arator*	Auckland, New Zealand	41, 63
Heterorhabditis sp. A Province, China	C8404	—	Hailing Is., Guangdong	41
Australia	D1	—	Darwin, Northern Territory,	41
	HQ 380	—	Yeppoon, Queensland, Australia	41

Heterorhabditis sp. B	Tetuan	*Cylas formicarius*	Artemisa, Cuba	66
	P₂M	*Pachneus litus*	Havana, Cuba	66
	HW79	—	The Netherlands	63
	Elgria	*Popillia japonica*	Ohio, U.S.	Present study
	HL81	Soil	The Netherlands	63

Figure 4. Infective-stage juveniles of *Heterorhabditis* spp. M = *H. megidis*; T = *H. zealandica*; B = *H. bacteriophora* strain HB1; H = *H. bacteriophora* strain NC1; A = *Heterorhabditis bacteriophora* strain Arg 1 (all presented at the same scale; bar = 102 μm).

infected insect cadavers in nature. First hermaphroditic generation usually followed by one or more amphimictic generations. Stylet absent. Head truncate or slightly rounded. Six distinct lips present which may be partially fused at base; each lip with a single labial papilla; two additional papillae at the base of each submedial lip; lateral lip with a single cephalic papilla and a circular amphidial opening. Cheilorhabdions present as a refractile ring in anterior portion of stoma. Posterior portion of stoma collapsed, with reduced pro-, meso-, and metarhabdions. Anterior portion of pharynx surrounds base of stoma. Procorpus of pharynx wide and cylindrical, with pharynx narrowing at isthmus and expanding into a distinct basal bulb with reduced valve plates. Nerve ring distinct, usually located near middle of isthmus in female and on basal bulb in male.

Females: Amphidelphic with median vulva and reflexed portion of ovaries often extending past the vulvar openings. Oviparous becoming ovoviviparous in later life. Hermaphroditic females with sperm in proximal portion of ovotestis and functional vulva. Amphimictic females with sperm in proximal portion of oviduct and vulva nonfunctional for egg passage (functional only for mating). Mated females often with copious deposit (copulation plug) surrounding vulvar opening. Tail pointed, usually with a postanal or anal swelling. Rectal glands present.

Males: Produced only during the amphimictic generation. Single, reflexed testis. Spicules paired and separate, nearly straight. Gubernaculum present. Bursa present, open, peloderan or weakly leptoderan, attended by a complement of nine genital papillae (Figure 3B).

Infective-stage juvenile: Third-stage juvenile often still inside its second-stage cuticle and narrower than corresponding parasitic juvenile. Cuticle with paired longitudinal double lines (ensheathing second-stage cuticle with numerous longitudinal ridges). Mouth and anus closed. Head with armature on the dorsal side (dorsal protrusion, hook, or spine) and in some cases, further modification on the subventral surfaces (callus, small spine, hook, or thickening) (Figure 3A). Pharynx and intestine collapsed. Tail pointed. Excretory pore posterior to nerve ring (Figure 3C, contrast with Figure 3D). Cells of symbiotic bacteria found throughout the lumen of the alimentary tract (Figure 5). Development of infective stage always into a hermaphroditic female.

The family contains the single genus *Heterorhabditis* with *H. bacteriophora* as the type species. Characters for the genus are similar to those outlined above for the family. A synopsis of the species of *Heterorhabditis* is presented below.

Heterorhabditis bacteriophora Poinar 1976 (Figures 3A-3C, 4A, 4B, 4H, 5, 6)

This species was collected from *Heliothis punctiger* in Brecon, Australia.[38] It is widely distributed and composed of many strains, each differing from one another in behavior and physiology. These strains can be identified by DNA fingerprinting and enzyme analysis.

An electrophoretic study by Akhurst[41] indicated three broad groupings for an assemblage of 23 geographic strains of *Heterorhabditis*. Thus far, only 3 species have been formally described. However, the Akhurst study revealed only a 10% dissimilarity between *H. bacteriophora* and *H. heliothidis*. On the basis of these results, coupled with DNA fingerprinting and morphological studies, it is concluded that the two species are conspecific. Descriptions of *bacteriophora* include strain HB1 from Australia,[38] strain NC1 from North Carolina,[39] and strain HP88 from Utah.[42]

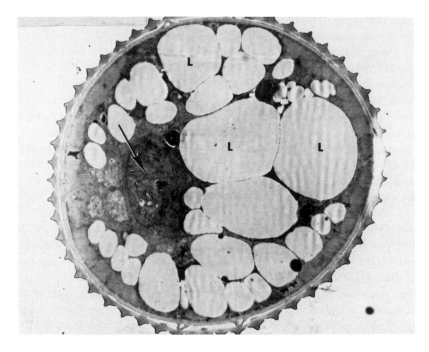

Figure 5

Figure 6

Diagnostic characters for *bacteriophora* include the length of the infective stage, the position of the pharynx base in the infective juvenile, and male tail morphology.

Geographical range. This species occurs in North and South America, Australia, and Europe.

Heterorhabditis zealandica Poinar syn. New Zealand population of *H. heliothidis* sensu Wouts[43] (Figure 4T)

This species was originally described as a New Zealand population of *H. heliothidis*;[43] however, morphological evidence indicates that it is a separate species. Also, the results of Akhurst's electrophoresis study show that this nematode (NZH) is separated from *H. bacteriophora* (NC) by 41% dissimilarity.[41] The infective juveniles of *H. zealandica* are larger than those of *H. bacteriophora*, and the distance from the head to the pharynx base is greater in *H. zealandica* than in *H. bacteriophora*.

The type specimens (Holotype = male) (Allotype = infective juvenile) are deposited in the National Museum in Wellington, New Zealand.

Geographical range. From data presented by Akhurst,[41] this species occurs in New Zealand, Australia, and Europe.

Heterorhabditis megidis Poinar, Jackson, and Klein[40] (Figure 4M)

This species was collected from diseased Japanese beetle larvae (*Popillia japonica*) in Ohio.[40] Unfortunately, it was not included in the study by Akhurst,[41] but DNA analysis and morphological features separate this species from the other two heterorhabditids.[40]

The length of the infective juvenile and the distance from the head to the excretory pore separate *H. megidis* from *H. bacteriophora* and *H. zealandica*. The males of *H. megidis* possess a weakly leptoderan (pseudopeloderan) bursa which differs from the peloderan bursa found in *H. bacteriophora* and *H. zealandica*.

Geographical range. H. megidis has only been reported from its original location in the state of Ohio (U.S.).[40]

Other *Heterorhabditis* spp.

Reference to three other heterorhabditids occurs in the literature. The first is *H. hambletoni* Pereira (described in the genus *Rhabditis*) from Brazil,[44] and

---

Figure 5 (Opposite, top). Electron micrograph (TEM) cross section through the midbody region of an infective juvenile of *H. bacteriophora* showing large drops of lipid storage material (L) in the lateral and dorsal hypodermal chords and *Xenorhabdus luminescens* bacteria (arrow) in the intestinal lumen. Note strongly ridged ensheathing second-stage cuticle and paired longitudinal striae on inner third stage cuticle (photo by R. Hess-Poinar).

Figure 6 (Opposite, bottom). Electron micrograph (TEM) longitudinal section through the two subventral teeth on the tip of an infective stage *H. bacteriophora* (photo by R. Hess-Poinar).

the second is *Heterorhabditis* sp. (Nematode 41088) from Saucier, Mississippi. The latter form was never officially published and probably represents a now lost strain of *H. bacteriophora*.[45] The third is *H. hoptha* (Turco) which was originally described in the genus *Neoaplectana*.[46] The descriptions of the above species are too incomplete to assign them to any species and they should be considered as nomen dubia.

Key to the Infective Juveniles of *Heterorhabditis*

Included here are all available measurements of different strains of the species to present normal variability within the species. Average refers to the arithmetic mean of 10 individuals. All measurements are in microns. Heterorhabditid infective juveniles are illustrated in Figure 4.

1.   Average length of infective stage from 570-644 (range = 512-671); distance from head to pharynx base 125 (100-139) — *H. bacteriophora* Poinar

1′.   Average length of infective stages from 685-768 (range = 570-800); distance from head to pharynx base 147 (135-160) — 2

2.   Total length 768 (range = 736-800); distance from head to pharynx base 155 (range = 147-160) — *H. megidis* Poinar, Jackson, and Klein

2′.   Total length 685 (range = 570-740); distance from head to pharynx base 140 (range = 135-147) — *H. zealandica* Poinar

Measurements of the infective juveniles of the above three species are presented in Table 5.

IV. BIOLOGY

Because of basic patterns of similarity between the Steinernematidae and Heterorhabditidae as a result of parallel evolution, the biology of both are covered collectively.

A. Infection

Infection by both *Steinernema* and *Heterorhabditis* is initiated by a third-stage juvenile which is morphologically and physiologically adapted to remain in the environment for a prolonged period (without taking nourishment) while waiting for an insect host. The infective stages are the only survival stage in the life cycle of these nematodes. Their role is to locate a suitable host, and their well-developed amphids and head papillae probably assist them in this goal. In the absence of biotic antagonists and natural disasters, the survival of the infective stage depends on the nematode species (physiological adaptations) and physical conditions in the environment (especially temperature and moisture). Morphological adaptations for this survival period include a compaction

TABLE 5. Collective Measurements of the Infective-Stage Juveniles of *Heterorhabditis* Species[a]

Character	*bacteriophora* Poinar	*zealandica* Poinar	*megidis* Poinar, Jackson, and Klein
Total length	588(512-671)	685(570-740)	768(736-800)
Greatest width	23(18-31)	27(22-30)	29(27-32)
Distance: head to excretory pore	103(87-110)	112(94-123)	131(123-142)
Distance: head to nerve ring	85(72-93)	100(90-107)	109(104-115)
Distance: head to pharynx base	125(100-139)	140(135-147)	155(147-160)
Length tail	98(83-112)	102(87-119)	119(112-128)
Ratio A[b]	25(17-30)	25(24-26)	26(23-28)
Ratio B[c]	4.5(4.0-5.1)	4.9(4.2-5.0)	5.0(4.6-5.9)
Ratio C[d]	6.2(5.5-7.0)	6.6(6.2-6.7)	6.5(6.1-6.9)
Ratio D[e]	0.84(0.76-0.92)	0.80(0.70-0.84)	0.85(0.81-0.91)
Ratio E[f]	1.12(1.03-1.30)	1.08(1.03-1.09)	1.10(1.03-1.20)
Ratio F[g]	0.25(0.22-0.36)	0.25(0.24-0.26)	0.25(0.23-0.28)

[a] All measurements in microns, and range is given in parentheses following the average value (N = 25).
[b] Length divided by width.
[c] Length divided by distance from head to pharynx base.
[d] Length divided by tail length.
[e] Distance from head to excretory pore divided by distance from head to pharynx base.
[f] Distance from head to excretory pore divided by tail length.
[g] Width divided by tail length.

or collapse of certain body tissues. For example, the alimentary tract is essentially nonfunctional because the walls of the intestine and pharynx have closed together, thereby greatly reducing the lumen of the digestive tract (Figure 5). The mouth and anus are also closed. Ultrastructural investigations reveal the presence of numerous mitochondria (probably for supplying energy for host searching) and large amounts of food reserves stored in the enlarged hypodermal glands and intestinal cells (Figure 5). Symbiotic bacteria (*Xenorhabdus* spp.), which play an important nutritional role inside the host, are found in the alimentary tract of the infective stage. In *Steinernema*, the great majority of the bacteria are found in the modified ventricular portion of the intestine; in *Heterorhabditis*, *Xenorhabdus* is found in this location but can also occur throughout the intestinal lumen and even in the pharyngeal lumen (Figure 5).

The behavior of the infective juveniles varies from species to species. Those of *S. carpocapsae* have a tendency to search for insects on the soil surface and can be seen standing on their tails waving back and forth or bending over in preparation to "jump" or bridge to another soil particle. The larger species of *Steinernema*, *S. glaseri*, and *S. anomali*, and the heterorhabditids tend to disperse through the soil in search of hosts.

During this period of soil migration, the infective juveniles are vulnerable to attack by invertebrates and microorganisms. Under laboratory conditions infective juveniles without the enclosing second-stage cuticle will survive as long as infective juveniles which are still ensheathed, but the sheath can serve as a means of protection. Although predators such as mites, collembola, etc. can ingest ensheathed and exsheathed forms, some nematophagous fungi such as *Hirsutella* will not adhere to ensheathed steinernematid nematodes.[47]

Once contact with a potential host is made, the infective juveniles have several alternative methods of entry. Those of *Steinernema* appear to enter the host's hemocoel through natural openings (mouth, anus, or spiracles). However, *Heterorhabditis* possesses a dorsal (and sometimes, two smaller subventral) tooth or hook which can be used to break the outer cuticle of an insect and allow the infective juveniles to enter (Figures 3A, 6).[48] This method of entry is successful with the smaller, more fragile insects but is not known to occur with larger insects.

B. Symbiotic Bacteria

Presumably in Steinernematidae and Heterorhabditidae, we have two phylogenetically different lines of rhabditids that have developed similar symbiotic association with closely related bacterial species.

As soon as the infective juveniles enter the host hemocoel, they initiate development. The alimentary tract becomes functional and cells of the symbiotic *Xenorhabdus* bacteria are released through the anus and start to multiply in the insect's hemocoel. These bacteria are consumed and digested by the developing nematodes.

Successful maturation and multiplication of the nematodes, culminating in infective juvenile formation, depend on a well-established population of *Xenorhabdus* bacteria in the insect's hemolymph. The bacteria occur in two major phases; the phase one variant is most ideal for nematode development, probably because it furnishes a good source of nourishment and produces an assortment of antibiotics which prohibit the establishment of other microorganisms.[49] However, for no obvious reason, phase one will convert to a phase two variant which neither supplies as much nutritional value nor the types or amount of antibiotics as phase one. A possible explanation for this conversion is related to the survival of the bacteria. A bacteriophage recently discovered from *X. luminescens* attacked phase one and not phase two cells.[50] The presence of this bacteriophage severely affects nematode production by destroying the phase one cells.

C. Development and Reproduction

One of the major differences between *Steinernema* and *Heterorhabditis* is their development subsequent to the infective stage. In *Steinernema*, the infective juveniles develop into amphimictic females or males but never hermaphrodites. In *Heterorhabditis*, each infective juvenile develops into a hermaphroditic female and never an amphimictic female or male. However, the second generation (progeny of the initial females formed from the infective stage) consists of amphimictic females and males in both genera (Figure 7). There are some unpublished accounts of second generation amphimictic females of *Heterorhabditis* being autotokous (producing progeny without males) but this is doubtful. It is possible to have a continuous line of *Heterorhabditis* hermaphrodites, but only when an infective juvenile is formed in each generation (in this case, amphimictic females and males never occur).

Spermatozoa in both *Steinernema* and *Heterorhabditis* are the amoeboid forms characteristic of the phylum. In *Heterorhabditis*, spermatogenesis occurs in two separate locations, in the ovotestes of the first generation hermaphrodites and in the testes of the second generation males.[51] Spermatogenesis appears to be similar in both forms and is initiated by spermatogonia which undergo mitosis and mature into primary spermatocytes. Each primary spermatocyte divides into two secondary spermatocytes and each of these divides into two spermatids. The spermatids lose their residual body, reorganize their fibrous bodies, develop a condensed nucleus and pseudopod, and become mature spermatozoa. The hermaphroditic females accumulate spermatozoa at the junction of the ovotestes and oviduct, whereas in the amphimictic female spermatozoa accumulated in the seminal receptacle between the uterus and oviduct.[52] Spermatozoan development in *Steinernema* is similar to that in the male of *Heterorhabditis*; however, individual spermatozoa are larger and they often group together in chains.[32]

Since no mating occurs with the hermaphrodites of *Heterorhabditis*, the vulva is only used for the exit of eggs from the reproductive tract. Usually only a portion of the eggs are oviposited by the hermaphrodite female, whereas the remainder hatch inside the uterus (ovoviviparous), and the juveniles destroy their mother as they develop. Where mating occurs with amphimictic females of *Heterorhabditis* and *Steinernema*, spermatozoon is passed through the vulva. In *Heterorhabditis*, the amphimictic females appear to be completely ovoviviparous and after mating the vulva is nonfunctional. All *Steinernema* females deposit some eggs in the initial stages (oviparous) but become ovoviviparous later in their development.

V. SPECIATION AND COEVOLUTION

If the steinernematids and heterorhabditids evolved with some of the non-insect arthropods such as the Diplopoda and Arachnida, they could have appeared back in the Devonian, some 375 million years ago. Experimental infection of some present day arachnids by these nematodes supports the

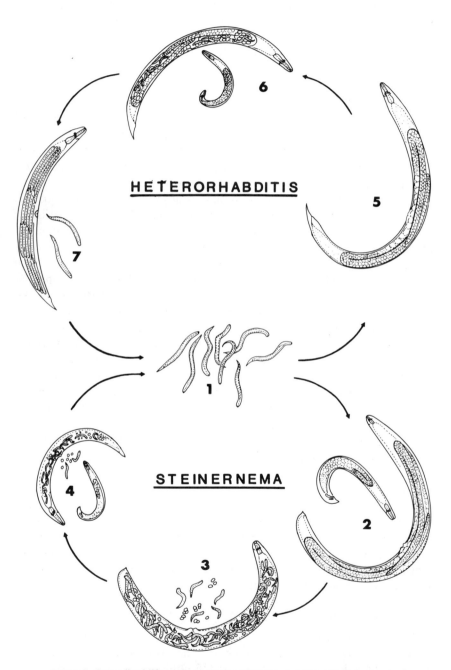

Figure 7. Generalized life cycle of species of *Heterorhabditis* and *Steinernema*.

possibility. On the other hand, if these nematodes evolved with the Insecta — in nature they have only been found parasitizing these arthropods — then they probably arose later, when insects were present. This could have still been as early as the Carboniferous (some 300 million years ago) when the orthopteroid groups made their first appearance (Protorthoptera, Blattodea, Caloneurodea, Orthoptera, and Palaeodictyoptera).[53]

Natural populations of *Steinernema* have been recovered from orthopteroid insects (Gryllotalpidae). It is not known whether this is a recent adaptation from forms that parasitized an Endopterygota or whether these are relic forms that evolved with the Orthoptera. Certainly the Orthoptera are the most primitive order of insects that have been found naturally infected with these rhabditoid nematodes. Moreover, together with related but now extinct orders, they were probably the first terrestrial insects physically large enough to support a population of these purely terrestrial nematodes. It should be noted that most populations of steinernematids and heterorhabditids have been recovered from terrestrial, holometabolous insects. If they evolved with the holometabolous forms, they could have a possible origin at some point in the Permian.

Due to recognition of basic differences between *Steinernema* and *Heterorhabditis* they are presumed to be derived from two distinct ancestral lineages. Though the two families are considered diphyletic, their similar physical environment (soil) and host choice (insects) have apparently resulted in the almost identical life histories of these nematodes.[54]

Geographical separation of the nematodes and adaptation to specific hosts would have led to adaptive radiation, and thus the formation of new species and strains. Nematode host seeking could have become focused on particular insect groups with the insect-produced attractants and their physical location in the soil serving as selective keys to host location. Further specializations could have occurred regarding the method of entering these hosts, and dealing with their possible defense mechanisms. The nematodes could have further adapted to be active at temperatures and moisture levels which support the host, to survive and move in a particular soil type, and even to avoid predatory or pathogenic organisms.

Two examples show host specialization in *Steinernema*. In the first, New Zealand (South Island) populations of *S. feltiae* occur at the bases of tussock grass, where they have become adapted to parasitize lepidopterous larvae (Noctuidae and Hepialidae) that feed on the roots of these grasses. These strains of *S. feltiae* are not effective against native scarabaeid larvae living in the same habitat. In addition, these nematode strains have become tolerant of low temperature and can infect and develop at 5-7°C.[55,56] Danish populations of this same nematode species have specialized in the parasitization of bibionid flies.[23] In the second example, *Steinernema scapterisci* is well adapted to parasitize mole crickets (Gryllotalpidae) in specific habitats in Uruguay, but develops poorly in *Galleria* larvae,[37] a host which normally supports the development of all *Steinernema* and *Heterorhabditis* species. Adaptation to a specific host group may be more normal than is commonly recognized.

VI. BIOGEOGRAPHY

Although not all of the species and strains of these rhabditoids have been discovered, some interesting biogeographic patterns are beginning to emerge. Distribution records show that both *Steinernema* and *Heterorhabditis* occur on all the continents with the exception of the Antarctic.

A major problem regarding their current distribution stems from determining whether a given population is present as a result of natural dispersal or human introduction. The dumping of soil used as ballast in the early sailing ships, the introduction of potted plants, or the previous field application of nematodes from foreign lands might explain how populations of *S. feltiae* and *S. carpocapsae* reached Australia and New Zealand. With such types of transport, hardy species (long-lived, resistant to temperature extremes and desiccation) would survive. Certainly *S. feltiae* would fall into this category. Once established, these nematodes could become further dispersed by adult insects. Glaser and Farrell[8] were the first to note that the infective stage of *S. glaseri* could be carried on and within adult insects (Japanese beetles) before the hosts perished from the infection. More recently, Timper et al.[57] showed that adult Lepidoptera (the beet armyworm) could also carry steinernematid nematodes. It is conceivable that this is how populations of *S. feltiae* became established on off-shore islands in Australia.

VII. SAFETY

Although *Steinernema* and *Heterorhabditis* are considered entomopathogenic nematodes because all natural infections have involved insects, they do have the ability to enter and kill noninsect invertebrates, at least in the laboratory.[58] The following basic steps have to be completed for the nematode to successfully complete its life cycle: (1) penetration into the body cavity of the potential host, (2) release of *Xenorhabdus* bacteria, (3) development to mature adults, and (4) mating and production of infective juveniles. These steps are normally completed in most insects, but various obstacles to their completion occur with many noninsect invertebrates.[58]

The effect of these nematodes on vertebrates is a greater concern. Of all vertebrates thus far challenged with infective stages of *Steinernema* and *Heterorhabditis* (see Table 6), only young tadpoles of frogs and toads were vulnerable. The infective stages of both genera penetrated into the coelom of the tadpole and attempted to develop. Although nematode maturation is rarely achieved, the tadpoles are killed by foreign bacteria introduced with the nematodes.

Recently, five bacterial isolates from humans displayed similarities with *Xenorhabdus luminescens*.[59] This curious discovery shows that terrestrial luminescent bacteria do occur in the soil environment. Recent data (molecular,

TABLE 6. Vertebrates Challenged with *Steinernema* and *Heterorhabditis* Infective Juveniles

Class	Host	Nematode	Method of application	Dose (no. nematodes)	Effect	Ref.
Pisces	*Lebistes reticulatus*	*S. carpocapsae*	in water	200,000	none	67
Reptilia	*Anolis marmoratus*	*S. carpocapsae*	oral	50,000	none	67
	Chelydra serpentina	*S. carpocapsae*	oral	3,000	none	68
		H. bacteriophora	oral	3,000	none	68
Amphibia	*Bufo marinus*	*S. carpocapsae*	in water	100/ml; 200/ml	mortality of young tadpoles	67
	Hyla regilla	*S. carpocapsae*	in water of young tadpoles	400/ml	mortality of	69
		H. bacteriophora	in water	100/ml; 200/ml; 400/ml	mortality of young tadpoles	69
	Xenopus laevis	*S. carpocapsae* 400/ml;	in water of young tadpoles	200/ml; 800/ml	mortality	69
		H. bacteriophora	in water	200/ml; 400/ml; 800/ml	mortality of young tadpoles	69
Aves	*Gallus gallus*	*S. carpocapsae*	oral	60,000	none	67
Mammalia	*Microtus pennsylvanicus*	*S. carpocapsae*	oral	200,000	none	70

TABLE 6 (continued). Vertebrates Challenged with *Steinernema* and *Heterorhabditis* Infective Juveniles

Class	Host	Nematode	Method of application	Dose (no. nematodes)	Effect	Ref.
	Rattus rattus	*S. carpocapsae*	oral	—	none	2
		S. glaseri	interperiton-eally	5-6,000	some nematode viability after 5 days	71
		S. carpocapsae	interperiton-eally	50,000	some nematode viability after 2 days	72
	mice	*S. carpocapsae*	oral	50,000	none	72
		S. carpocapsae	subcutan-eously	1000	no development	73
		H. bacteriophora	subcutan-eously	1000	no development	69
	rats	*S. glaseri*	orally, abdominally	—	none	74
	rabbits	*S. glaseri*	orally, abdominally, orbital cavity	—	none	75
	Rhesus monkey	*S. glaseri*	orally, abdominally, respiratory tract	—	none	76

mice				
S. carpocapsae (over 25 days)	oral	62,500	no development	67
S. carpocapsae	oral	1000; 10,000	no development	77
S. feltiae	oral	1000; 10,000	no development	77
S. glaseri	oral	1000; 10,000	no development	77
H. bacteriophora	oral	1000; 10,000	no development	77
S. carpocapsae	subcutaneously	20,000; 100,000	ulcers in skin, no development	77
S. feltiae	subcutaneously	20,000; 100,000	no development	77
S. carpocapsae	interperitoneally	20,000; 100,000	no development	77
S. feltiae	interperitoneally	20,000; 100,000	no development	77

cultural, and physiological) indicate that despite the resemblance of the luminescent human isolates to the nematode symbiont, they certainly belong to a separate species, if not a separate genus of bacteria.

VIII. CONCLUSION

Now that steinernematids and heterorhabditids have demonstrated their potential for insect control, the number of researchers studying them will increase considerably in the next decade. In working with these nematodes, it is important to record and report the species and strain used and to maintain records of species introduced in areas where they are not considered native.

REFERENCES

1. **Poinar, G. O., Jr.,** *Entomogenous Nematodes,* E. J. Brill, Leiden, 1975.
2. **Poinar, G. O., Jr.,** *Nematodes for Biological Control of Insects,* CRC Press, Boca Raton, FL, 1979.
3. **Steiner, G.,** *Aplectana kraussei* n. sp., einer in der Blattwespe *Lyda* sp. parasitierende Nematodenform, nebst Bemerkungen über das Seitenorgan der parasitischen Nematoden, *Zentralbl. Bakteriol. Parasitenkd. Infektionskr. Hyg. Abt.,* 2, 59, 14, 1923.
4. **Steiner, G.,** *Neoaplectana glaseri,* n.g., n. sp. (Oxyuridae), a new nemic parasite of the Japanese beetle (*Popillia japonica* Newm.), *J. Wash. Acad. Sci.,* 19, 436, 1929.
5. **Glaser, R. W.,** The cultivation of a nematode parasite of an insect, *Science,* 73, 614, 1931.
6. **Glaser, R. W.,** The bacterial-free culture of a nematode parasite, *Proc. Soc. Exp. Biol. Med.,* 43, 512, 1940.
7. **Glaser, R. W.,** A pathogenic nematode of the Japanese beetle, *J. Parasitol.,* 18, 199, 1932.
8. **Glaser, R. W., and Farrell, C. C.,** Field experiments with the Japanese beetle and its nematode parasite, *J. N. Y. Entomol. Soc.,* 43, 345, 1935.
9. **Glaser, R. W., McCoy, E. E., and Girth, H. B.,** The biology and economic importance of a nematode parasitic in insects, *J. Parasitol.,* 26, 479, 1940.
10. **Wouts, W. M., Mrácek, Z., Gerdin, S., and Bedding, R. A.,** *Neoaplectana* Steiner, 1929 a junior synonym of *Steinernema* Travassos, 1927 (Nematoda, Rhabditida), *Syst. Parasitol.,* 4, 147, 1982.
11. **Filipjev, I. N.,** Miscellanea Nematologica 1. Eine neue Art der Gattung *Neoaplectana* Steiner nebst Bermerkungen über die systematische Stellung der letzteren, *Mag. Parasitol. Instit. Zool. Acad. USSR,* 4, 229, 1934.
12. **Kozodoi, E. M., Voronov, D. A., and Spiridonov, S. E.,** New data on taxonomic studies of *Neoaplectana feltiae* (Nematoda, Rhabditida), *Zool. Zhur.,* 66, 980, 1987.
13. **Stanuszek, S.,** *Neoaplectana feltiae* (Filipjev, 1934) - a facultative parasite of the caterpillars of Agrotinae in Poland, *Proc. 9th Int. Symp. Eur. Soc. Nematol.,* Warsaw, August 1967, 1970, 355.
14. **Stanuszek, S.,** *Neoaplectana feltiae* complex (Nematoda: Rhabditoidea, Steinernematidae) its taxonomic position within the genus *Neoaplectana* and intraspecific structure, *Zesz. Probl. Postepów Nauk Roln.,* 154, 331, 1974.
15. **Stanuszek, S.,** *Neoaplectana feltiae pieridarum,* n. ecotype (Nematoda: Rhabditoidea, Steinernematidae) - a parasite of *Pieris brassicae* L. and *Mamestra brassicae* L. in Poland, Morphology and biology, *Zesz. Probl. Postepów Nauk Roln.,* 154, 361, 1974.
16. **Poinar, G. O., Jr.,** Examination of the neoaplectanid species *feltiae* Filipjev, *carpocapsae* Weiser and *bibionis* Bovien (Nematoda: Rhabditida), *Rev. Nématol.,* 12, 375, 1989.

17. **Sha, C-Y.**, A comparative analysis of esterase of the insect parasitic nematodes of the genus *Neoaplectana*, *Acta Zootax. Sin.*, 10, 246, 1985.

18. **Kozodoi, E. M., and Spiridonov, S. E.**, Cuticular ridges on lateral fields of invasive larvae of *Neoaplectana* (Nematoda: Steinernematidae), *Folia Parasitol.* (Praha), 35, 359, 1988.

19. **Mrácek, Z.**, *Steinernema kraussei*, a parasite of the body cavity of the sawfly, *Cephaleia abietis*, in Czechoslovakia, *J. Invertebr. Pathol.*, 30, 87, 1977.

20. **Wouts, W. M.**, Biology, life cycle and re-description of *Neoaplectana bibionis* Bovien, 1937 (Nematoda: Steinernematidae), *J. Nematol.*, 12, 62, 1980.

21. **Poinar, G. O., Jr.**, Generation polymorphism in *Neoaplectana glaseri* Steiner (Steinernematidae: Nematoda), re-described from *Strigoderma arboricola* (Fab.) (Scarabaeidae: Coleoptera) in North Carolina, *Nematologica*, 24, 105, 1978.

22. **Artyukhovsky, A. K.**, *Neoaplectana arenaria* nov. sp. (Steinernematidae, Nematoda) agent of a nematode disease in the May beetle in the Voronez Province, *Tr. Voronezh. Gos. Zapov.*, 15, 94, 1967.

23. **Bovien, P.**, Some types of association between nematodes and insects. *Vidensk. Medd. Dan. Naturahist. Foren. Khobenhavn*, 101, 1, 1937.

24. **Poinar, G. O., Jr.**, Re-description of *Neoaplectana affinis* Bovien (Rhabditida: Steinernematidae), *Rev. Nématol.*, 11, 143, 1988.

25. **Weiser, J.**, *Neoaplectana carpocapsae* n. sp. (Anguillulata, Steinernematinae), novy cizopasnik housenik obalece jablecného, *Carpocapsa pomonella* L., *Vestn. Cesk. Spol. Zool.*, 19, 44, 1955.

26. **Poinar, G. O., Jr.**, Description and taxonomic position of the DD-136 nematode (Steinernematidae, Rhabditoidea) and its relationship to *Neoaplectana carpocapsae* Weiser, *Proc. Helminthol. Soc. Wash.*, 34, 199, 1967.

27. **Veremchuk, G. V.**, A new species of *Neoaplectana* (Rhabditida: Steinernematidae), pathogenic to insects, *Parazitologiya*, 3, 249, 1969.

28. **Poinar, G. O., Jr., and Veremchuk, G. V.**, A new strain of nematodes pathogenic to insects and the geographical distribution of *Neoaplectana carpocapsae* Weiser (Rhabditida, Steinernematidae), *Zool. Zhur.*, 49, 966, 1970.

29. **Kozodoi, E.**, A new entomopathogenic nematode *Neoaplectana anomali* sp. n. (Rhabditida, Steinernematidae) and observations on its biology, *Zool. Zhur.*, 63, 1605, 1984.

30. **Poinar, G. O., Jr., and Kozodoi, E. M.**, *Neoaplectana glaseri* and *N. anomali*: sibling species or parallelism?, *Rev. Nematol.*, 11, 13, 1988.

31. **Poinar, G. O., Jr.**, *Neoaplectana intermedia* n. sp. (Steinernematidae: Nematoda) from South Carolina, *Rev. Nématol.*, 8, 321, 1985.

32. **Hess, R. T., and Poinar, G. O., Jr.**, Sperm development in the nematode *Neoaplectana intermedia* (Steinernematidae: Rhabditida), *J. Submicrosc. Cytol. Pathol.*, 21, 543, 1989.

33. **Doucet, M. M. A. de**, A new species of *Neoaplectana* Steiner, 1929 (Nematoda: Steinernematidae) from Córdoba, Argentina, *Rev. Nématol.*, 9, 317, 1986.

34. **Poinar, G. O., Jr., Mrácek, Z., and Doucet, M. M. A. de**, A re-examination of *Neoaplectana rara* Doucet, 1986 (Steinernematidae: Rhabditida), *Rev. Nématol.*, 11, 447, 1988.

35. **Mamiya, Y.**, *Steinernema kushidai* n. sp. (Nematoda: Steinernematidae) associated with scarabaeid beetle larvae from Shizuoka, Japan, *Appl. Entomol. Zool.*, 23, 313, 1988.

36. **Nguyen, K. B., and Smart, G. C., Jr.**, *Steinernema scapterisci* n. sp. (Steinernematidae: Nematoda), *J. Nematol.*, 22, 187, 1990.

37. **Nguyen, K. B., and Smart, G. C., Jr.**, A new steinernematid nematode from Uruguay, *J. Nematol.*, 20, 651, 1988.

38. **Poinar, G. O., Jr.**, Description and biology of a new insect parasitic rhabditoid, *Heterorhabditis bacteriophora* n. gen. n. sp. (Rhabditida; Heterorhabditidae N. Fam.), *Nematologica*, 21, 463, 1975.

39. **Khan, A., Brooks, W. W., and Hirschmann, H.**, *Chromonema heliothidis* n. gen., n. sp. (Steinernematidae, Nematoda), a parasite of *Heliothis zea* (Noctuidae, Lepidoptera), and other insects, *J. Nematol.*, 8, 159, 1976.

40. **Poinar, G. O., Jr., Jackson, T., and Klein, M.,** *Heterorhabditis megidis* sp. n. (Heterorhabditidae: Rhabditida), parasitic in the Japanese beetle, *Popillia japonica* (Scarabaeidae: Coleoptera), in Ohio, *Proc. Helminthol. Soc. Wash.*, 54, 53, 1987.

41. **Akhurst, R. J.,** Use of starch gel electrophoresis in the taxonomy of the genus *Heterorhabditis* (Nematoda: Heterorhabditidae), *Nematologica*, 33, 1, 1987.

42. **Poinar, G. O., Jr., and Georgis, R.,** Description and field application of the HP88 strain of *Heterorhabditis bacteriophora*, *Rev. Nématol.*, 1990, in press.

43. **Wouts, W. M.,** The biology and life cycle of a New Zealand population of *Heterorhabditis heliothidis* (Heterorhabditidae), *Nematologica*, 25, 191, 1979.

44. **Pereira, C.,** *Rhabditis hambletoni* n. sp. nema apparentemente semiparasito da "broca do algodoeiro" (*Gasterocercodes brasiliensis*), *Arch. Inst. Biol.* (Sao Paulo), 8, 25, 1937.

45. **Littig, K. S., and Swain, R. B.,** Studies on nematode 41088, a nematode parasite of the white-fringed beetles, White-fringed Beetle Investigations, unpubl. rept., Gulfport, 1943.

46. **Turco, C. P.,** *Neoaplectana hoptha*, sp. n. (Neoaplectanidae, Nematoda), a parasite of the Japanese beetle, *Popillia japonica* Newm., *Proc. Helminthol. Soc. Wash.*, 37, 119, 1970.

47. **Timper, P., and Kaya, H. K.,** Role of the second-stage cuticle of entomogenous nematodes in preventing infection by nematophagous fungi, *J. Invertebr. Pathol.*, 54, 314, 1989.

48. **Bedding, R. A., and Molyneux, A. S.,** Penetration of insect cuticle by infective juveniles of *Heterorhabditis* spp. (Heterorhabditidae: Nematoda), *Nematologica*, 28, 354, 1982.

49. **Akhurst, R. J.,** Morphological and functional dimorphism in *Xenorhabdus* spp., bacteria symbiotically associated with the insect pathogenic nematodes, *Neoaplectana* and *Heterorhabditis*, *J. Gen. Microbiol.*, 121, 303, 1980.

50. **Poinar, G. O., Jr., Hess, R. T., Lanier, W., Kinney, S., and White, J. H.,** Preliminary observations of a bacteriophage infecting *Xenorhabdus luminescens* (Enterobacteriaceae), *Experientia*, 45, 191, 1989.

51. **Poinar, G. O., Jr., and Hess, R.,** Spermatogenesis in the insect-parasitic nematode, *Heterorhabditis bacteriophora* Poinar (Heterorhabditidae: Rhabditida), *Rev. Nématol.*, 8, 357, 1985.

52. **Hess, R. T., and Poinar, G. O., Jr.,** Ultrastructure of the genital ducts and sperm behavior in the insect parasitic nematode, *Heterorhabditis bacteriophora* Poinar (Heterorhabditidae: Rhabditida), *Rev. Nématol.*, 9, 141, 1986.

53. **Carpenter, F. M., and Burnham, L.,** The geological record of insects, *Annu. Rev. Earth Planet. Sci.*, 13, 297, 1985.

54. **Poinar, G. O., Jr.,** *The Natural History of Nematodes*, Prentice Hall, Englewood Cliffs, NJ, 1983.

55. **Wright, P. J., and Jackson, T. A.,** Low temperature activity and infectivity of a parasitic nematode against porina and grass grub larvae, *Proc. 41st. N.Z. Weed and Pest Control Conference*, 138, 1988.

56. **Wright, P. J.,** Selection of entomogenous nematodes from the families Steinernematidae and Heterorhabditidae (Nematoda) to control grass grub and Porina larvae in pasture, Ph.D. thesis, Lincoln College, 1989.

57. **Timper, P., Kaya, H. K., and Gaugler, R.,** Dispersal of the entomogenous nematode *Steinernema feltiae* (Rhabditida: Steinernematidae) by infected adult insects, *Environ. Entomol.*, 17, 546, 1988.

58. **Poinar, G. O., Jr.,** Non-insect hosts for the entomogenous rhabditoid nematodes *Neoaplectana* (Steinernematidae) and *Heterorhabditis* (Heterorhabditidae), *Rev. Nématol.*, 12, 423, 1989.

59. **Farmer, III., J. J., Jorgensen, J. H., Grimont, P. A. D., Akhurst, R. J., Poinar, G. O., Jr., Ageron, E., Pierce, G. V., Smith, J. A., Carter, G. P., Wilson, K. L., and Hickman-Brenner, F. W.,** *Xenorhabdus luminescens* (DNA hybridization group 5) from human clinical specimens, *J. Clin. Microbiol.*, 27, 1594, 1989.

60. **Poinar, G. O., Jr.,** Recognition of *Neoaplectana* species (Steinernematidae: Rhabditida), *Proc. Helminthol. Soc. Wash.*, 53, 121, 1986.

61. Mrácek, Z., Gut, J., and Gerdin, S., *Neoaplectana bibionis* Bovien, 1937, an obligate parasite of insects isolated from forest soil in Czechoslovakia, *Folia Parasitol. (Praha)*, 29, 139, 1982.

62. Pizano, M. A., Aguillera, M. M., Monteiro, A. R., and Ferroy, L. C. C. B., Incidencia de *Neoaplectana glaseri* Steiner, 1929 (Nematoda: Steinernematidae) parasitando ovo de *Migdolus fryanus* (Westwood, 1863) (Col: Cerambycidae), *Soc. Brasil. Nematol. 9th Reun. Piracicaba*, 1, 1985.

63. Curran, J., Chromosome numbers of *Steinernema* and *Heterorhabditis* species, *Rev. Nematol.*, 12, 145, 1989.

64. Doucet, M. M. A. de, and Poinar, G. O., Jr., Estudio del ciclo de vida de una ipoblacion de *Heterorhabditis* sp. proveniente de Rio Cuarto, Provincia de Cordoba, *Rev. UNRS*, 2, 145, 1989.

65. Creighton, C. S., and Fassuliotis, G., *Heterorhabditis* sp. (Nematoda: Heterorhabditidae): a nematode parasite isolated from the banded cucumber beetle *Diabrotica balteata*, *J. Nematol.*, 17, 150, 1985.

66. Hernandez, E. M., and Mrácek, Z., *Heterorhabditis heliothidis*, a parasite of insect pest in Cuba, *Folia Parasitol. (Praha)*, 31, 11, 1984.

67. Kermarrec, A., and Mauleon, H., Nocuite potentielle du nematode entomoparasite *Neoaplectana carpocapsae* Weiser pour le crapaud *Bufo marinus* aux Antilles, *Meded. Fac. Landbouw. Rijksuniv. Gent.*, 50, 831, 1985.

68. Poinar, G. O., Jr., and Miller, R. W., unpublished data, 1989.

69. Poinar, G. O., Jr., and Thomas, G. M., Infection of frog tadpoles (Amphibia) by insect parasitic nematodes (Rhabditida), *Experientia*, 44, 528, 1988.

70. Schmiege, D. C., The feasibility of using a neoaplectanid nematode for control of some forest insect pests, *J. Econ. Entomol.*, 56, 427, 1963.

71. Jackson, G. J., and Bradbury, P. C., Cuticular fine structure and molting of *Neoaplectana glaseri* (Nematoda), after prolonged contact with rat peritoneal exudate, *J. Parasitol.*, 56, 108, 1970.

72. Gaugler, R., and Boush, G. M., Nonsusceptibility of rats to the entomogenous nematode, *Neoaplectana carpocapsae*, *Environ. Entomol.*, 8, 658, 1979.

73. Poinar, G. O., Jr., Thomas, G. M., Presser, S. B., and Hardy, J. L., Inoculation of entomogenous nematodes, *Neoaplectana* and *Heterorhabditis* and their associated bacteria, *Xenorhabdus* spp. into chicks and mice, *Environ. Entomol.*, 11, 137, 1982.

74. Wang, J. X., Qui, L. H., and Liu, Z. M., The safety of the nematode, *Neoaplectana glaseri* Steiner, to vertebrates. I. A test on rats, *Nat. Enemies Insects*, 5, 240, 1983.

75. Wang, J. X., and Liu, Z. M., The safety of the nematode, *Neoaplectana glaseri* Steiner to vertebrates. II. A test on rabbits, *Nat. Enemies Insects*, 5, 240, 1983.

76. Wang, J. X., Chen, Q. S., and Huang, J. T., The safety of the nematode *Neoaplectana glaseri* Steiner to vertebrates. III. A test on monkeys, *Mucaca mulatta*, *Nat. Enemies Insects*, 6, 41, 1984.

77. Kobayashi, M., Okano, H., and Kirihara, S., The toxicity of steinernematid and heterorhabditid to the male mice, in *Recent Advances in Biological Control of Insect Pests by Entomogenous Nematodes in Japan*, Ishibashi, N., Ed., Ministry of Education, Japan, Grant No. 5986005, 1987, 153.

3. Molecular Techniques in Taxonomy

John Curran

I. INTRODUCTION

The rapid development of DNA-based taxonomic techniques has provided a fresh impetus to the use of molecular techniques in systematics in the 1980s, and these technologies are being increasingly applied to the resolution of taxonomic problems in economically important nematode groups.[1] The aims of this review are to (1) identify taxonomic problems which can be most effectively resolved by molecular techniques, (2) summarize the available data and literature, and (3) evaluate current molecular methods and identify which techniques would practically answer these taxonomic problems.

II. TAXONOMIC PROBLEMS IN ENTOMOPATHOGENIC NEMATODES

In the Rhabditoidea, entomopathogenic nematodes form the two families Steinernematidae and Heterorhabditidae, each family with one genus: *Steinernema* and *Heterorhabditis*, respectively. Within *Steinernema*, species have been recognized by morphological differences, supplemented by cross-breeding tests, and despite some problems of misidentification and nomenclature, there has been little difficulty in species recognition. In contrast, within *Heterorhabditis* it is arduous to obtain cross-breeding data, and there has been difficulty in defining taxonomically useful morphological characters. Although the taxonomy of *Heterorhabditis* spp. will likely be founded on morphological studies, such approaches as protein electrophoresis and DNA sequence analysis can provide supplementary information useful for delimiting species before conducting morphological analysis and constructing of phylogenies and derived classification systems.

The major role for molecular techniques in entomopathogenic nematode taxonomy is the identification of sibling species, subspecies, and other intraspecific groupings. Besides contributing to understanding the biology and evolution of entomopathogenic nematodes, this role may become of critical importance in identifying species and isolates for registration, quarantine, and proprietary protection purposes. The ability to identify individual infective-stage nematodes would also be of considerable benefit (e.g., in ecological studies following the fate of inundatively or inoculatively released isolates of entomopathogenic nematodes).

III. MOLECULAR TECHNIQUES AND NEMATODE TAXONOMY

The taxonomy of the genus *Steinernema* (=*Neoaplectana*) and several nominal species has been in a state of flux. Without knowing the precise origins and culture history of the various isolates referred to in the literature, it is difficult to cross-reference the various designations. I cannot offer a resolution of the matter in this review, but it does highlight the potential problems that may arise from misidentification of nematode isolates and lack of a universal nomenclature system for isolates. To avoid further confusion and provide ready reference back to the original publications, the species and isolate designations given are those of the original authors, followed in parentheses by Poinar's[2] revised nomenclature.

A. Protein Electrophoresis

The first application of molecular techniques to the study of variation in entomopathogenic nematodes was an examination of isozyme differences between different cultures and life stages of *Neoaplectana glaseri* (*Steinernema glaseri*) and *N. carpocapsae* (*S. carpocapsae*).[3] This study demonstrated that there were qualitative and quantitative differences in the esterases and alkaline phosphatase of *N. glaseri* (*S. glaseri*) and *N. carpocapsae* (*S. carpocapsae*). Tests for alkaline phosphatase and lactic dehydrogenase were negative for both species. No qualitative isozyme differences were detected between different life stages or could be attributed to different culture media, but this may reflect the limited number of enzyme systems studied. There were some quantitative differences in acid phosphatase activity between life stages of *N. glaseri* (*S. glaseri*) with infective-stage juveniles having lower activity than a mixed population of adults and developing juveniles.

The taxonomic application of protein techniques to entomopathogenic nematodes was not pursued until over 20 years later, when a comparison of esterases of nematodes of the genus *Neoaplectana* (*Steinernema*) was published.[4] Qualitative differences were noted between the esterase patterns of *N. bibionis* (*S. feltiae*), and *N. glaseri* (*S. glaseri*) and a group of three nematode isolates: *N. feltiae* (*S. carpocapsae*) and two isolates of *N. carpocapsae* (*S. carpocapsae*) designated DD-136 and DD-136-1-8. The common esterase pattern shared by these nematodes was taken as evidence supporting Stanusek's view[5] that *N. feltiae* (*S. carpocapsae*) was a strain of *N. carpocapsae* (*S. carpocapsae*). Alternatively, the *N. feltiae* (*S. carpocapsae*) found by Stanusek could have been in fact a reisolate of *N. carpocapsae* (*S. carpocapsae*) (as discussed below).[6] Differences in esterase pattern were noted between juvenile and adult stages within *N. bibionis* (*S. feltiae*), *N. feltiae* (*S. carpocapsae*), and *N. carpocapsae* (*S. carpocapsae*) (Figure 1).

Kozodoi et al.[7] combined analysis of total protein patterns, isozyme patterns, morphology, and cross-breeding experiments to elucidate the relationships between ten isolates of *Neoaplectana* (*Steinernema*). Cross-breeding identified four genetically separate groups, which included the following named isolates: (1) *N. anomali* (*Steinernema anomali*), (2) *N. carpocapsae* (*S. carpocapsae*), (3) *N. feltiae* (*S. feltiae*), *N. bibionis* (*S. feltiae*), *Neoaplectana*

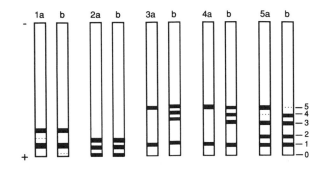

Figure 1. Comparison of adult and larval esterase patterns of five isolates of *Steinernema*, following disc gel electrophoresis (redrawn from Sha, C-Y.[4]). 1. *N. bibionis* (*S. feltiae*), 2. *N. glaseri* (*S. glaseri*), 3. *N. feltiae* (*S. carpocapsae*), 4. *N. carpocapsae* (*S. carpocapsae*) DD-136-1-8, 5. *N. carpocapsae* (*S. carpocapsae*) DD-136; a = juvenile, b = adult.

sp. 1 (*S. feltiae*), and (4) *Neoaplectana* sp. 2 (*Steinernema* sp. 2). Examination of total protein patterns by density gradient polyacrylamide gel electrophoresis placed the nematode isolates in the same groups, as did isozyme patterns for nonspecific esterases and alkaline phosphatase (Figure 2a-c). There was marked ontogenetic variation in alkaline phosphatase of *N. anomali* (*S. anomali*). In this study the *Neoaplectana* (*Steinernema*) isolates had previously been separated into the same four groups on the basis of morphology. The confirmation of isolate identity by three independent genetic means validates the diagnostic morphological characters used and adds considerable weight to their taxonomic utility. These data also raised a number of questions regarding species designation of laboratory isolates of *Neoaplectana* (*Steinernema*). It supported Poinar's[6] view that Stanusek had reisolated *N. carpocapsae* (*S. carpocapsae*) not *N. feltiae* (*S. feltiae*) and that isolates designated *N. bibionis* (*S. feltiae*) and *N. feltiae* (*S. feltiae*) were freely interbreeding and therefore conspecific. The resolution of this nomenclature problem has been addressed by Poinar.[2]

Differences in total proteins, malate dehydrogenase, esterases, and acid phosphatase have been reported between *N. glaseri* (*S. glaseri*) and *N. anomali* (*S. anomali*) and could be used to separate these morphologically very similar species. There was no variation in these characters between two isolates of *N. anomali* (*S. anomali*) from different geographical locations.[8]

In a study aimed at assigning *Heterorhabditis* isolates to groups prior to morphological analysis, Akhurst examined the isozyme patterns for ten enzymes of 23 *Heterorhabditis* isolates.[9] The potential problems of ontogenetic variation noted in all analyses of *Steinernema* isozymes[3,4,7] were avoided by examining the same ontogenetic stage for each isolate, the infective juvenile. This detailed analysis of isozyme patterns allowed the construction of a dissimilarity matrix and cluster analysis which assigned the 23 isolates to 3-8 groupings (=species) of *Heterorhabditis* (Figure 3).

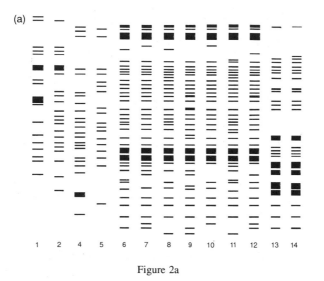

Figure 2a

Figure 2. Total protein, nonspecific esterase, and alkaline phosphatase isozyme patterns of *Steinernema* spp. (redrawn from Kozodoi et al.[7]). (a) Total protein pattern following electrophoresis in a 3-30% concentration gradient polyacrylamide gel and staining with SPV250 stain; (b, opposite page, top) nonspecific esterases; (c, opposite page, middle) alkaline phosphatase. Lane 1. *Galleria mellonella*, 2. females of *N. anomali* (*S. anomali*), 3. males of *N. anomali* (*S. anomali*), 4. infective-stage juveniles of *N. anomali* (*S. anomali*), 5. *Neoaplectana* (*Steinernema*) sp. 2, 6. *Neoaplectana* (*Steinernema*) sp. 1 (daughter culture), 7. *Neoaplectana* (*Steinernema*) sp. 1, 8. *N. feltiae* (*S. feltiae*) from Latvia, 9. *N. feltiae* (*S. feltiae*) from the Transcarpathian Region, 10. *N. feltiae* (*S. feltiae*) from Ryazan Province, 11. *N. bibionis* (*S. feltiae*), 12. *N. feltiae* (*S. feltiae*) from the Czechoslavic Socialist Republic, 13. *N. carpocapsae* (*S. carpocapsae*) from the Primorsk region, 14. *N. carpocapsae* (*S. carpocapsae*) Agriotos. In (a), Lane 3 was not included in original data.

Figure 3 (Opposite page, bottom). Dendogram derived from dissimilarity matrix (Manhattan distance) based on isozyme data for *Heterorhabditis* isolates (from Akhurst[9]). NC* = includes *Heterorhabditis* isolates NC 14, NC 69, NC 127, NC 162, NC 200, NC 323, NC 405, and NC 476; HB1 = *H. bacteriophora*; NC1 = *H. heliothidis* (*H. bacteriophora*); other isolate designations represent *Heterorhabditis* isolates from around the world.

B. Immunological Techniques

Gel diffusion and immunoelectrophoresis techniques, which utilized antisera raised against nematode homogenates injected into rabbits, were capable of discriminating three nematode isolates: *N. glaseri* (*S. glaseri*), *N. carpocapsae* (*S. carpocapsae*), and *N. dutkyi* DD-136 (*S. carpocapsae*).[10]

C. DNA Sequence Analysis

DNA sequence divergence between species of entomopathogenic nematodes was first studied by visualization (in ethidium bromide stained agarose gels) of repetitive DNA restriction fragment length differences (RFLDs) in

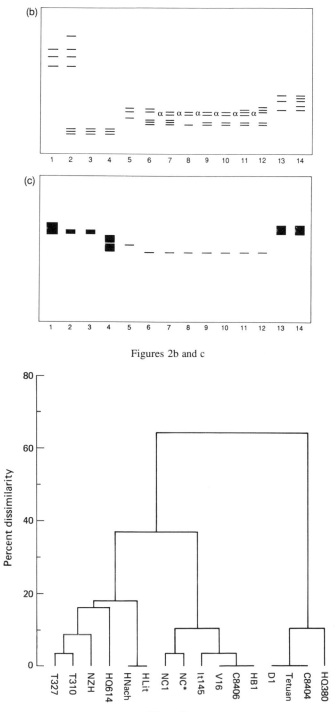

Figures 2b and c

Figure 3

total genomic DNA of *S. glaseri, S. feltiae (S. carpocapsae), Steinernema* spp., and *Heterorhabditis* isolates.[11] This approach was extended to ten isolates of *Steinernema* spp. (including *S. glaseri, S. bibionis [S. feltiae]*, and *S. carpocapsae*). Each isolate was correctly assigned to species group by detection of RFLDs in repetitive DNA. The use of a heterologous ribosomal DNA (rDNA) probe from the free-living nematode *Caenorhabditis elegans*[12] identified some of these diagnostic repetitive DNA bands as part of the rDNA repeat. These studies established that RFLDs in EcoRI digested genomic DNA, in particular rDNA, were of taxonomic value in *Steinernema*.[13,14] The same approach provided supplementary data for the recognition of *S. intermedia* as a new species.[15]

Subsequent screening of randomly selected plasmid clones of *S. carpocapsae* EcoRI genomic DNA fragments identified a species-specific DNA probe which, in Southern blots of DNA from all ten isolates, only hybridized to the genomic DNA of the three *S. carpocapsae* isolates.[1,13] In addition, when labelled total DNA from *S. carpocapsae* (Simon Fraser University isolate #44) was used as a probe against Southern blots of EcoRI digested genomic DNA from the same ten *Steinernema* isolates, intense labelling of the three *S. carpocapsae* isolates (Simon Fraser University #6,44,123) occurred with only low level labelling of the repetitive DNA bands of the other isolates (labelling was predominantly confined to the rDNA bands).

DNA sequence analysis has been applied to *Heterorhabditis* isolates from North America,[12] and three genotypic groups based on RFLDs in repetitive DNA and rDNA, were identified. An analysis of rDNA in ten isohermaphrodite lines derived from *Heterorhabditis* NC447 failed to detect any variation in restriction fragment length pattern.[13] In the absence of supporting morphological or cross-breeding data, it was not possible to determine if these groupings of *Heterorhabditis* isolates represent species or intraspecific forms. Similar, but not yet completed, analyses of *Heterorhabditis* isolates from around the world have identified other genotypic groupings of *Heterorhabditis*,[16] which, in general, coincide with published isozyme data.[9]

This inability to assign entomopathogenic nematodes, on the basis of DNA genotypes, to a given taxonomic level without supporting data such as cross-breeding, is a major gap in our knowledge of the use of RFLDs and other DNA sequence based procedures. To determine the taxonomic level at which different degrees of sequence divergence operate requires an indepth knowledge of inter- and intraspecific variation in DNA sequence within *Steinernema* and *Heterorhabditis*.

The ongoing research of Reid and Hominick[17] is making a major contribution to this area. RFLDs have been determined in rDNA between *S. feltiae (S. carpocapsae)* and *S. bibionis (S. feltiae)*, and total genomic DNA from *S. bibionis (S. feltiae)* can be used as a diagnostic probe to distinguish *S. bibionis (S. feltiae)* from *S. feltiae (S. carpocapsae)* and *Heterorhabditis* isolates. Importantly, an indepth analysis of sequence divergence in over 50 natural isolates of *S. bibionis* (*S. feltiae*), coupled with cross-breeding experiments, has been initiated. Preliminary data on 25 isolates have documented the levels of intraspecific sequence divergence in rDNA.

In an equally effective approach to determining the level of intraspecific sequence divergence, White[18] has analyzed DNA from several inbred lines established from a single isolate of *S. carpocapsae*. Progeny of some of these inbred lines showed mutant phenotypes and inbreeding depression or recessive lethals. Further evidence for heterozygosity was obtained by analyzing DNA from the different generations of these lines. Use of a heterologous probe MSP (the major sperm protein in *C. elegans*) contributed evidence for heterozygosity. A reduction or loss of EcoRI restriction fragments which cross hybridized to MSP was detected in three out of eight full sibling inbred lines. These three lines could not be maintained beyond the 4th or 5th generation. This demonstration of heterozygosity establishes the potential for intraisolate variation within *S. carpocapsae*. Ongoing research will further characterize this and other genes in isolates of *S. carpocapsae* with the goal of using such genes for inter- and intraspecific comparisons of entomopathogenic nematodes. The studies of Reid and Hominick[17] and White,[18] and hopefully similar studies on intraspecific variation in other species, will form the solid base needed for the establishment of molecular procedures as a taxonomic tool for the classification and identification of intraspecific forms of entomopathogenic nematodes.

IV. METHODOLOGIES

Each of the three molecular methods that have been applied to the taxonomy of entomopathogenic nematodes (i.e., protein electrophoresis, immunology, and DNA sequence analysis) has certain benefits and limitations for resolving taxonomic problems. The following section assesses the comparative utility of these techniques, the taxonomic level at which each operates, and the most profitable future avenues of research.

A. Assessment of Current Approaches

The taxonomic value of protein electrophoresis has been extensively reviewed;[19,20] it is of little use above the level of genus nor is it useful for separating organisms at or below the level of subspecies (unless unusual levels of divergence are present). Good agreement between species determinations based on morphology and protein analysis usually occurs. Therefore, the main value of protein electrophoresis is in the identification of sibling species. A major constraint on this approach is the need to control for environmental and ontogenetic variation in protein expression. Protein analysis may play a future role in determining the population genetics of entomopathogenic nematodes, but as a taxonomic tool for the identification of these nematodes, it probably has limited potential below the level of species.

The use of immunological techniques in nematode taxonomy, in general, has received little attention. This is due in part to the early recognition that phenotypic variation can reduce the discriminatory power of antibody-based technology with the additional complication of possible cross-reactivity occurring between different antibodies and antigens. A further constraint is the degree of antigen conservation between closely related taxa. Monoclonal antibodies potentially provide a more powerful immunological approach, but there are scant data in this area, and this technology currently offers little promise at higher taxonomic levels or below the species level.

DNA sequence analysis offers the greatest flexibility of the molecular techniques currently available. The direct examination of an organism's genotype by analysis of DNA sequence is a taxonomically powerful tool that can operate at all taxonomic levels (e.g., from insights into the evolutionary relationships between pro- and eukaryotic organisms[21] to differentiation of infrasubspecific groupings of nematodes[1,12,22]). The DNA-based taxonomic approach has several distinct advantages over the previously described taxonomic techniques. It examines the genotype of the nematode directly and so avoids the diagnostic problems associated with phenotypic variation. Furthermore, greater discrimination is possible than with previous techniques because the entire genome is examined, whereas other molecular methods indirectly assay the often highly conserved coding sequences which are only 20-25% of the genome. The remaining 75-80% of the genome, which can be analyzed only by this DNA approach, contains many highly variable sequences. This DNA sequence variability provides a large pool of variation which can serve as useful diagnostic characters for separation of taxa and infrasubspecific categories.

Detailed protocols of DNA sequence analysis can be found in numerous molecular biology reference manuals (e.g., Maniatis et al.[23]), and the general concepts and application of this technology to nematode taxonomy have been recently reviewed.[1] The basic approach relies on the digestion of DNA with restriction endonucleases. Because the distribution of a particular restriction endonuclease's recognition site is determined by the nucleotide sequence (genotype) of the isolate, the size distribution of the DNA restriction fragments generated is characteristic for that genotype. A logical progression ensues from the detection of RFLDs in repetitive DNA as visualized by examination of ethidium bromide stained agarose gels, to the detection of RFLDs in repetitive DNA and low or single copy DNA sequences by hybridization of heterologous or homologous cloned DNA fragments to Southern blots. Labelled total genomic DNA can be used as a species-specific hybridization probe, as can certain randomly cloned DNA fragments.[1] RFLDs have been detected between isolates within species of entomopathogenic nematodes,[1,17,18] but to date no isolate-specific hybridization probes are available for positive or negative determinations in dot-blot procedures.

B. Future Developments

DNA sequence analysis is the method of choice for, and may be the only method capable of, discriminating between infrasubspecific categories of entomopathogenic nematodes. The DNA analyses conducted on entomopathogenic nematodes have examined both sequence divergence in total DNA (i.e., genomic and mitochondrial DNA [mtDNA]) and, more frequently, RFLDs in repetitive and rDNA. Intraspecific variation has been detected by these methods. Mitochondrial DNA sequence data have been used extensively in the taxonomy of other organisms and has provided useful diagnostic characters below the level of species.[24] Because of the higher rates of evolution in mtDNA as compared with genomic DNA,[25] mtDNA sequence data are thought to be more likely to display taxonomically useful variation at lower taxonomic levels.[26] However, research on the molecular variability of the lizard *Sceloporus grammicus* has demonstrated a high rate of rDNA divergence, relative to the degree of allozyme and mtDNA divergence.[27] It would be useful to examine mtDNA divergence in entomopathogenic nematodes and assess its comparative taxonomic value. This could be readily achieved by hybridization of labelled cloned mtDNA fragments to Southern blots of total DNA. Restriction fragments of *C. elegans* mtDNA have been cloned and used to distinguish intraspecific isolates of *C. elegans* by detection of RFLDs.[28] Such heterologous probes or the mtDNA cloned from a representative entomopathogenic nematode could be similarly employed.

Molecular taxonomic techniques have centered on the determination of natural variation within a nematode grouping and the correlation of a certain degree of variation to a particular taxonomic level. Inter- and intraisolate variation in DNA sequence has been demonstrated in entomopathogenic nematodes. Given that it is ultimately possible to detect a single nucleotide change in the millions of base pairs of a nematode's genome, the assignment to a particular taxonomic grouping below species level will depend upon the acceptance of an arbitrary degree of sequence divergence as representative of a given grouping. This approach, acceptable in systematics, may not meet the requirements of proprietary or patent right protection for a sequence uniquely identifying an isolate. For instance, in the case of a randomly chosen DNA sequence it would be necessary to define the nature and extent of its variation for all isolates prior to establishing it as uniquely identifying a given isolate. This process of validation would be difficult in the case of entomopathogenic nematodes. New isolates can be easily obtained, thus increasing the number of required comparisons. Moreover, genetic transfer of identifying sequences into or out of different isolates can be easily accomplished by cross-breeding, unless the critical sequence defined an essential gene(s) of a unique advantageous phenotype.

Instead of using naturally occurring sequence divergence, identification could be based upon a unique "man-made" sequence inserted into the genome by genetic engineering (e.g., chromosomal transformation by injection of cloned DNA into oocyte nuclei). Such a unique sequence or tag should have no harmful phenotype and should be heritable, preferably integrated into a chromosome. Procedures which routinely achieve the integration of exotic DNA into the genome of *C. elegans* could be readily transferred to entomopathogenic nematodes. This approach, however, results in a genetically engineered organism, a major drawback because of ensuing additional requirements in the product registration process for the product. Also, to avoid removal of the tag by genetic means, the unique sequence should be intimately linked with essential advantageous genes or have several copies dispersed throughout the genome.

DNA-DNA hybridization techniques using Southern or dot blots can readily detect a 5% sequence divergence between DNA samples. Either the presence of this level of sequence divergence in a given naturally occurring DNA fragment or constructed DNA fragments would allow the use of species and isolate specific probes. This technology is restricted by the lower limit to the amount of DNA needed for detection. Insufficient DNA is present in an individual infective-stage entomopathogenic nematode to be visualized by current labelling technology.[1] For taxonomic purposes the ease of mass culture of entomopathogenic nematodes would guarantee a virtually unlimited quantity available. Therefore, the minimum amount of DNA could be readily obtained.

For ecological and population studies, the identification of individual infective-stage nematodes would have many benefits. It is technologically possible for the DNA extracted from a single infective-stage nematode to be amplified, using polymerase chain reaction (PCR) technology, to give sufficient DNA for detection by DNA-DNA hybridization techniques. Unfortunately, the cost and laborious procedures involved are likely to make PCR technology too expensive for routine use in ecological studies. An alternative technology would be the use of *in situ* hybridization techniques to detect a unique mRNA, if this was expressed in sufficient quantity. This technique has been used to visualize mRNA of moderately expressed genes within individual *C. elegans*. Given the generally higher rate of sequence conservation in coding regions, it is unlikely that sufficient sequence divergence would be found between naturally occurring mRNAs of different isolates; the integration of an exogenous expressible gene would be required. Along similar lines, the integration and expression of genes such as β-galactosidase or β-glucuronidase in the appropriate null background, would allow the detection and identification of nematodes by the action of these enzymes on chromogenic substrates. For example, in *C. elegans*, individual nematodes containing the introduced β-galactosidase or β-glucuronidase genes can be identified by their blue or red color, respectively, following incubation in the appropriate enzyme substrate.[29,30]

While labelling infective-stage nematodes prior to their release may prove cost-effective for certain types of ecological study, an alternative approach based on the differential staining of entomopathogenic nematode isolates with dyes has been developed.[31] The procedure simply involves the incorporation of 0.5% w/w of dye into the culture medium before inoculation with *Steinernema* spp. It results in the production of stained infective-stage nematodes. Seven dyes have been tested including Sudan II and III which stain the intestine red. When tested for any undesirable side-effects, Sudan II did not alter the production (i.e., yield per flask) or infectivity of the progeny. Moreover, the color remained distinguishable until shortly after penetration of the insect host. Further evaluation of this technique is warranted because this method offers great potential for certain ecological studies on inundatively or inoculatively released entomopathogenic nematodes.

V. CONCLUSIONS

Classical morphological methods have been successfully applied to the description, identification, and classification of entomopathogenic nematodes to genus and species. In *Steinernema,* cross-breeding provides an invaluable genetic proof of identity which should be incorporated into any taxonomic study. Cross-breeding is difficult in *Heterorhabditis* and molecular methods can provide valuable additional information for species delimitation. Nevertheless, morphological studies will continue to form the essential core of entomopathogenic nematode taxonomy.

In the immediate future, molecular methods will become indispensable for distinguishing between isolates of entomopathogenic nematodes below the level of species. This level of discrimination will be necessary to meet the possible requirements of quarantine and registration authorities for isolate identification, and to provide markers for verification of proprietary rights to patented organisms. DNA sequence analysis is the method of choice for, and may be the only method capable of, discriminating between such infrasubspecific categories of entomopathogenic nematodes.

REFERENCES

1. **Curran, J., and Webster, J. M.,** Identification of nematodes using restriction fragment length differences and species-specific DNA probes, *Can. J. Plant Pathol.,* 9, 162, 1987.
2. **Poinar, G. O., Jr.,** Biology and taxonomy of Steinernematidae and Heterorhabditidae, in *Entomopathogenic Nematodes in Biological Control,* Gaugler, R., and Kaya, H. K., Eds., CRC Press, Boca Raton, FL, 1989, chap. 2.
3. **Herman, I. W., and Jackson, G. J.,** Zymograms of the parasitic nematodes, *Neoaplectana glaseri* and *N. carpocapsae,* grown axenically, *J. Parasitol.,* 49, 392, 1963.
4. **Sha, C-Y.,** A comparative analysis of esterase of the insect parasitic nematodes of the genus *Neoaplectana, Acta Zootax. Sin.,* 10, 246, 1985.
5. **Stanusek, S.,** *Neoaplectana feltiae* complex (Nematode: Rhabditoidea, Steinernematidae): its taxonomic position within the genus *Neoaplectana* and intraspecific structure, *Zesz. Probl. Postepow. Nauk Roln.,* 154, 331, 1974.
6. **Poinar, G. O., Jr.,** On the nomenclature of the genus *Neoaplectana* Steiner 1929 (Steinernematidae: Rhabditida) and the species *N. carpocapsae* Weiser, 1955, *Rev. Nematol.,* 7, 199, 1984.
7. **Kozodoi, E. M., Voronov, D. A., and Spiridinov, S. E.,** The use of an electrophoretic method to determine the species association of *Neoaplectana* specimens, *Trudy Gel'minthologicheskoi laboratorii (Voprosy biotsenologii gel'mintov),* 34, 55, 1986.

8. **Poinar, G. O., Jr., and Kozodoi, E. M.**, *Neoaplectana glaseri* and *N. anomali*: sibling species or parallelism?, *Rev. Nematol.*, 11, 13, 1988.
9. **Akhurst, R. J.**, Use of starch gel electrophoresis in the taxonomy of the genus *Heterorhabditis* (Nematoda: Heterorhabditidae), *Nematologica*, 33, 1, 1987.
10. **Jackson, G. J.**, Differentiation of three species of *Neoaplectana* (Nematoda: Rhabditida), grown axenically, *Parasitology*, 55, 571, 1965.
11. **Curran, J., Baillie, D. L., and Webster, J. M.**, Use of restriction fragment length differences in genomic DNA to identify nematode species, *Parasitology*, 90, 137, 1985.
12. **Curran, J., and Webster, J. M.**, Genotypic analysis of *Heterorhabditis* isolates from North Carolina, USA, *J. Nematol.*, 21, 140, 1989.
13. **Curran, J.**, unpublished data, 1984.
14. **Murray, K. A.**, unpublished data, 1983.
15. **Poinar, G. O., Jr.**, *Neoaplectana intermedia* n. sp. (Steinernematidae: Nematoda) from South Carolina, *Rev. Nematol.*, 8, 321, 1985.
16. **Curran, J.**, unpublished data, 1988.
17. **Reid, A., and Hominick, W. M.**, unpublished data, 1989.
18. **White, J. H.**, unpublished data, 1989.
19. **Hussey, R. S.**, Biochemical systematics of nematodes - a review, *Helminthol. Abstr.*, 48, 141, 1979.
20. **Platzer, E. G.**, Potential use of protein patterns and DNA nucleotide sequences in nematode taxonomy, in *Plant Parasitic Nematodes*, Vol. III, Zuckerman, B. M., and Rohde, R. A., Eds., Academic Press, New York, 1981, 3.
21. **Vahidi, H., Curran, J., Nelson, D. W., Webster, J. M., McClure, M. A., and Honda, B. M.**, Unusual sequences, homologous to 5s RNA, in ribosomal DNA repeats of the nematode *Meloidogyne arenaria*, *J. Mol. Evol.*, 27, 222, 1988.
22. **Curran, J., McClure, M. A., and Webster, J. M.**, Genotypic analysis of *Meloidogyne* populations by detection of restriction fragment length differences in total DNA, *J. Nematol.*, 18, 83, 1986.
23. **Maniatis, T., Fritsch, E. J., and Sambrook, J.**, *Molecular Cloning, a Laboratory Manual*, Cold Spring Harbor, New York, 1982.
24. **Birley, A. J., and Croft, J. H.**, Mitochondrial DNAs and phylogenetic relationships, in Dutta, S. K., Ed., *DNA Systematics*, Vol. I, *Evolution*, CRC Press, Boca Raton, FL, 1986, 107.
25. **Brown, W. M.**, The mitochondrial genome of animals, in *Molecular Evolutionary Genetics*, McIntyre, R. J., Ed., Plenum, New York, 1985, 93.
26. **Avise, J. C., Arnold, J., Ball, R. M., Bermingham, E., Lamb, T., Niegel, J. E., Reeb, C. A., and Saunders, N.C.**, Intraspecific phylogeography: the mitochondrial DNA bridge between population genetics and systematics, *Annu. Rev. Ecol. Syst.*, 18, 489, 1987.
27. **Sites, J. W., Jr., and Davis, S. K.**, Phylogenetic relationships and molecular variability within and among six chromosome races of *Sceloporus grammicus* (Saiura, Iguanidae), based on nuclear and mitochondrial markers, *Evolution*, 43, 296, 1989.
28. **Beckenbach, K., Baillie, D. L., and Rose, A. M.**, unpublished data, 1988.
29. **Fire, A.**, Integrative transformation of *Caenorhabditis elegans*, *EMBO J.*, 5, 2673, 1986.
30. **Jefferson, R., A., Klass, M., Wolf, N., and Hirsh, D.**, Expression of chimeric genes in *Caenorhabditis elegans*, *J. Mol. Biol.*, 193, 41, 1987.
31. **Yang, H., and Jian, H.**, Labelling living entomopathogenic nematodes with stains, *Chinese J. Biol. Cont.*, 4, 59, 1988.

4. Biology and Taxonomy of *Xenorhabdus*

R. J. Akhurst and N. E. Boemare

I. INTRODUCTION

All known species of nematodes of the families Steinernematidae and Heterorhabditidae are symbiotically associated with bacteria of the genus *Xenorhabdus*.[1-8] The increasing interest in *Xenorhabdus* reflects growing awareness that the bacterial partner plays a significant role in nematode/bacterium associations beyond mass production of the nematode vector. Appreciation of the significance of this role is an integral part of the development of entomopathogenic nematodes as a viable means of insect pest control.

II. BIOLOGY

The interaction between nematode and bacterium has been shown to have many facets. *Xenorhabdus* spp. apparently do not survive well in soil or water[9] and are not pathogenic for insects when ingested.[10,11] Steinernematid and heterorhabditid nematodes provide protection for *Xenorhabdus* outside the insect host and a means of transmission from cadaver to the hemocoel of a new host. In addition to transporting the bacterium to a new host, the nematodes provide protection from some host defense mechanisms.[12] The bacterial contribution to the association is the provision of nutrients for the nematodes. Axenic nematodes are unable to reproduce in axenic insects and require bacterial activity to produce suitable nutrient conditions.[13,14] Although the nematodes can reproduce in *Galleria mellonella* infected with bacteria other than *Xenorhabdus*, they do not reproduce as prolifically as when *Xenorhabdus* dominates the bacterial flora of the cadaver.[13]

Insects infected by nematodes are also subject to secondary invasion by other microorganisms which might modify conditions within the cadaver to the detriment of the nematodes' capacity to reproduce. Contamination of the insect cadaver is minimized initially by the phagocytic activity of the insect hemolymph.[15,16] This short initial protection is followed by that provided by the various antimicrobial agents produced by the *Xenorhabdus* spp., which inhibit a wide range of bacteria, yeasts, and fungi.[17] The agents so far identified include defective phages, reported from *X. nematophilus* and *X. luminescens*,[18] which may be bacteriocins that inhibit *Xenorhabdus* introduced by other nematode species. Other antimicrobials detected in *Xenorhabdus* are chemical compounds. Paul et al.[19] found four indole derivatives from strain R, recorded as *X. nematophilus*, and two *trans*-stilbene derivatives from *X. luminescens* strain Hb. Examination of *Xenorhabdus bovienii* (originally identified as *X. nematophilus*) yielded three antimicrobial compounds in another class, designated Xenorhabdins.[20] Both *X. nematophilus* and *X. luminescens*, as well as strain Q1 of an undescribed species,[21] produce yet another class of antimicrobials, water-soluble compounds designated Xenocoumacins.[22] The variety of antimicrobial agents produced by each species of *Xenorhabdus* suggests that antimicrobial activity is very important for the nematode/bacterium associations.

The significance of luminescence in *X. luminescens* has not been determined. One hypothesis[23] suggested that the glowing cadaver might attract other insects that could be infected as the nematodes emerged from the cadaver, simplifying the problem of locating a new host. However, luminescence is greatly reduced or even absent by the time of infective juvenile emergence. Moreover, light does not penetrate soil very effectively and so could only act as a trap rather than as a lure. An alternative hypothesis is that the bioluminescence may be inhibitory. Bioluminescence is at its peak when the nematodes are not infective and are most vulnerable to the effects of saprophagic invertebrates invading the cadaver. These saprophages, being soil-dwellers, might be expected to be photophobic and so avoid the luminescent cadaver. The nonluminescent strain of *X. luminescens*[24] may be useful for testing these hypotheses.

A. Pathogenicity

Most *Xenorhabdus* spp. are highly pathogenic when introduced into the hemocoel of *Galleria* larvae, the LD_{50} usually being less than 50 cells.[14,25] However, the data obtained from *Galleria* larvae may be misleading. *X. nematophilus* is much less pathogenic for the pupae of another lepidopteran, *Hyalophora cecropia* (LD_{50} = 500 cells),[12] and not at all pathogenic for the dipteran *Chironomus*.[26] The pathogenicity of *Xenorhabdus* for various insect species has not been investigated extensively.

The pathogenicity of *Xenorhabdus* spp. depends on their entry into the host hemocoel and ability to multiply in the hemolymph in spite of the host's defense response. *X. nematophilus* is not pathogenic for *Chironomus* because the highly effective humoral encapsulation system of that species prevents its establishment.[26] In some insect species, however, living *Xenorhabdus* cells are not recognized as nonself[27] or are not killed by the hemocytes[28] and are able to multiply and kill the host. In some insects, the nematodes are encapsulated and killed when few enter the host. This encapsulation does not necessarily protect the host which may be killed either by *Xenorhabdus* released by the nematodes before encapsulation is complete, by other bacteria carried into the host on the external surface of the nematode,[29,30] or by the nematode toxin.

Xenorhabdus pathogenicity can also depend on interaction with its nematode associate. The nematode may produce a toxin[31,32] or a factor that destroys the inducible enzymatic defense response of the insect.[12] The significance of this interaction is most clearly illustrated by the *Steinernema glaseri/Xen-*

orhabdus poinarii combination which is highly pathogenic for *Galleria mellonella* larvae, although neither *S. glaseri* nor *X. poinarii* alone is pathogenic for *G. mellonella*.[4] This type of interaction may also be important for other nematode/bacterium associations, including *S. carpocapsae/X. nematophilus*, where the target insect is not sensitive to the nematode toxin or for those associations in which the nematode does not produce a toxin.

Growth of *Xenorhabdus* is accompanied by the production of exo-[33,34] and endotoxins.[16,35,36] The exotoxin activity is probably linked to the exoenzymatic functions found in *Xenorhabdus* (protease, lecithinase, lipase) and thought to be insect toxins in other bacteria.[37] The endotoxins are lipopolysaccharide components of the bacterial cell wall, a common feature of Gram negative bacteria.[16,38] It is probable that insects of some species are killed by exotoxin activity whereas others are more susceptible to the endotoxins or nematode toxin.

Xenorhabdus spp. and their associated nematodes form nonspecific entomopathogenic combinations. To function nonspecifically they must be able to avoid or overcome the variety of defense mechanisms (external and internal) of a wide range of insects. They must also have several toxin strategies to ensure that they can kill that wide range of insects.

B. Nematode/Bacterial Specificity

Taxonomic studies confirm that each species of entomopathogenic nematode has a specific natural association with only one *Xenorhabdus* species (though a *Xenorhabdus* sp. may be associated with more than one nematode species, Table 1).[21] The specificity of association between nematode and bacterium operates on two levels: the provision of essential nutrients for the nematode by the bacterium and the retention of the bacterium within the intestine of the nonfeeding infective juvenile nematode.

TABLE 1. *Xenorhabdus* Species and Associated Nematodes

Xenorhabdus species	Associated nematode species	Ref.
X. nematophilus	*Steinernema carpocapsae*	1, 3, 45, 56
X. luminescens[a]	All *Heterorhabditis* spp.	1, 3, 7, 45, 56
X. bovienii	*S. feltiae (=bibionis)*	3, 45, 56
	S. kraussei	2, 45, 56
	S. affinis	45
	S. intermedia	6, 56
	undescribed *Steinernema* sp. F3	45, 56
	undescribed *Steinernema* sp. F9	45

TABLE 1 (continued). *Xenorhabdus* Species and Associated Nematodes

Xenorhabdus species	Associated nematode species	Ref.
X. poinarii	*S. glaseri*	3, 4, 56
	Steinernema sp. NC513	4, 55
X. beddingii	undescribed *Steinernema* sp. M	3, 5, 45, 56
	undescribed *Steinernema* sp. N	3, 5
Xenorhabdus sp.	*S. rara*	8, 56
Xenorhabdus sp.	*S. anomali*	56

[a]   *X. luminescens* is a multispecies taxon.[56,59]

1. Nutrients

The requirement by the nematode of bacterial-produced nutrients does not impose a high level of specificity; *S. carpocapsae* can reproduce in culture with some species of bacteria other than *Xenorhabdus*.[39] However, there is some degree of specificity involved because no steinernematid or heterorhabditid can be cultured monoxenically with all other *Xenorhabdus* spp.[40] Although *Steinernema feltiae* (=*bibionis*) and an undescribed *Steinernema* sp. were successfully cultured monoxenically with the symbiont of any of five other *Steinernema* spp., they could not be cultured with *X. luminescens*. *S. carpocapsae*, *S. glaseri*, and another undescribed *Steinernema* sp. were even more limited and could not even utilize the symbionts of some other *Steinernema* spp. The best nutrient conditions for the nematodes are not necessarily produced by its natural symbiont. *S. glaseri* reproduced more rapidly on lipid fortified agar when cultured with an *X. bovienii* isolate than with *X. nematophilus* or its natural symbiont, *X. poinarii*.[41]

Within the species *X. luminescens* there is variation in the production of essential nutrients for *Heterorhabditis* spp. Each *X. luminescens* isolate tested supported monoxenic *in vitro* culture of only some strains of *Heterorhabditis* (Table 2).[33]

2. Retention

The transmission of *Xenorhabdus* is determined at a higher level of specificity. Although *Steinernema* spp. can be cultured with bacteria other than their natural symbionts, they do not retain non-*Xenorhabdus* within the intestine of the infective juvenile and are limited in their ability to retain the symbiont of another *Steinernema* sp.[40] Specificity is very high in the *S. carpocapsae/X. nematophilus* association; infective juveniles do not contain the symbiont of any other species.[40] In contrast, *S. feltiae* (=*bibionis*) *and S. glaseri* infective

TABLE 2. Monoxenic Culture of Various *Xenorhabdus luminescens* (Phase One) and *Heterorhabditis* Isolates[33]

Heterorhabditis strains[70]	*Xenorhabdus luminescens* isolates						
	Hb	**Hl**	**NC19**	**D1**	**Q614**	**NZ**	**T310**
Hb	+[a]	+	+	-[b]	-	+	+
Hl	+	+	+	-	-	-	-
NC19	+	+	+	-	-	+	+
NC69	+	+	+	-	-	+	+
NC200	+	+	+	-	-	+	+
D1	-	-	-	+	-	-	-
Q614	-	-	-	-	+	+	-
NZ	-	-	+	-	+	+	+
T310	-	-	+	-	-	+	+

[a] + = Culture sustained through at least three subcultures.
[b] - = Culture could not be sustained for two subcultures.

juveniles carry the symbionts of some other *Steinernema* spp., though usually not as efficiently as their own.[40-42]

S. glaseri is inefficient in retaining its natural symbiont within the infective stage.[40-42] Whereas the retention rate for *S. carpocapsae* and *S. feltiae* (=*bibionis*) infective juveniles for their natural symbionts exceeds 90%, the average is only 50% for *S. glaseri*. No morphological basis for the difference has been detected[43] and it may reflect an imbalance between the nematode and bacterial populations.[42] A similarly low retention rate has been shown for an undescribed *Steinernema* sp. and *X. beddingii*, its natural symbiont.[40] These low retention rates may be indicative of less highly evolved nematode/bacterium relationships than that between *S. carpocapsae* and *X. nematophilus*.

C. Phase Variation

Xenorhabdus isolated from the infective-stage nematode produce dye adsorbing colonies.[1] However, when *in vitro* cultures of *Xenorhabdus*, either stationary phase axenic cultures or monoxenic cultures with nematodes, are sampled some nonadsorbing colonies can be detected.[44] The adsorbing and nonadsorbing variants were initially designated primary and secondary form, respectively. However, it has now been shown that reversion from secondary to primary occurs not only in *X. nematophilus*[44] but in other *Xenorhabdus* as well,[45,46] identifying this as a phase variation. Consequently, the primary and secondary variants are referred to as phase one and phase two, respectively.[45]

Reversion from secondary to primary has not yet been detected in any *X. luminescens* strain. The expression of variation may differ for these bacteria, which are taxonomically distinct from other *Xenorhabdus* spp. (see Section III). However, for consistency the variants will be referred to as phases one and two for all *Xenorhabdus* spp. If *X. luminescens* is reclassified to a separate genus and reversion from secondary to primary cannot be demonstrated, then the variation should be referred to as form variation for the new genus.

Phase one variants of *X. nematophilus, X. luminescens, X. bovienii,* and *X. beddingii* differ from phase two in dye adsorption, pigmentation, production of antimicrobials and lecithinase,[45] and presence of proteinaceous inclusion bodies.[47,48] In addition to these general differences, there are other phase differences that have so far been reported for only one or more species.[45] For example, phase one *X. nematophilus* also differs from phase two in cell size,[14] internal morphology, interaction with hemocytes,[28,49] and its fimbriae,[50] and phase one *X. luminescens* differs from phase two in being sensitive to a bacteriophage.[51]

Although both phases are equally pathogenic for *Galleria mellonella* larvae, the phase one variant produces better conditions for nematode reproduction, an important consideration in commercial production of the nematodes. The relative effectiveness of the phases in supporting nematode reproduction in monoxenic *in vitro* culture varies between species. There was no significant difference between the phases for the culture of *S. feltiae* (=*bibionis*) with *X. bovienii*, but cultures of *Heterorhabditis* spp. with phase one *X. luminescens* usually produced three times more infective-stage nematodes than those with phase two.[52] For steinernematids, the difference was even greater *in vivo*.[25] More than seven times as many *S. feltiae* (=*bibionis*) and *S. carpocapsae* per *Galleria* larva were produced when the infective nematodes carried phase one. The difference *in vivo* may be the combined effects of nutritional differences between the phases and antibiotic production by phase one. It is not unusual for insects infected with nematodes bearing only phase two *Xenorhabdus* to be so badly contaminated by invading bacteria or fungi that the nematodes are unable to complete development. However, the difference may also be nutritional, because the insect is poor in growth factors required by the nematodes.

X. poinarii differs from other species in having no close association between dye adsorption and antimicrobial activity. Neither character correlates with improved yield in monoxenic *in vitro* culture but antimicrobial activity appears to be advantageous *in vivo*.[4]

Although the infective-stage juvenile is capable of retaining phase two of its *Xenorhabdus* symbiont within its intestine, nematodes collected from nature almost invariably contain only phase one cells. Even where 80% of *X. nematophilus* was in phase two at the time of *S. carpocapsae* infective juvenile formation, infective juveniles carried only phase one cells.[44] There is no evidence for phase two to phase one reversion within the infective juvenile, suggesting that phase one is significantly more effective in colonizing the intestinal vesicle of the infective-stage nematode.

Clearly, there is a significant advantage for the nematode to be associated with phase one rather than phase two *Xenorhabdus*. However, the basis for the existence of phase two has yet to be explained. Poinar[53] has proposed that phase change may be a mechanism for avoiding the elimination of *Xenorhabdus* from a cadaver containing a bacteriophage that lyses phase one but not phase two.[51] However, phase variation occurs in the stationary phase and so would perhaps occur too late to allow *Xenorhabdus* to escape the lytic activity of a phage. In addition, we have only detected phase one *Xenorhabdus* in field-collected nematodes and have only recovered phase two from infective juveniles cultured monoxenically with phase two (i.e., no phase two to one reversion within the infective juvenile). The discovery or production of a stable phase one *Xenorhabdus* sp. will facilitate investigation of the significance of phase variation as well as improving the efficiency of commercial production of the nematodes.

III. TAXONOMY

Clarification of the taxonomy of *Xenorhabdus* serves two utilitarian purposes: providing labels to report data obtained with different *Xenorhabdus* isolates so that meaningful comparisons can be made, and providing the basis for a system of identification that may be required for government registration purposes. This latter consideration becomes more urgent now that isolates identified as *X. luminescens* have been obtained from human clinical specimens.[54] Although these clinical isolates clearly constitute a new species,[54] they may cause some concern to registration authorities until they are formally distinguished from the symbionts of *Heterorhabditis* spp. This section examines approaches to understanding taxonomic relationships among the bacterial symbionts of entomopathogenic nematodes and to similar bacterial groups.

A. Current Status

Xenorhabdus are Gram negative, facultatively anaerobic rods, classified within the family Enterobacteriaceae. Five species have been described: *X. nematophilus, X. luminescens, X. bovienii, X. poinarii*, and *X. beddingii* (Table 1).[1,21] Preliminary studies indicate that the symbionts of some undescribed and recently described species constitute new species of *Xenorhabdus*.[56]

B. Classical Taxonomy

The classical approach of comparing biochemical and cultural characteristics is the one used most widely for studying the taxonomy of *Xenorhabdus*.

There have been some significant differences in the characteristics reported for *Xenorhabdus* (e.g., fermentation of some sugars, protease activity). Although in some cases the differences have been due to the use of different isolates, significant differences have also been due to variation in the methods or conditions used to test for these characteristics. Boemare and Akhurst[45] found that results obtained for citrate utilization and gelatin liquefaction varied with the type of medium used. Temperature has been shown to affect the predominant fimbrial type[50] and the type of hemolysis.[54] Major differences in data on fermentation of carbohydrates seem to have been due to the pH indicator chosen. The interpretation of weak responses has also generated differences; a weak response may indicate a low level of activity or may be indicative of a phase change occurring during the conduct of the test.[21] Numerical analysis of the data indicated that treating weak responses as negative rather than positive produced a more logical taxonomic structure.[21]

Analysis of data should take into consideration phase-related characters.[21] Cluster analysis of a data set involving both phases of 21 isolates tested for 240 characters showed a strong tendency for isolates to cluster first by phase. Elimination of five characters identified as phase-related (pigmentation, BTB adsorption, neutral red adsorption, antimicrobial activity, lecithinase) reduced this tendency but produced some anomalous groupings. When the data for the two phases of each isolate were combined, and the isolate scored "+" for a character if either phase was positive for that character, the groupings of isolates reflected the species groups of their nematode symbionts (Figure 1).

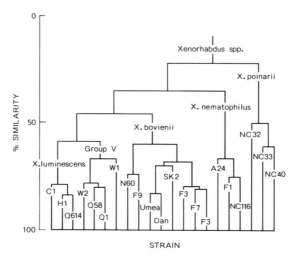

Figure 1. Taxonomic relationships of *Xenorhabdus* suggested by numerical analysis of 45 variable characters of both phases of 20 isolates. The data for the two phases of each strain were combined and scored "-" only where both phases were negative for that character. Redrawn from Akhurst and Boemare.[21]

Some characteristics may be encoded on plasmid-borne genes. Plasmids have been identified in *Xenorhabdus* spp.[48,51] but no role has yet been determined. Some of the variation between isolates detected by phentoypic methods may be due to the presence or absence of plasmids and consequently give an exaggerated estimate of the difference between isolates.

C. DNA/DNA Hybridization

DNA relatedness has been widely used to examine intrageneric relationships among bacteria. DNA relatedness in excess of 60-70% is taken to be indicative of conspecificity; 20-60% DNA relatedness indicates that the isolates are from closely related species.[57,58] It is recommended that a DNA homology group should not be described as a distinct species unless it can also be differentiated by some phenotypic character.[58]

Grimont et al.[59] found three DNA relatedness groups which corresponded to *X. luminescens, X. nematophilus*, and a group associated with *S. feltiae* (=*bibionis*) and several unnamed species of nematode. The levels of relatedness suggested that the last two groups are species within the same genus. However, their relatedness to the *X. luminescens* group indicated that this group should be considered a separate genus. Within the *X. luminescens* group, there were four subgroups indicative of distinct species. Farmer et al.[54] found that the "clinical" isolates identified as *X. luminescens* formed a fifth DNA homology group, indicating that they represent a separate species. Grimont et al.[59] and Farmer et al.[54] indicated that they had too few isolates to produce reliable phenotypic descriptions of the DNA homology groups as new species.

Analysis of our DNA homology data[56] generally supports the taxonomic conclusions drawn from phenotypic data[21,45] and the DNA homology data of Grimont et al.[59] The type strains of the five described species of *Xenorhabdus* fall into separate groups within which DNA homology exceeds 70%. DNA homology between the groups containing the type strains of *X. nematophilus, X. bovienii, X. poinarii, and X. beddingii* ranged from 25% to 36%. The DNA relatedness of the *X. luminescens* isolates to the other *Xenorhabdus* isolates was less than 25%.

The DNAs of isolates Q1 and W1, which clustered with *X. beddingii* on the basis of phenotypic characters (Figure 1), had less than 50% homology with that of Q58, the type strain of *X. beddingii*. Differences were also found among the *X. luminescens* symbionts of *H. bacteriophora* (Hb), *H. bacteriophora* (=*heliothidis*) NC19 (=C1), and *H. megidis*, indicating that these bacteria belong to two separate species.

The homologies of DNA from *Xenorhabdus* associated with *S. anomali, S. rara*, a *Steinernema* sp. from Argentina, and one from China, with those from the type strains of the *Xenorhabdus* species, excluding *X. luminescens*, ranged from 31% to 43%. This suggests that the bacterial symbionts of these nematodes represent at least three new species (their relatedness to each other has not been fully tested). The DNA homology of the symbiont of *S. intermedia* and three *X. bovienii* strains (T228, F3, SK2[3,45]) varied from 61% to 84%, indicating that they belong to the same species.

No significant differences between phases one and two of any isolate were detected by DNA/DNA hybridization.

D. 16S rRNA Oligonucleotide Cataloguing

Comparison of 16S rRNA sequences by oligonucleotide cataloguing[60] provides a means of assessing higher taxonomic relationships that cannot be examined by DNA/DNA homology.

Ehlers et al.[61] catalogued the oligonucleotides of 16S rRNA of the type strains of *X. luminescens* and *X. nematophilus*, and of *X. luminescens* ATCC29304, the symbiont of *H. bacteriophora* (=*heliothidis*) NC19. Comparison of this data with the rRNA catalogues of nearly 200 Gram negative eubacteria did not clarify the position of *Xenorhabdus* in relation to the family Enterobacteriaceae. The dendrogram generated from analysis of the data indicated that *Xenorhabdus* is more dissimilar to other Enterobacteriaceae than *Proteus*, previously considered to have the least relatedness to the rest of the family.[62] However, this is based on data from only three *Xenorhabdus* strains and may be misleading because of group-size dependency in the sorting strategy. It is notable that the similarity coefficient calculated from these data for *X. nematophilus/E. coli* (0.70) is identical to that for *X. nematophilus/X. luminescens* ATCC29304. The *Xenorhabdus* isolates also contain almost all of the enterobacteria-specific oligonucleotides. Ehlers et al.[61] noted that whether *Xenorhabdus* should be seen as a separate family or be retained within the Enterobacteriaceae is a moot point because of its production of the enterobacterial common antigen,[63] phenotypic similarities, and possession of enterobacteria-specific oligonucleotides.

Later work revealed a variable domain within the 16S rRNA which shows significant differences to other Enterobacteriaceae and differences within *Xenorhabdus*.[64] This led to the cloning of 16-20 nucleotide sequences for species-specific identification of *X. luminescens* (except strain HL81), *X. nematophilus*, *X. bovienii*, *X. poinarii*, and *X. beddingii*. Probes that can be used to distinguish the *X. luminescens* symbionts of *H. bacteriophora* Hb and *H. bacteriophora* (=*heliothidis*) NC19 from each other have also been cloned.[64]

E. Immunology

Determination of immunological distance between homologous proteins by quantitative microcomplement fixation provides a measure of percent amino acid sequence difference. Studies based on glutamine synthetase and superoxide dismutase[64,66] indicated that *X. luminescens* was more closely related to terrestial than to luminous and nonluminous marine Gram negative eubacteria. *X. luminescens* was also more similar to other Enterobacteriaceae than to Vibrionaceae, but was on the outer edge of the Enterobacteriaceae cluster.

Slight intraspecific antigenic differences between the DD-136, Agriotos, and Mexican strains of *X. nematophilus* have been detected by immunodiffusion, agglutination, and imunofluorescence techniques.[67] These tests also showed that *X. poinarii* and *X. bovienii* were substantially different antigenically from the DD-136 strain of *X. nematophilus*.

F. Electrophoresis

Electrophoretic techniques can be useful for distinguishing species and subspecific groups. Hotchkin and Kaya[68] compared total protein and isozyme profiles of both phases of *Xenorhabdus* associated with several *Steinernema* and *Heterorhabditis* species. Total protein profiles on one-dimensional polyacrylamide gels revealed little intraspecific variation, except in *X. luminescens*. There was no difference in the profiles of the symbionts of *H. bacteriophora* Hb, *H. bacteriophora* (=*heliothidis*) NC19, phase one of the Netherlands, or phase two of the Victorville strains. However, the profiles of the phase two and phase one of the Netherlands and Victorville strains, respectively, were appreciably different from each other and the other *X. luminescens* isolates. The other four species were distinguished from each other but there was little intraspecific variation. The profile of the *S. kraussei* symbiont was interpreted as being more similar to the *X. beddingii* profile than to the *X. bovienii* profile.[68]

Isozyme analysis of four enzymes confirmed the similarities and differences between the *X. luminescens* isolates detected by total protein profile,[68] and showed differences between the remaining four species. Although the symbiont of *S. kraussei* differed slightly from *X. bovienii*, it was clearly different from *X. beddingii*.

Restriction fragment length differences (RFLDs) in genomic DNA can be used to detect interspecific differences between *Xenorhabdus* species. Intraspecific differences between the type strain of *X. nematophilus*, the symbiont of *S. carpocapsae* DD-136, and the Agriotos and Plougastel (=Breton)[45] strains were detected by hybridizing a labelled probe to Southern blots of genomic DNA.[46]

Intraspecific differences have also been detected in the electrophoretic profiles of lipopolysaccharides of the Breton strain of *X. nematophilus* and those of the Mexican and DD-136 strains.[38]

The usefulness of electrophoresis in the taxonomy of *Xenorhabdus* has yet to be tested by extensive examination of intraspecific variation. Its greatest value will probably lie in the detection of genetic polymorphisms.

G. Future Directions

There remain important taxonomic questions to be answered. Examination of the relationship of bacteria symbiotically associated with *Heterorhabditis* to those associated with the Steinernematidae shows that isolates currently identified as *X. luminescens* should be divided into several taxonomic groups, probably at the species level, and perhaps be assigned to a new genus. However, it would be preferable that an extensive evaluation of DNA/DNA homologies among *Xenorhabdus* be completed to determine whether there are two or more groups at the generic level or a continuum of DNA/DNA homologies. It is also necessary to complement this with a phenotypic study so that any new taxonomic groups can be described adequately. Before describing a new species to accommodate the symbiont of a newly discovered species of nematode, one should examine several isolates of the bacterium to provide a measure of intraspecific variation. The study should also include representative, preferably type, strains of existing species as controls to reduce the problem of the choice of test conditions affecting the result.

There are significant differences among *X. luminescens* isolates in their association with nematodes (Table 2). However, subdivision of *X. luminescens* may not be a simple matter. Numerical analysis of phenotypic data (59 isolates, 45 characters) produced two groups from the 13 *X. luminescens* isolates, with phase one of the Polish strain in one group and phase two in the other.[25] Combining the data of phase one and two isolates[21] may resolve this problem, although it is desirable that phase two isolates be derived from phase one at the outset of the study to minimize the effects of between isolate variation.

The taxonomic relationships between the symbionts of *S. feltiae* (=*bibionis*), *S. kraussei*, *S. affinis*, *S. intermedia*, and some unidentified *Steinernema* spp. are uncertain. Although differences between some of these bacteria have been detected by comparing phenotypes[21,45] and protein profiles[68] and by the use of 16S rRNA probes[64] and DNA/DNA hybridization,[56] there are insufficient data on intraspecific variation to assess whether these reflect differences at the species level. In assessing their relationships, one should consider the nematode/bacterium interaction. *S. feltiae* (=*bibionis*) reproduced well in monoxenic *in vitro* culture with the symbiont of *S. kraussei* and was almost as efficient in retaining it in the infective-stage juvenile intestinal vesicle as its own.[40] However, *S. feltiae* (=*bibionis*) did not associate as effectively with two other strains that are phenotypically similar to *X. bovienii*; it did not retain the symbiont of *S. affinis* (Dan)[45] within the infective-stage intestine and could not be cultured with the symbiont of *Steinernema* sp. F3 (Figure 1).[69]

It may soon be necessary for government insecticide registration and, perhaps for proprietary rights, that *Xenorhabdus* be distinguished at the strain level because the nematode/bacterium association is unlikely to be broken when the organisms are used on a large scale for insect pest control. Lipolysaccharide profiles or isozyme analysis may be useful for distinguishing strains when intraspecific variation has been explored. However, DNA techniques are effective for identification at the species and subspecies levels and are generally easier to use.

IV. CONCLUSIONS

The interaction between nematode and bacterium has not been fully explored. The nature of essential nutrients provided by the bacterium, the significance and underlying mechanism(s) of phase change and the consequences of a stable phase one, the pathogenicity of *Xenorhabdus* for a range of insects, and the mechanism(s) determining retention of *Xenorhabdus* within the infective juvenile intestine have yet to be determined. Moreover, most studies of the role of *Xenorhabdus* have been addressed using *S. carpocapsae* and *X. nematophi-*

lus as a model. Interactions between other *Steinernema* spp. and their symbionts do not always follow this model[4] and require separate examination. This is particularly true for *Heterorhabditis*/X. *luminescens*, a different family of nematode and probably a different genus of bacterium from the model.

ACKNOWLEDGMENTS

The authors thank Drs. M. Brehélin, L. Drif, R.-U. Ehlers, and C. J. Thomas for generously providing us with their unpublished data.

REFERENCES

1. **Thomas, G. M., and Poinar, G. O., Jr.**, *Xenorhabdus* gen. nov., a genus of entomopathogenic nematophilic bacteria of the family Enterobacteriaceae, *Int. J. Syst. Bacteriol.*, 29, 352, 1979.

2. **Akhurst, R. J.**, A *Xenorhabdus* sp. (Eubacteriales: Enterobacteriaceae) symbiotically associated with *Steinernema kraussei*, *Rev. Nématol.*, 5, 277, 1982.

3. **Akhurst, R. J.**, Taxonomic study of *Xenorhabdus*, a genus of bacteria symbiotically associated with insect pathogenic nematodes, *Int. J. Syst. Bacteriol.*, 33, 38, 1983.

4. **Akhurst, R. J.**, *Xenorhabdus poinarii*: its interaction with insect pathogenic nematodes, *System. Appl. Microbiol.*, 8, 142, 1986.

5. **Akhurst, R. J.**, *Xenorhabdus nematophilus* subsp. *beddingii* (Enterobacteriaceae) a new subspecies of bacteria mutualistically associated with entomopathogenic nematodes, *Int. J. Syst. Bacteriol.*, 36, 454, 1986.

6. **Poinar, G. O., Jr.**, *Neoaplectana intermedia* n. sp. (Steinernematidae: Nematoda) from South Carolina, *Rev. Nématol.*, 8, 321, 1985.

7. **Poinar, G. O., Jr., Jackson, T., and Klein, M.**, *Heterorhabditis megidis* sp. n. (Heterorhabditidae: Rhabditida), parasitic in the Japanese beetle, *Popillia japonica* (Scarabaeidae: Coleoptera), in Ohio, *Proc. Helminthol. Soc. Wash.*, 54, 53, 1987.

8. **Poinar, G. O., Jr., Mrácek, Z., and Doucet, M. M. A.**, A re-examination of *Neoaplectana rara* Doucet, 1986 (Steinernematidae: Rhabditida), *Rev. Nématol.*, 11, 447, 1988.

9. **Poinar, G. O., Jr.**, *Nematodes for Biological Control of Insect Pests*, CRC Press, Boca Raton, 1979, 143.

10. **Poinar, G. O., Jr., and Thomas, G. M.**, The nature of *Achromobacter nematophilus* as an insect pathogen, *J. Invertebr. Pathol.*, 9, 510, 1967.

11. **Milstead, J. E.**, *Heterorhabditis bacteriophora* as a vector for introducing its associated bacterium into the hemocoel of *Galleria mellonella* larvae, *J. Invertebr. Pathol.*, 33, 324, 1979.

12. **Götz, P., Boman, A., and Boman, H. G.**, Interactions between insect immunity and an insect-pathogenic nematode with symbiotic bacteria, *Proc. Roy. Soc., Ser. B*, 212, 333, 1981.

13. **Poinar, G. O., Jr., and Thomas, G. M.**, Significance of *Achromobacter nematophilus* Poinar and Thomas (Achromobacteraceae: Eubacteriales) in the development of the nematode, DD-136 (*Neoaplectana* sp., Steinernematidae), *Parasitology*, 56, 385, 1966.

14. **Boemare, N.**, Recherches sur les Complexes Némato-bactériéns Entomopathogèns: Étude Bactériologique, Gnotobiologique et Physiologique du Mode d'Action Parasitaire de *Steinernema carpocapsae* Weiser (Rhabitida: Steinernematidae), Ph.D. thesis, Université des Sciences, Montpellier, 1983.

15. **Brehélin, M., and Boemare, N.,** Immune recognition in insects: conflicting effects of autologous plasma and serum, *J. Comp. Physiol. B*, 17, 759, 1988.
16. **Dunphy, G. B., and Webster, J. M.,** Virulence mechanisms of *Heterorhabditis heliothidis* and its bacterial associate, *Xenorhabdus luminescens*, in non-immune larvae of the greater wax moth, *Galleria mellonella, Int. J. Parasitol.*, 18, 729, 1988.
17. **Akhurst, R. J.,** Antibiotic activity of *Xenorhabdus* spp., bacteria symbiotically associated with insect pathogenic nematodes of the families Heterorhabditidae and Steinernematidae, *J. Gen. Microbiol.*, 128, 3061, 1981.
18. **Poinar, G. O., Jr., Hess, R., and Thomas, G. M.,** Isolation of defective bacteriophages from *Xenorhabdus* spp. (Enterobacteriaceae), *IRCS Med. Sci.*, 8, 141, 1980.
19. **Paul, V. J., Frautschy, S., Fenical, W., and Nealson, K. H.,** Antibiotics in microbial ecology. Isolation and structure assignment of several new antibacterial compounds from the insect-symbiotic bacteria *Xenorhabdus* spp., *J. Chem. Ecol.*, 7, 589, 1981.
20. **Rhodes, S. H., Lyons, G. R., Gregson, R. P., Akhurst, R. J., and Lacey, M. J.,** Xenorhabdin antibiotics, Can. Patent 1214130, U.S. Patent 4672130, 1984.
21. **Akhurst, R. J., and Boemare, N. E.,** A numerical taxonomy study of the genus *Xenorhabdus* (Enterobacteriaceae) and proposed elevation of the subspecies of *X. nematophilus* to species, *J. Gen. Microbiol.*, 134, 1835, 1988.
22. **Gregson, R. P., and McInerny, B. V.,** Xenocoumacins, Int. Patent Appl. PCT/AU83/00156, 1986.
23. **Poinar, G.O., Jr., Thomas, G.M., Haygood, M., and Nealson, K.H.,** Growth and luminescence of the symbiotic bacteria associated with the terrestrial nematode *Heterorhabditis bacteriophora, Soil Biol. Biochem.*, 12, 5, 1980.
24. **Akhurst, R.J., and Boemare, N.E.,** A non-luminescent strain of *Xenorhabdus luminescens* (Enterobacteriaceae), *J. Gen. Microbiol.*, 132, 1917, 1986.
25. **Akhurst, R. J.,** Bacterial Symbionts of Insect Pathogenic Nematodes of the Families Steinernematidae and Heterorhabditidae, Ph.D. thesis, University of Tasmania, Hobart, 1982.
26. **Götz, P., and Boman, H. G.,** Insect immunity, in *Comprehensive Insect Physiology, Biochemistry and Pharmacology*, Vol. 3, Kerkut, G. A., and Gilbert, L. J., Eds, Pergamon Press, Oxford, 1985, 453.
27. **Drif, L., Brehélin, M., and Boemare, N.,** unpublished data, 1989.
28. **Dunphy, G. B., and Webster, J. M.,** Interaction of *Xenorhabdus nematophilus* subsp. *nematophilus* with the haemolymph of *Galleria mellonella, J. Insect Physiol.*, 30, 883, 1984.
29. **Welch, H. E., and Bronskill, J. F.,** Parasitism of mosquito larvae by the nematode, DD-136 (Nematoda: Neoaplectanidae), *Can. J. Zool.*, 40, 1263, 1962.
30. **Poinar, G. O., Jr., and Kaul, H. N.,** Parasitism of the mosquito *Culex pipiens* by the nematode *Heterorhabditis bacteriophora, J. Invertebr. Pathol.*, 39, 382, 1982.
31. **Boemare, N., Laumond, C., and Luciani, J.,** Mise en évidence d'une toxicogenèse provoquée par le Nématode axénique entomophage *Neoaplectana carpocapsae* Weiser chez l'Insecte axénique *Galleria mellonella* L., *C.R. Acad. Sci. Paris*, 295, 543, 1982.
32. **Burman, M.,** *Neoaplectana carpocapsae*: toxin production by axenic insect parasitic nematodes, *Nematologica*, 28, 62, 1982.
33. **Akhurst, R. J.,** unpublished data, 1984.
34. **Bowen, D. J., Barman, M. A. E., Beckage, N. E., and Ensign, J. C.,** Extracellular insecticidal activity of *Xenorhabdus luminescens* strain NC-19, in *Proc. XVIII Int. Congr. Entomol.*, Vancouver, 1988, 256.
35. **Seryczynska, H.,** Toxicity of *Achromobacter nematophilus* Poinar et Thomas cell suspensions against *Leptinotarsa decemlineata* Say, *Bull. Acad. Pol. Sci.*, 23, 347, 1975.
36. **Kamionek, M.,** Effect of heat-killed cells of *Achromobacter nematophilus* Poinar et Thomas, and the fraction (endotoxin) isolated from them on *Galleria mellonella* L. caterpillars, *Bull. Acad. Pol. Sci.*, 23, 277, 1975.

37. **Lysenko, O.,** Principles of pathogenesis of insect bacterial diseases as exemplified by the nonsporeforming bacteria, in *Pathogenesis of Invertebrate Microbial Diseases*, Davidson, E. W., Ed., Allenheld/Osmun, Totowa, 1981, chap. 6.

38. **Dunphy, G. B., and Webster, J. M.,** Lipopolysaccharides of *Xenorhabdus nematophilus* (Enterobacteriaceae) and their haemocyte toxicity in non-immune *Galleria mellonella* (Insecta: Lepidoptera) larvae, *J. Gen. Microbiol.,* 134, 1017, 1988.

39. **Boemare, N., Bonifassi, E., Laumond, C., and Luciani, J.,** Etude expérimentale de l'action pathogène du nématode *Neoaplectana carpocapsae* Weiser; recherches gnotobiologiques chez l'insecte *Galleria mellonella, Agronomie*, 3, 407, 1983.

40. **Akhurst, R. J.,** *Neoaplectana* species: specificity of association with bacteria of the genus *Xenorhabdus, Exp. Parasitol.,* 55, 258, 1983.

41. **Dunphy, G. B, Rutherford, T. A., and Webster, J. M.,** Growth and virulence of *Steinernema glaseri* influenced by different subspecies of *Xenorhabdus nematophilus, J. Nematol.,* 17, 476, 1985.

42. **Akhurst, R. J.,** The nematode/bacterium complex, *Steinernema glaseri*/*Xenorhabdus nematophilus* subsp. *poinarii,* pathogenic for root-feeding scarab larvae, in *Proc. 4th Australasian Conf. Grassland Invertebr. Ecol.,* Chapman, R. B., Ed., Caxton Press, Lincoln, 1985, 262.

43. **Bird, A. F., and Akhurst, R. J.,** The nature of the intestinal vesicle in nematodes of the family Steinernematidae, *Int. J. Parasitol.,* 13, 599, 1983.

44. **Akhurst, R. J.,** Morphological and functional dimorphism in *Xenorhabdus* spp., bacteria symbiotically associated with the insect pathogenic nematodes *Neoaplectana* and *Heterorhabditis, J. Gen. Microbiol.,* 121, 303, 1980.

45. **Boemare, N. E., and Akhurst, R. J.,** Biochemical and physiological characterization of colony form variants in *Xenorhabdus* spp. (Enterobacteriaceae), *J. Gen. Microbiol.,* 134, 751, 1988.

46. **Akhurst, R. J., and Boemare, N. E.,** unpublished data, 1988.

47. **Boemare, N., Louis, C., and Kuhl, G.,** Etude ultrastructurale des cristaux chez *Xenorhabdus* spp., bacteries inféodées aux nématodes entomophages Steinernematidae et Heterorhabditidae, *C. R. Sci. Soc. Biol.,* 177, 107, 1983.

48. **Couche, G. A., Lehrbach, P. R., Forage, R. G., Cooney, G. C., Smith, D. R., and Gregson, R. P.,** Occurrence of intracellular inclusions and plasmids in *Xenorhabdus* spp., *J. Gen. Microbiol.,* 133, 967, 1987.

49. **Brehélin, M. A., and Boemare, N. E.,** Recognition of particulate material by haemocytes in *Locusta migratoria, Dev. Comp. Immunol.,* 10, 639, 1986.

50. **Thomas, C. J.,** personal communication, 1989.

51. **Poinar, G. O., Jr., Hess, R. T., Lanier, W., Kinney, S., and White, J. H.,** Preliminary observations of a bacteriophage infecting *Xenorhabdus luminescens* (Enterobacteriaceae), *Experientia*, 45, 191, 1989.

52. **Akhurst, R. J.,** unpublished data, 1980.

53. **Poinar, G. O., Jr.,** personal communication, 1989.

54. **Farmer, J. J., III, Jorgensen, J. H., Grimont, P. A. D., Akhurst, R. J., Poinar, G. O., Jr., Ageron, E., Pierce, G. V., Smith, J. A., Carter, G. P., Wilson, K. L., and Hickman-Brenner, F. W.,** *Xenorhabdus luminescens* (DNA hybridization group 5) from human clinical specimens, *J. Clin. Microbiol.,* 27, 1594, 1989.

55. **Curran, J.,** Chromosome numbers of *Steinernema* and *Heterorhabditis* species, *Rev. Nématol.,* 12, 145, 1989.

56. **Akhurst, R. J., Boemare, N. E., and Mourant, R. G.,** unpublished data, 1989.

57. **Johnson, J. L.,** Nucleic acids in bacterial classification, in *Bergey's Manual of Systematic Bacteriology*, Vol.1, Krieg, N. R., Ed., Williams and Wilkins, Baltimore, 1985, 8.

58. **Wayne, L. G., Brenner, D. J., Colwell, R. R., Grimont, P. A. D., Kandler, O., Krichevsky, M. I., Moore, L. H., Moore, W. E. C., Murray, R. G. E., Stackebrandt, E., Starr, M. P., and Trüper, H. G.,** Report of the *ad hoc* committee on reconciliation of approaches to bacterial systematics, *Int. J. Syst. Bacteriol.,* 37, 463, 1987.

59. **Grimont, P. A. D., Steigerwalt, A. G., Boemare, N., Hickman-Brenner, F. W., Deval, C., Grimont, F., and Brenner, D. J.,** Deoxyribonucleic acid relatedness and phenotypic study of the genus *Xenorhabdus, Int. J. Syst. Bacteriol.,* 34, 378, 1984.

60. **Fox, G. E., Pechman, K. R., and Woese, C. R.,** Comparative cataloguing of 16S ribosomal ribonucleic acid: molecular approach to procaryotic systematics, *Int. J. Syst. Bacteriol.,* 27, 44, 1977.

61. **Ehlers, R-U., Wyss, U., and Stackebrandt, E.,** 16S rRNA cataloguing and the phylogenetic position of the genus *Xenorhabdus, System. Appl. Microbiol.,* 10, 121, 1988.

62. **Brenner, D. J.,** Family Enterobacteriaceae, in *Bergey's Manual of Systematic Bacteriology,* Vol. 1, Krieg, N. R., Ed., Williams and Wilkins, Baltimore, 1985, 506.

63. **Ramia, S., Neter, E., and Brenner, D. J.,** Production of enterobacterial common antigen as an aid to classification of newly identified species of the families Enterobacteriaceae and Vibrionaceae, *Int. J. Syst. Bacteriol.,* 32, 395, 1982.

64. **Ehlers, R-U.,** personal communication, 1989.

65. **Baumann, L., and Baumann, P.,** Immunological relationships of glutamine synthetases from marine and terrestial enterobacteria, *Curr. Microbiol.,* 3, 191, 1980.

66. **Baumann, L., Bang, S. S., and Baumann, P.,** Study of relationships among species of *Vibrio, Photobacterium,* and terrestrial enterobacteria by an immunological comparison of glutamine synthetase and superoxide dismutase, *Curr. Microbiol.,* 4, 133, 1980.

67. **Poinar, G. O., Jr.,** *Nematodes for Biological Control of Insects,* CRC Press, Boca Raton, 1979, 199.

68. **Hotchkin, P. G., and Kaya, H. K.,** Electrophoresis of soluble proteins from two species of *Xenorhabdus,* bacteria mutualistically associated with the nematodes *Steinernema* spp. and *Heterorhabditis* spp., *J. Gen. Microbiol.,* 130, 2725, 1984.

69. **Akhurst, R. J., and Boemare, N. E.,** unpublished data, 1984.

70. **Akhurst, R. J.,** Use of starch gel electrophoresis in the taxonomy of the genus *Heterorhabditis* (Nematoda: Heterorhabditidae), *Nematologica,* 33, 1, 1987.

Ecology

5.   Soil Ecology

Harry K. Kaya

I. INTRODUCTION

Soil, the natural habitat for entomopathogenic nematodes, varies greatly in chemical composition and physical structure. It is a dynamic system in a continual state of flux. Combined with its physical, biological, and chemical complexity, this dynamic state makes soil a difficult medium in which to conduct quantitative research. Despite the complexity and variety of this environment, entomopathogenic nematodes have been recovered from soils throughout the world; their distribution may primarily be limited by the availability of susceptible hosts.

Initial studies with an entomopathogenic nematode were conducted in the 1930s to test *Steinernema glaseri* as a biological control agent of Japanese beetle grubs in soil.[1] In contrast, when *S. carpocapsae* was isolated in the 1950s, it was used principally as a biological insecticide against above ground insects.[2] Excellent control was obtained against insects in moist, cryptic habitats, but many failures occurred against foliage-feeding insects.[3] The success in cryptic habitats and misdirected use of this nematode as a biological insecticide against foliage-feeding insects probably accounted for the delay in intensive soil studies. The lack of researchers and resources in insect nematology also contributed significantly to this delay. During the 1980s, as applications made against foliage-feeding insects generally proved to be ineffective because of rapid desiccation and inactivation by sunlight, a shift to usage in their natural habitat against soil insects occurred. These trials had greater success, particularly against root weevils, but field evaluations of efficacy against soil insect pests other than root weevils produced inconsistent results.[3,4] The sources of variation included application methods, nematode species and strains, host stage and defense mechanisms, and the biotic and abiotic soil environments. Thus, Gaugler[3] stated that "We lack definitive information on the fate of nematodes introduced into the soil, on factors regulating their population dynamics, on optimal conditions for epizootic initiation, and on the ecological barriers to infection."

Our knowledge of entomopathogenic nematode behavior in the soil environment is limited. If these nematodes are to be exploited in their natural habitat, we need to understand their interactions in the soil environment. This knowledge would form the basis for further studies on population dynamics and epizootiology of nematode diseases in soil. Ultimately, this information will allow us to optimize the control potential of entomopathogenic nematodes against soil insects.

II. SOIL ENVIRONMENT

Phytonematologists have long recognized the importance of the soil environment on plant-parasitic nematodes so a number of reviews elucidating soil-nematode interactions are available.[5-7] More recently, entomologists have focused their attention on soil insect ecology, particularly for pests of agricultural importance.[8] Their work will also provide important background information for scientists working with entomopathogenic nematodes. Conducting further ecological studies concentrating on the interactions of steinernematids and heterorhabditids will require a basic understanding of the soil environment. Detailed information on the technical aspects of soil is available in textbooks in soil science,[9-11] and only a brief summary is provided here as background for this chapter.

Soil texture, a basic concept in soil science, is determined by the ratio of particle types, which are divided into three arbitrary fractions based on size: sand, silt, and clay. According to the classification of the U.S. Department of Agriculture, sand ranges from 2.0-0.05 mm in diameter, silt between 0.05 and 0.002 mm, and clay <0.002 mm. Textural class names such as sand, sandy loam, loam, sandy clay, and silty clay describe the approximate ratios of these particle types.

Wallace[5] states that the principal soil factors affecting nematodes are pore size, water, aeration, temperature, and the chemistry of the soil solution. The pore space of a soil is occupied by air and water as well as nematodes and other soil organisms. The amount of pore space is determined, in part, by the arrangement of the solid particles,[9] and is thus dependent on soil texture. Sandy soil have large pore spaces but less total pore space than loam or clay soils. Soil depth, overall compaction, and cultivation techniques also influence porosity.

Although many factors affect nematode activity in soil, moisture is considered central. The many methods for measuring soil water fall into two general categories, those calculating the moisture content and those determining the soil water potential. The moisture content is an expression of the grams of water per 100 g of dry soil. Water potential is a measure of the forces holding water in the soil. Adsorption of water molecules to the soil particles, capillary forces, and osmotic pressure from solutes contribute to the total water potential.[12] Soil moisture content is easier to measure, but in terms of comparing the biological availability of water in different soil types, water potential is a better measurement. Its use should be encouraged so that the results of different investigations can be compared.

Soil aeration depends on oxygen consumption and carbon dioxide production. Oxygen is replaced by diffusion from the atmosphere into the soil pores; carbon dioxide diffuses from the soil pores to the atmosphere. For gaseous movement to occur, there must be a gradient of both gases. The gradient is established by biological and chemical activities in the soil, and the rate of diffusion is controlled by the physical properties of the soil. As nematodes live in the water film within the soil pores, their oxygen diffuses from the pore space into the water film, and their waste carbon dioxide diffuses from the water into the pore space.

Soil temperature is determined to a large extent by factors which control transfer of heat in and out of soil. Wet soil has a greater conductance and smaller rise in temperature than dry soil with the same input of heat at the surface. Thus, solar heat penetrates deeper in wet soil but produces a smaller rise in temperature than in dry soil. Wet or dry, soil undergoes slower and smaller temperature fluctuations than the atmosphere. The deeper layers are more buffered than the surface, which tends to heat and cool rapidly along with the atmosphere and under the influence of direct sunlight. During the day, higher surface temperature allows for heat flow downward; at night, when the soil surface cools, heat flows upward.

The chemistry of the soil solution affects nematodes through such factors as pH and osmotic pressure. In most soils, pH ranges from 4 to 8 and probably has little effect on nematode activity. In fact, survival and pathogenicity of *S. glaseri* and *S. carpocapsae* decreased only slightly as soil pH decreased from pH 8 to pH 4 but were adversely affected at pH 10.[13] Chemicals, however, may act directly as orientation stimuli for nematodes, and steinernematids do respond to various ions and chemicals.[14]

Plants directly and indirectly affect nematodes in many ways. They create a moisture gradient in the rhizosphere. Plant roots also respire, thus reducing oxygen and increasing the carbon dioxide in the surrounding soil, and creating gradients of oxygen and carbon dioxide close to the roots. In addition, plants influence soil temperature by intercepting solar radiation.

Plants are central to the food web. Microbes, invertebrates, and vertebrates utilize all plant parts as a food source, and exudates from the roots affect the soil microflora. The invertebrate and vertebrate herbivores provide a food source for predators and parasites and produce waste products which serve as food for saprophytes. All these organisms eventually die and provide additional food for saprophytes. Most of these activities occur in the soil and add to the complexity of the soil environment. With seasonal climatic variation and human activities come further complications, making it difficult to interpret cause and effect. Simplistic generalizations do not easily explain the myriad of interacting abiotic and biotic factors in natural and cultivated soils. Consequently, research under controlled conditions has been conducted to present hypotheses to explain activities of entomopathogenic nematodes in the soil environment.

III. DISPERSAL

Factors affecting the movement of steinernematid and heterorhabditid nematodes in soil are examined in this section. Unfortunately, in many cases, the experiments discussed here put an emphasis on the dispersal capability of the nematode species at a constant temperature and moisture level rather than in a particular soil type.

Movement can be active or passive. In the former, dispersal of nematodes is through their own locomotion and in the latter, dispersal is through the action of another agent.

A. Active Dispersal

When the infective juveniles of steinernematids or heterorhabditids are placed on a soil surface, the majority of them remain near the point of placement.[15-19] The exception is *S. glaseri*.[19] In general, only a small proportion of the steinernematid or heterorhabditid populations disperse. Those that do disperse are found 4 to 90 cm from the point of placement (Tables 1 and 2). When infective juveniles are placed in the middle of a soil column, significantly more *Heterorhabditis bacteriophora*[18] and *S. carpocapsae*[16] move upwards than downwards, whereas significantly more *S. glaseri* move downwards than upwards.[17] In a limited, unreplicated field study, *S. carpocapsae* applied in a furrow was recovered 46 cm away from the placement site within 2 weeks.[20] Rain occurred 11 days after application and may account, in part, for the rapid dispersal through field soil.

Soil texture and the presence of a host also affect dispersal (Tables 1 and 2).

TABLE 1. Vertical Dispersal of *Steinernema carpocapsae, S. glaseri, S. feltiae (=S. bibionis)*, and *Heterorhabditis bacteriophora* in Soil

Soil texture (%)			Time in soil (days)	Host present	Dispersal distance (cm)[a]		Ref.
sand	silt	clay			downwards	upwards	
S. carpocapsae							
95	-	<5[b]	30	-	15	45	19
100	0	0	5	-	10[c]	-	16
100	0	0	5	+	10[c]	-	16
100	0	0	5	-	10[c]	-	16
80	10	10	5	-	10[c]	-	16
80	10	10	5	+	10[c]	-	16
65	15	20	5	-	6	-	16
65	15	20	5	+	10	-	16
42	24	34	5	-	6[c]	-	16
42	24	34	5	+	6[c]	-	16
100	0	0	2	-	10	-	15
100	0	0	7	+	15[d]	15[d]	59

TABLE 1 (continued). Vertical Dispersal of *Steinernema carpocapsae, S. glaseri, S. feltiae (=S. bibionis),* and *Heterorhabditis bacteriophora* in Soil

Soil texture (%)			Time in soil (days)	Host present	Dispersal distance (cm)[a]		Ref.
sand	silt	clay			downwards	upwards	
				S. glaseri			
95	-	<5[b]	30	-	90	90	19
100	0	0	5	-	14[c]	-	17
100	0	0	5	+	14[c]	-	17
80	10	10	5	-	14[c]	-	17
80	10	10	5	+	14[c]	-	17
65	15	20	5	-	8	-	17
65	15	20	5	+	10	-	17
42	24	34	5	-	6[e]	-	17
42	24	34	5	+	6[e]	-	17
100	0	0	7	+	35[d]	30[d]	59
				S. feltiae			
100	0	0	7	+	25[d]	20[d]	59
				H. bacteriophora			
95	-	<5[b]	30	-	0	45	19
80	10	10	5	-	10[c]	-	19
80	10	10	5	+	10[c]	-	18
100	0	0	7	+	35[d]	35[d]	59

[a] The nematode movement to the greatest distance away from the placement site.
[b] Combination of silt and clay.
[c] A significant difference in nematode distribution between host present and absent occurred within the same soil texture. Higher nematode numbers were recovered from soil with the host present.
[d] Dispersal distance was assessed by infection of host.
[e] No significant difference in nematode distribution.

TABLE 2. Horizontal Dispersal of *Steinernema carpocapsae, S. glaseri, S. feltiae (=S. bibionis)* and *H. bacteriophora* in Soil

| Soil texture (%) | | | | | | |
sand	silt	clay	Time in soil (days)	Host present	Dispersal distance (cm)[a]	Ref.
			S. carpocapsae			
95	-	<5[b]	30	-	15	19
100	0	0	2	-	14	15
100	0	0	7	+	25[c]	59
40	33	26	14	-	46[c]	20
			S. glaseri			
95	-	<5[b]	30	-	90	19
100	0	0	7	+	30[c]	59
			S. feltiae			
100	0	0	7	+	30[c]	59
			H. bacteriophora			
95	-	<5[b]	30	-	45	19
100	0	0	7	+	30[c]	59

[a]   Nematode movement is recorded for the greatest distance away from the placement site.
[b]   Combination of silt and clay.
[c]   Dispersal distance was assessed by infection of host.
[d]   Study was conducted in the field.

In heavy clay soils, nematode movement is impaired and the presence of a host does not increase infective juvenile dispersal.[16,17] Thus, porosity affects nematode movement, with less dispersal occurring as the percentage of silt and clay increases in the soil.[16,17]

With the exception of *S. glaseri*, most entomopathogenic nematodes do not become active dispersers in the presence or absence of hosts. Although the presence of a host significantly increases the number of dispersing infective juveniles, the majority are still found near the placement site.[16-18] Ishibashi and

Kondo[21] suggested that *S. carpocapsae* enters a quiescent state following application to the soil. Gaugler et al.,[22] in studies to improve host-finding ability, suggested that only a small proportion of *S. carpocapsae* became aggressive host seekers (dispersers); the larger proportion conserved their energy and presumably waited for the host to come to them or became aggressive only in close proximity to a host. The heterorhabditids also appeared to not move as readily in soil as compared with *S. glaseri*, but recent studies indicated that five heterorhabditids placed on a soil surface were sufficient to infect a high proportion of hosts buried in soil.[23]

Laboratory studies in petri dishes have shown that these nematodes respond positively to a number of chemical and physical cues produced by insects.[2] Despite these studies, no definitive experiments have been conducted to unequivocally demonstrate that the nematodes respond to kairomones in the soil. Heterorhabditids and *S. glaseri* appear to initiate random movement in soil and, when they are in close proximity to a host, may utilize chemical and physical cues for host finding.

The effects of moisture and temperature on *S. glaseri* dispersal in bark compost in the presence of a host have been studied.[24] Under dry conditions (defined as 25% maximum water capacity ratio) the percentage of nematodes dispersing increased from 0 to 50% as the temperature rose from 5°C to 25°C. Soil moisture levels >50% reduced nematode dispersal. Dry conditions and low temperatures inhibited nematode movement, because of the nonavailability of a water film in which to move and temperature-induced sluggishness. Excess water reduced nematode movement due to anoxia and slippage.[5]

B. Passive Dispersal

Numerous means of passive dispersal are known for nematodes. Entomogenous nematodes can be dispersed by water, wind,[25] and human activity, but dispersal by these agents, especially wind, has not been adequately documented with steinernematids and heterorhabditids. Though both active and passive dispersal were probably involved, irrigation after application of steinernematids and heterorhabditids as biological insecticides against soil-inhabiting insects was shown to enhance efficacy.[26] This suggested that water dispersed the nematodes downward through thatch and the soil pores.

The dispersal of steinernematid by small invertebrates functioning as phoretic hosts has been observed under laboratory conditions.[27] For *S. carpocapsae*, eight species of mesostigmatid mites and one species of oribatid mite served as phoretic hosts, some species being better hosts than others. The nematodes escaped consumption by nematophagous mites by orienting perpendicular to a substrate and bridging onto the dorsum of a searching mite. Over time, the nematodes were tightly packed on the dorsum of the mites. Although eight of nine mite species fed on the nematodes, they still served as phoretic hosts. Phoretic hosts may be important dispersal agents of steinernematids in the soil environment, taking them greater distances than if they moved on their own energy. No phoretic relationship between a heterorhabidi-tid and mites was observed, and a different behavioral strategy (i.e., active downward movement as opposed to nictating behavior) may account for the lack of a phoretic relationship.

Nematode-infected hosts, which live about 48 hr before they die, may also serve as a means to disperse nematodes in the soil. Infected larvae[28] may move laterally and downward in soil, and infected adults[29] may fly several meters before dying and establishing new foci of infection.

IV. SURVIVAL

Studies on factors affecting nematode survival or persistence in soil have focused on moisture, temperature, and more recently, parasites and predators. A major criticism of many of these studies is that they have been conducted with sterilized or pasteurized soil. However, in defense of this approach, baseline data are required to understand survival in soil. Once these data are obtained, the impact of biotic or other factors can be evaluated. One aspect of survival is the ability of the nematode to recycle, which is discussed in Section VI.

A. Soil Texture

Soil texture affects survival of the steinernematids *S. carpocapsae* and *S. glaseri*.[30] When infective juveniles were placed in sterilized sand, sandy loam, clay loam, or clay soils for 16 weeks, the lowest survival was seen in the clay soil for both nematode species. Survival of *S. carpocapsae* and *S. glaseri* was best in the sandy loam and sand soil, respectively. Generally, higher clay content resulted in lower nematode survival, and survival of *S. glaseri* was lower than *S. carpocapsae* in all soil textures. Pathogenicity studies paralleled results obtained for survival. As pore space size and aeration are reduced in clay soils, these two parameters probably have a major influence on survival. The larger, motile *S. glaseri* would have expended more energy supplies during dispersal in the smaller pore spaces than the inactive *S. carpocapsae*. Low oxygen levels would have also been more detrimental to the active *S. glaseri*. This explanation covers the clay soils, but reasons for the lower survival of *S. glaseri* in the sandy soil remains unknown. Perhaps *S. glaseri* abrades its cuticle during movement in the sand, resulting in lower survival.

B. Moisture and Temperature

S. carpocapsae can survive some degree of desiccation provided it occurs slowly.[31] Nematodes placed in moist soil survived for 20 days when the soil was slowly dried at 70% relative humidity (rh).[32] When *S. carpocapsae* and *S. glaseri* were placed in a sandy loam soil at different soil moistures, both species survived best at low soil moistures (2 and 4%, respectively).[33] In simulated soil conditions, 90% of *S. carpocapsae* placed in humidity chambers were alive after 12 days at 79.5% rh.[34] Because steinernematids survive at relatively low soil moisture, moisture levels normally occurring in natural soils may not be a limiting factor (i.e., nematodes survive but host finding and infectivity may be reduced).

Temperature alone had a direct effect on steinernematid and heterorhabditid survival in sand at 7% moisture (water potential of -0.03 bars).[35] More than 90% of *S. glaseri* survived for 32 weeks at 15°C in the absence of an insect host. At lower and higher temperatures, survival decreased; overall better survival was observed at the cooler temperatures. *S. carpocapsae* was recovered at a high proportion (ca. 60%) after 32 weeks at 10°C. At the other temperatures, nematode survival was <10% after ca. 6 weeks. *H. bacteriophora* showed trends similar to those of *S. carpocapsae* but at an overall lower survival rate, whereas *Heterorhabiditis* sp. D1 did not survive beyond 8 weeks at any temperatures except 15°C. Molyneux[35] observed that within 2 weeks at 23° and 28°C, the heterorhabditids and *S. carpocapsae* appeared transparent and lethargic, whereas *S. glaseri* became transparent after 8 weeks at 28°C. Moreover, *S. glaseri* were frequently observed in an immobile, coiled position, a condition associated with enhanced survival.[31] Coiling was not observed with the heterorhabditids or *S. carpocapsae*.

The poor survival of heterorhabditids and *S. feltiae* (=*S. bibionis*) at the higher temperatures was probably related to their relatively high motility and respiration which would have quickly depleted food reserves.[36] At lower temperatures, these activities were minimal and increased their survival, but *Heterorhabditis* sp. D1 did not survive well at 10°C, probably because this nematode is adapted to tropical conditions. The excellent survival of *S. glaseri* was attributed to its ability to become inactive, a strategy also used by some plant-parasitic nematodes.

In a sandy loam soil at 12% moisture, survival of *S. carpocapsae* was greater at 5 to 15°C than at 35°C.[33] In contrast, survival of *S. glaseri* was greater at 15 to 35°C than at 5°C. The reason for the temperature-related difference in survival for *S. carpocapsae* and *S. glaseri* was ascribed to their possible origins. The temperate origin of *S. carpocapsae* may favor its survival at low temperatures, and the tropical or subtropical origin of *S. glaseri* may favor its survival at high temperatures.[33]

Temperatures above 30°C tend to inhibit nematode development in a host.[37,38] Temperatures above 35° over an extended period of time are detrimental to infective juveniles.[39] Generally, these temperatures are not experienced by naturally occurring nematodes in the field unless the host cadaver is near the soil surface. Deeper in the soil, the infective juveniles are buffered from environmental extremes. As many species and strains of steinernematids and heterorhabditids have been isolated from temperate soils, cooler temperatures are not detrimental to survival. Moreover, the nematodes, because of their motility, may move below the frost line to survive. Studies by Molyneux[36] strongly suggest that the "native" home of the nematode species and strains determines their ability to tolerate temperature extremes.

Only limited studies have been conducted to evaluate the combined effects of moisture and temperature on survival. In tests with *S. carpocapsae*, temperature was the single factor that affected nematode survival in a silty loam soil.[40] Survival (>75%) was greatest at 30°C at 10, 20, and 30% soil moisture over 1 or 2 weeks. At temperatures above 30°C, survival decreased significantly as temperature increased at the three moisture levels. At temperatures below 30°C survival generally decreased. Although laboratory data suggest that *S. carpocapsae* can survive warm temperatures (40°C) occurring near the soil surface for at least 2 weeks when moisture levels are adequate,[40] in the field nematodes exposed on the unprotected soil surface probably desiccate and die; they are also exposed to the direct detrimental effects of high temperature and ultraviolet radiation.[41]

Obvious differences in the data are apparent among the various researchers. Yet, some trends are emerging and we can speculate on the survival strategies of these entomopathogenic nematodes. The dispersal studies showed that *S. glaseri* is a very motile species compared with *S. carpocapsae*; however, the highly motile species is often observed to persist longer than the less motile species. The following hypothesis is proposed to explain these differences. The larger *S. glaseri* possesses abundant storage reserves, and actively searches for hosts (abundant food reserves may not be correlated with overall survival because a larger nematode is expected to have more reserves). If no host is found and the storage material reaches a critical point, the nematode coils, enters an inactive state and only becomes active when a suitable host is nearby. The smaller *S. carpocapsae* has less storage reserves and is less active, but because it cannot or does not coil, it moves until its reserves are exhausted and then dies. Heterorhabditids, smaller than *S. glaseri* and more active than *S. carpocapsae* and noncoilers, also die quickly. Thus, coiling may be a key survival strategy for some nematodes.[31]

C. Aeration

In nonsoil laboratory studies, *S. carpocapsae* can survive with oxygen tension as low as 0.5% of saturation at 20°C.[42] In another study, the infective juveniles apparently survived only when this amount of oxygen was bubbled in water.[43] Lindegren et al.[43] demonstrated that the respiratory rate of *S. carpocapsae* was temperature dependent and that the nematode died at low oxygen concentrations. In sandy loam soil, survival of *S. glaseri* and *S. carpocapsae* decreased as oxygen concentrations decreased from 20 to 1%.[13] *S. carpocapsae* survived better than *S. glaseri* at oxygen concentrations of 1, 5, and 10% after 2 weeks. After this time period, both species showed a rapid decline in survival even at 20% oxygen concentration. Although oxygen is essential for nematode survival, other factors such as moisture, temperature, and soil type also play important roles. Oxygen becomes a limiting factor in clay soils, water-saturated soils, or soils with high organic content.

D. Pesticides

A number of studies have shown that *S. carpocapsae* and *H. bacteriophora* can survive exposure to various kinds of chemical pesticides.[2] In some cases, a given pesticide was nematostatic, but its removal through washing resulted in "normal" behavior. When *S. carpocapsae* was exposed to fenamiphos, no effect on survival or infectivity was observed when the infective juveniles were washed from the treated sand after 4 days.[44] Presumably, long-term exposure to pesticides will cause nematode mortality.

E. Antagonists

Antagonist is a plant pathological term which refers to biological control agents with the potential to interfere with the life processes of an organism. Natural enemies, an entomological term, refers to parasitoids, predators, and pathogens. The term "antagonist", which encompasses antibiosis, competition, parasites, pathogens, and predators, is widely accepted in the biological control literature of plant-parasitic nematodes and will be used in this chapter. Many antagonists of nematodes exist and a number of reviews on the subject, particularly with plant-parasitic nematodes, are available.[45-49]

Survival of *S. carpocapsae* and *S. glaseri* was greater in sterilized and nonsterilized bark compost and sterilized sandy soil than in nonsterile sandy soil.[50] Kaya et al.[51] also observed better survival of *S. carpocapsae* in sterilized soil as compared with nonsterilized soil. The dearth of micro- and macroorganisms in the sterilized soils probably accounted for the enhanced survival of the steinernematids. In the case of the nonsterilized compost, decomposing organic material in a compost pile can reach temperatures of 60°C, accounting for the poor soil biota. This pasteurization probably explains the enhanced persistence. The direct cause for the reduced survival of the steinernematids in nonsterile soil is unknown, and a number of factors could contribute to these results. These factors would range from the production of toxic metabolites by microorganisms to predation by invertebrates.

Antagonists, particularly bacteria, fungi, and predatory invertebrates, reduce plant-parasitic nematode populations, and steinernematids and heterorhabditids are susceptible to a number of these same antagonists. Under laboratory conditions, the entomopathogenic nematodes are infected by nematophagous fungi[52,53] and microsporidians,[54] trapped by nematode-trapping fungi,[55] and preyed upon by mononchid and dorylaimid nematodes,[56] mites,[27,56] a collembolan species,[56] and a tardigrade species.[56] These observations suggest that some of these antagonists may be significant mortality factors in the soil.

With the great array of antagonists capable of reducing nematode numbers, there is considerable pressure for evolution of strategies to avoid predation and infection. Thus, the retention of the second-stage cuticle of infective juveniles may have evolved as a survival mechanism. Small mesostigmatid mites, which puncture (or mangle) their prey and ingest fluids, had difficulty attacking the larger infective juveniles, suggesting that the second-stage cuticle might be an effective defense structure.[27] Exactly how the second-stage cuticle protected the nematode from mite attack is not clear. Conceivably, the loose cuticle becomes entangled within the mites' appendages and discourages further feeding activity.

The second-stage cuticle has been shown to have a protective function against other antagonists. Steinernematids trapped by nematode-trapping fungi can escape by slipping out of their second-stage cuticle.[55] Infective juveniles with the second-stage cuticle were not susceptible to infection by the nemato-phagous fungi, *Drechmeria coniospora* and *Hirsutella rhossiliensis*.[52] Conidial attachment was observed on the second-stage cuticle of heterorhabditids in soil[57] and in petri dishes,[52] but no infection occurred. Removal of the second-stage cuticle resulted in conidial attachment to the third-stage cuticle and infection. It is tempting to argue that the heterorhabditids, because they retain their second-stage cuticle, are refractory to nematophagous fungi and have developed a strategy of actively searching for their host, whereas steinernemat-ids, because they lose their second-stage cuticle, are highly susceptible to fungal infection and have developed the strategy of inactivity, wait for the host to approach them. Although this argument appears attractive, *S. glaseri*, an active disperser, loses its cuticle readily and is susceptible to fungal infection.

Jaffee et al.[58] studied the impact of *Hirsutella rhossiliensis* on field populations of the plant-parasitic nematode, *Criconomella xenoplax*. This fungus is apparently weakly density dependent and has little impact on *Criconoemella* populations. Whether this situation will hold true for steinernematids and heterorhabditids remains to be seen.

Competition between entomopathogenic nematode species or between these nematodes and other organisms has been documented. When a steinerne-matid species and *H. bacteriophora* were placed in equal numbers on soil and allowed to compete for the same lepidopteran hosts, the steinernematids infected a greater proportion of proximally located hosts than *H. bacteriophora*; *H. bacteriophora* infected a greater proportion of distally located hosts than the steinernematids.[59] *H. bacteriophora*'s greater motility accounts, in part, for its ability to infect hosts located farthest from the point of nematode application. Motility, per se, is not the only factor to be considered, because *S. glaseri* shows greater motility than *H. bacteriophora* and infected more hosts located proximally than distally. Similarly, the steinernematids infected a greater proportion of hosts located proximal to nematode placement than *H. bacteriophora*. *H. bacteriophora*, in most cases, probably reached the host before the steinernematid because of its greater motility. The steinernematids, being intrinsically superior, prevented *H. bacteriophora* from colonizing the host. In a few instances when a dual infection occurred in the same host, neither nematode species survived.

In most field soils, only one species of steinernematid or heterorhabditid seem to occur, but on some occasions a heterorhabditid and a steinernematid have been isolated from the same soil.[60-62] Unless these nematode species are separated temporally or spatially, some form of competition can be expected.

Competition studies with other insect pathogens in the same hosts have been summarized by Kaya and Burlando[63] and Barbercheck and Kaya.[64] Briefly, baculoviruses do not affect nematode development, but interference with nematode development occurs in hosts infected with *Bacillus thuringiensis.*[63] On the other hand, *S. carpocapsae* and its mutualistic bacterium inhibit *Beauveria bassiana* development.[64] Although none of these interactions was studied in the soil, they probably occur because *B. thuringiensis*-killed, fungal-killed, and virus-killed insects are found in or on soil.

Steinernematids can be antagonists to plant-parasitic nematodes.[65] Inundative application of *S. glaseri* prevented the establishment of root-knot nematode juveniles in tomato roots. Apparently, *S. glaseri* accumulated around the growing tips of the roots in response to carbon dioxide and prevented root penetration by 2-week-old root-knot juveniles. The addition of *S. glaseri* or *S. carpocapsae* also reduced populations of native plant-parasitic and free-living nematodes in sandy soil.[50] Free-living nematodes surpassed their initial density levels after 8 weeks, but the plant-parasitic nematodes remained at a low level. The reason for the initial decrease in nematode numbers is unknown, although the authors speculated that the competition for space or habitat in the soil may have been a factor.

V. HOST-FINDING AND INFECTIVITY

Host-finding and infectivity in the soil have not been studied extensively. The relationship between presence of a host and dispersal was discussed in Section III.A.

A. Soil Texture and Moisture

Soil texture influenced the ability of *S. glaseri* and *Heterorhabditis* sp. D1 to infect their hosts.[66] Infection was less in a clay loam soil with high clay content compared with a sandy soil. Similar observations with *S. carpocapsae* and *H. bacteriophora* were made by Choo[67] and Geden et al.[68]

Moisture affected infectivity of the host in clay loam, loamy sand, and fine sand.[66] At low water potential ($>$-1.0 bars) in the fine sand or clay loam soil, or very high water potentials ($\leq$-0.01 bars) in the clay loam, no nematode infection occurred. In loamy sand, nematode infection was obtained at water potentials between -0.01 and -100 bars (very wet to very dry), and in the fine sand, nematode infection was observed from water potentials of about -0.003 bars (near saturation) to -0.4 bars. *S. glaseri*, the larger nematode, was less effective in the clay loam with its small pore diameters than the smaller heterorhabditid species. Thus, nematode species, soil texture, and moisture interact to drastically affect the nematodes' ability to find and infect a host. The possibility for nematode infection is better over a wider range of water potentials in a sandy soil containing some silt and clay than in a clay soil. In all cases, water-saturated soil is detrimental to nematodes because oxygen in the soil is limiting, and active movement is impaired. Moreover, these conditions are detrimental for most insects, resulting in nonavailability of hosts for the nematode.

B. Temperature and Moisture

At a water potential of -0.03 bars, *S. feltiae* (=*S. bibionis*) could infect hosts in soil at temperatures varying from 2° to 30°C, whereas heterorhabditids were able to infect hosts from 7° to 35°C.[35,36] However, the temperature range for infection of a given nematode species or strain depended on its "native" home (Table 3). For example, *S. feltiae* (=*S. bibionis*) strain N60, whose native home temperature ranges from 7-20°C, was very active at low temperatures and infected >70% of its host from 2-20°C. In contrast, *Heterorhabditis* sp. D1, whose native home temperature ranges from 26 to 29°C, infected >70% of its host from 20-33°C.

C. Hosts

Behavioral, morphological, and physiological defense strategies of target insects may affect the nematodes' ability to infect the host in the soil environment.[3,69] Gaugler[3] provides an excellent discussion on strategies evolved by soil insects to avoid or minimize infection by steinernematid and heterorhabditid nematodes. Behavioral defenses included walling off infected individual to avoid contamination (termites), a high defecation rate to reduce infection via the anus (scarabaeids), low carbon dioxide output or release of carbon dioxide in bursts to minimize chemical cues (lepidopterous pupae), and boring into roots to escape infective juveniles in the soil (rootworms), or preening to remove infective juveniles (scarabaeids). Morphological defenses or barriers to infection involved pupal cells of dense soil particles, silken cocoons, sieve plates over spiracles, tightly closed spiracles, and hard body surfaces. Physiological defenses included encapsulation of invading nematodes by rootworms[70] and humoral response against the mutualistic bacteria.[71] Gaugler[3] concluded that known defensive strategies against nematode infection were skewed toward behavioral encounter measures, especially to escape detection or penetration, rather than physiological defenses or phenological separation.

An interesting observation concerning the mole cricket and *S. carpocapsae* was recently made in Brazil.[72] Nematode-trapping fungi in the genera *Arthrobotrys* and *Dactylaria* colonized the cuticle of the mole cricket, apparently trapping infective juveniles of *S. carpocapsae* and protecting the mole cricket. To verify this observation, mole crickets with and without nematode-trapping fungi were exposed to *S. carpocapsae*. Mole crickets without the nematode-trapping fungi had significantly higher mortality (63%) than those with the nematode-trapping fungi (13%). The authors concluded that this association between the fungi and mole cricket was facultative mutualism or protocooperation. This conclusion, however, is tentative because of the low

TABLE 3. Infectivity of *Heterorhabditis* spp. and *Steinernema* spp. to *Lucilia cuprina* Over a Range of Temperatures at a Water Potential of -0.03 Bars in Sand[36]

Nematode species (strain)	"Native" home temperature °C		Temperature range of infectivity °C		Temperature range >70% infectivity	
	low	high	low	high	low	high
Heterorhabditis sp. (D1)	26	29	12	35	20	33
H. zealandica (T327)[a]	7	16	7	31	10	25
Steinernema feltiae (T335)[b]	8	16	4	30	5	25
S. feltiae (N60)[b]	7	20	2	30	2	20

a Originally referred to as *H. heliothidis*.
b *Steinernema feltiae* (=*S. bibionis*).

numbers of test insects, inadequate controls, and the lack of direct observations of trapped nematodes on the mole cricket cuticle.

D. Species and Strains

Major differences in infectivity exist between steinernematid and heterorhabditid species and their various strains. The literature on field and laboratory tests is replete with the differential response of these nematodes to a given insect host. Differences, for example, in infectivity at different temperatures between two heterorhabditids and two *S. feltiae* (=*S. bibionis*) strains were cited earlier in Section V.B.[36,66] The differences observed were attributed, in part, to the "native" home of the nematode. For another example, a *Heterorhabditis* sp. Tasmanian strain had a lower LD_{50} than *H. bacteriophora* NC or HP88 against the western spotted cucumber beetle, *Diabrotica undecimpunctata*.[73] The reasons for these differences remain unclear and may be related to nematode behavior, quality of the nematodes, and adaptation to a given host (attraction to the host or ability to overcome defense mechanisms).

VI. RECYCLING

Ecologically, steinernematids and heterorhabditids are obligate pathogens of insects. In order to persist, they need to reproduce or recycle within a host. As these nematodes occur naturally in soil, recycling is obviously not a problem. However, the population dynamics of these nematodes in the soil environment have not been adequately studied, and methods to determine population fluctuations temporally in soil need to be developed.

A. Inundative Release

Field application of steinernematids and heterorhabditids has stressed inundative releases to reduce pest populations to levels below the economic threshold level. This insecticidal approach requires nematode survival or persistence for a sufficient period to obtain efficacious results. Although recycling is highly desirable to minimize subsequent applications, it is not a prerequisite for this control tactic. In many field applications, soil samples are routinely taken to examine for nematode persistence, but whether the nematodes recycled or merely persisted remains unknown. Thus, *H. bacteriophora* and *S. carpocapsae* applied against Japanese beetle grubs reduced the population at 28 days post-treatment; examination of the same plots without further application in the following generation of grubs showed that the nematodes were present, often with high mortality due to nematodes.[4] Recycling of *H. bacteriophora* and *S. carpocapsae* probably occurred in the grubs in the treated plots, but no direct evidence for this event was provided.

H. bacteriophora were applied to a loamy sand in containers with grape ivy plants in the following treatments: nematodes alone, nematodes plus hosts, no nematodes, and no nematodes plus hosts.[74] Using the *Galleria* trap method to detect the nematode,[75] *H. bacteriophora* declined to a low level within 48 days in the nematode alone plot and was detected occasionally up to 150 days posttreatment. In the nematode plus host plot, *H. bacteriophora* showed a similar decline but the nematode recycled in the host and was recovered for nearly a year (Figure 1). In loam soil in the field, *H. bacteriophora* also showed a rapid decline within 2 weeks, and then increased and persisted for 14 months (Figure 2).[76] Although no insects were added to the soil, caterpillars, ants, and earwigs were observed and probably served as alternate hosts for the nematode.

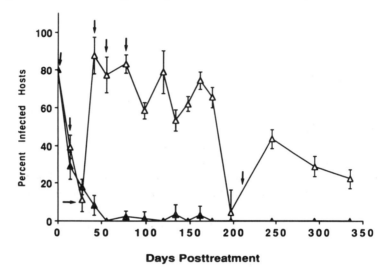

Figure 1. Recycling of *Heterorhabditis bacteriophora* in nursery soil with (Δ) and without (▲) an alternate host (*Galleria mellonella*) present (bars = ± standard error; arrows = when *Galleria* added to soil).

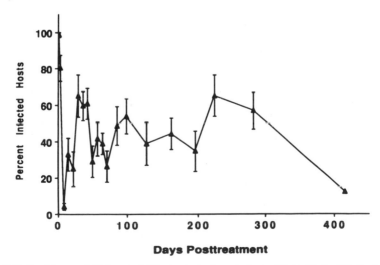

Figure 2. Persistence of *Heterorhabditis bacteriophora* in loam soil in Davis, CA (bars = ± standard error).

B. Inoculative Release

Inoculative releases of nematodes have not been attempted. This approach is worthy of experimentation and should be tried in locations having the following criteria: (1) a soil pest or a complex of soil pests should be present throughout most of the year, (2) the pest(s) preferentially should have a high economic threshold level and should be moderately susceptible to the nematode so that some pests survive to reinfest the soil, and (3) soil conditions should be favorable for nematode survival (e.g., moist, sandy soil, moderate temperatures, and few antagonists). These criteria are fulfilled by a number of habitats, including forests, pastures, orchards, and turf. Hopefully, the nematodes will respond in a density-dependent manner to the pest problem.

C. Natural Occurrence

Steinernematids and heterorhabditids are widespread and have been isolated from every inhabited continent and many islands.[61,77] Distribution studies utilizing the baiting technique[75] have shown that these nematodes are widely distributed in Ireland,[78] England,[62] Czechoslovakia,[79,80] U.S. (i.e., Florida[60] and North Carolina,[81]), Hungary,[82] Italy,[83] and Australia.[61]

In spite of the natural occurrence of these nematodes, very few long-term studies on population dynamics have been conducted, another indication of the difficulty of conducting research in the soil environment. In Florida citrus groves, steinernematids and heterorhabditids were recovered from soil most often from May to November.[60] Rainfall, moisture content, soil type, and pH were not correlated with nematode infection, although there was a positive correlation with soil temperature. Lower nematode parasitism was recorded in the winter when soil temperatures were <15°C. In Czechoslovakian forests, the steinernematid, *S. kraussei*, was recovered from humus (3-7 cm) and organic-mineral (5-10 cm) layers of soil where the host insect pupated, but not from the litter (2-3 cm).[79] This nematode was isolated from the soil throughout much of the year, although lower numbers were recovered during the summer than during spring or autumn. *S. kraussei*, a cool temperature nematode, was apparently adversely affected by the higher soil temperatures during the summer.[80]

In the field, steinernematids and heterorhabditids are faced with a vast number of antagonists and unfavorable abiotic conditions. Nematodes in sterile soil can often survive for an extended period of time, but the probability for long-term nematode survival in nonsterile soil is greatly diminished. Factors favoring nematode survival include adequate moisture or slow dehydration of soil, low temperatures, absence of antagonists, favorable soil texture (i.e., sandy soils), and continued presence of hosts.

Epizootics of nematode disease probably occur regularly in soil,[84] but because these epizootics are difficult to observe, they go unnoticed and unrecorded.[25] This area remains a virgin field of study.

VII. CONCLUSIONS

Understanding the soil environment in relation to the activities of entomopathogenic nematodes is a basic step toward managing soil pests and initiating studies in epizootiology of nematode diseases. Although these nematodes are prime candidates for biological control of a number of soil pests, their use as biological insecticides has often resulted in inconsistent control. These inconsistencies have been attributed to abiotic factors—in particular temperature extremes, ultraviolet radiation, soil texture, or overabundant or insufficient moisture levels—and to the nematode themselves for their lack of motility or poor quality due to production or storage conditions. Other factors which may contribute to the inconsistent results include host defense mechanisms and the presence of nematode antagonists. For the nematodes to be successful, soil conditions, abiotically and biotically, must be favorable for their motility and infectivity. Using the nematodes inundatively does not require that they survive or persist in the soil for any great length of time, especially if the pest is univoltine or has a synchronous life cycle. On the other hand, nematode survival (i.e., recycling and persistence) is virtually essential for pests with an asynchronous life cycle or for initiating epizootics.

Soil is diverse chemically and biologically, even over a small area, making it difficult to elucidate cause and effect in the field. Consequently, most studies have been conducted under controlled conditions to demonstrate the impact of various abiotic and biotic factors on nematode behavior and survival. Examination of certain nematode characteristics in the laboratory has enabled us to determine which nematode may have the greatest potential for success in the field in a given soil environment against a particular pest. For example, the greater motility of heterorhabditids in soil compared with *S. carpocapsae* has been clearly established. Assuming equal susceptibility of a host to heterorhabditids or steinernematids under laboratory conditions, the heterorhabditids would be the nematode of choice for biological control. However, the knowledge that *S. carpocapsae* is less motile has led to the development of application equipment which places this nematode in close proximity to a host[85] and to the selection of a better host-finding strain.[22] Laboratory studies have delineated that clay soils adversely affect nematode motility and infectivity, thus implying that greater success can be attained against pests in sandy soils. Clearly, as more information about the ecology of these nematodes and their ability to infect insect hosts in the soil environment are obtained, their effectiveness in biological control programs and our ability to manipulate them to initiate epizootics should be greatly enhanced.

REFERENCES

1. **Glaser, R. W., and Farrell, C. C.,** Field experiments with the Japanese beetle and its nematode parasite, *J. N. Y. Entomol. Soc.*, 43, 345, 1935.

2. **Kaya, H. K.**, Entomogenous nematodes for insect control in IPM systems, in *Biological Control in Agricultural IPM Systems*, Hoy, M. A., and Herzog, D. C., Eds., Academic Press, New York, 1985, 283.
3. **Gaugler, R.**, Ecological considerations in the biological control of soil-inhabiting insects with entomopathogenic nematodes, *Agric. Ecosys. Environ.*, 24, 351, 1988.
4. **Klein, M. G.**, Field efficacy against soil insects, in *Entomopathogenic Nematodes in Biological Control*, Gaugler, R., and Kaya, H. K., Eds., CRC Press, Boca Raton, FL, 1990, chap. 10.
5. **Wallace, H. R.**, *The Biology of Plant Parasitic Nematodes*, Edward Arnold Ltd., London, 1963.
6. **Wallace, H. R.**, Environment and plant health: a nematological perception, *Annu. Rev. Phytopathol.*, 27, 59, 1989.
7. **Vrain, T. C.**, Role of soil water in population dynamics of nematodes, in *Plant Disease Epidemiology*, Vol. I, Leonard, K. J., and Fry, W. E., Eds., Macmillan, New York, 1986, 101.
8. **Villani, M. G., and Wright, R. J.**, Environmental influences on soil macroarthropod behavior in agricultural systems, *Annu. Rev. Entomol.*, 35, 249, 1990.
9. **Brady, N. C.**, *The Nature and Properties of Soil*, 9th ed., Macmillan, New York, 1984.
10. **Donahue, R. L., Miller, R. W., and Shickluna, J. C.**, *Soils: An Introduction to Soils and Plant Growth*, 5th ed., Prentice-Hall, Englewood Cliffs, NJ, 1983.
11. **Foth, H. D.**, *Fundamentals of Soil Science*, 6th ed., John Wiley & Sons, New York, 1978.
12. **Kramer, P. J.**, *Plant and Soil Water Relationships: A Modern Synthesis*, McGraw-Hill, New York, 1969.
13. **Kung, S. P., Gaugler, R., and Kaya, H. K.**, Influence of soil pH and oxygen on entomopathogenic nematode persistence, *J. Nematol.*, 1990, in press.
14. **Pye, A. E., and Burman, M.**, *Neoaplectana carpocapsae*: nematode accumulations on chemical and bacterial gradients, *Exp. Parasitol.*, 51, 13, 1981.
15. **Moyle, P. L., and Kaya, H. K.**, Dispersal and infectivity of the entomogenous nematode, *Neoaplectana carpocapsae* Weiser (Rhabditida: Steinernematidae), in sand, *J. Nematol.*, 13, 295, 1981.
16. **Georgis, R., and Poinar, G. O., Jr.**, Effect of soil texture on the distribution and infectivity of *Neoaplectana carpocapsae* (Nematoda: Steinernematidae), *J. Nematol.*, 15, 308, 1983.
17. **Georgis, R., and Poinar, G. O., Jr.**, Effect of soil texture on the distribution and infectivity of *Neoaplectana glaseri* (Nematoda: Steinernematidae), *J. Nematol.*, 15, 219, 1983.
18. **Georgis, R., and Poinar, G. O., Jr.**, Vertical migration of *Heterorhabditis bacteriophora* and *H. heliothidis* (Nematoda: Heterorhabditidae) in sandy loam soil, *J. Nematol.*, 15, 652, 1983.
19. **Schroeder, W. J., and Beavers, J. B.**, Movement of the entomogenous nematodes of the families Heterorhabditidae and Steinernematidae in soil, *J. Nematol.*, 19, 257, 1987.
20. **Poinar, G. O., Jr., and Hom, A.**, Survival and horizontal movement of infective stage *Neoaplectana carpocapsae* in the field, *J. Nematol.*, 18, 34, 1986.
21. **Ishibashi, N., and Kondo, E.**, A possible quiescence of the applied entomogenous nematode, *Steinernema feltiae*, in soil, *Jpn. J. Nematol.*, 16, 66, 1986.
22. **Gaugler, R., McGuire, T., and Campbell, J.**, Genetic variability among strains of the entomopathogenic nematode *Steinernema feltiae*, *J. Nematol.*, 21, 247, 1989.
23. **Choo, H. Y., Kaya, H. K., Burlando, T. M., and Gaugler, R.**, Entomopathogenic nematodes: host-finding ability in the presence of plant roots, *Environ. Entomol.*, 18, 1136, 1989.
24. **Haraguchi, N., Kawasaki, M., and Shibata, M.**, Effects of temperature and moisture on the migration, infectivity and survival of the entomogenous nematode (*Steinernema glaseri*) in bark compost, in *Recent Advances in Biological Control of Insect Pests* by *Entomogenous Nematodes in Japan*, Ishibashi, N., Ed., Ministry of Education, Japan, Grant No. 59860005, 1987, 17.

25. **Kaya, H. K.**, Diseases caused by nematodes, in *Epizootiology of Insect Diseases*, Fuxa, J. R., and Tanada, Y., Eds., John Wiley and Sons, New York, 1987, 453.

26. **Shetlar, D. J., Suleman, P. E., and Georgis, R.**, Irrigation and use of entomogenous nematodes, *Neoaplectana* spp. and *Heterorhabditis heliothidis* (Rhabditida: Steinernematidae and Heterorhabditidae), for control of Japanese beetle (Coleoptera: Scarabaeidae) grubs in turfgrass, *J. Econ. Entomol.*, 81, 1318, 1988.

27. **Epsky, N. D., Walter, D. E., and Capinera, J. L.**, Potential role of nematophagous microarthropods as biotic mortality factors of entomogenous nematodes (Rhabditida: Steinernematidae, Heterorhabditidae), *J. Econ. Entomol.*, 81, 821, 1988.

28. **Molyneux, A. S., Bedding, R. A., and Akhurst, R. J.**, Susceptibility of larvae of the sheep blowfly *Lucilia cuprina* to various *Heterorhabditis* spp., *Neoaplectana* spp., and an undescribed steinernematid (Nematoda), *J. Invertebr. Pathol.*, 42, 1, 1983.

29. **Timper, P., Kaya, H. K., and Gaugler, R.**, Dispersal of the entomogenous nematode *Steinernema feltiae* (Rhabditida: Steinernematidae) by infected adult insects, *Environ. Entomol.*, 17, 546, 1988.

30. **Kung, S. P., Gaugler, R., and Kaya, H. K.**, Soil type and entomopathogenic nematode persistence, *J. Invertebr. Pathol.*, 1990, in press.

31. **Womersley, C. Z.**, Dehydration survival and anhydrobiotic potential, in *Entomopathogenic Nematodes in Biological Control*, Gaugler, R., and Kaya, H. K., Eds., CRC Press, Boca Raton, FL, 1990, chap. 6.

32. **Moore, G. E.**, The bionomics of an insect-parasitic nematode, *J. Kansas Entomol. Soc.*, 38, 101, 1965.

33. **Kung, S. P., Gaugler, R., and Kaya, H. K.**, Effects of soil temperature, moisture, and relative humidity on entomopathogenic nematode persistence, *J. Invertebr. Pathol.*, 1990, in press.

34. **Simons, W. R., and Poinar, G. O., Jr.**, The ability of *Neoaplectana carpocapsae* (Steinernematidae: Nematodea) to survive extended periods of desiccation, *J. Invertebr. Pathol.*, 22, 228, 1973.

35. **Molyneux, A. S.**, Survival of infective juveniles of *Heterorhabditis* spp., and *Steinernema* spp. (Nematoda: Rhabditida) at various temperatures and their subsequent infectivity for insects, *Rev. Nematol.*, 8, 165, 1985.

36. **Molyneux, A. S.**, The influence of temperature on the infectivity of heterorhabditid and steinernematid nematodes for larvae of the sheep blowfly, *Lucilia cuprina*, in *Proc. 4th Aust. Appl. Entomol. Res. Conf., Adelaide*, Bailey, P., and Swincer, D., Eds., 1984, 344.

37. **Kaya, H. K.**, Development of the DD-136 strain of *Neoaplectana carpocapsae* at constant temperatures, *J. Nematol.*, 9, 346, 1977.

38. **Milstead, J. E.**, Influence of temperature and dosage on mortality of seventh instar larvae of *Galleria mellonella* (Insecta: Lepidoptera) caused by *Heterorhabditis bacteriophora* (Nematoda: Rhabditoidea) and its bacterial associate *Xenorhabdus luminescens*, *Nematologica*, 27, 167, 1981.

39. **Schmiege, D. C.**, The feasibility of using a neoaplectanid nematode for control of some forest insect pests, *J. Econ. Entomol.*, 56, 427, 1963.

40. **Gray, P. A., and Johnson, D. T.**, Survival of the nematode *Neoaplectana carpocapsae* in relation to soil temperature, moisture and time, *J. Georgia Entomol. Soc.*, 18, 454, 1983.

41. **Gaugler, R., and Boush, G. M.**, Effects of ultraviolet radiation and sunlight on the entomogenous nematode, *Neoaplectana carpocapsae*, *J. Invertebr. Pathol.*, 32, 291, 1978.

42. **Burman, M., and Pye, A. E.**, *Neoaplectana carpocapsae*: respiration of infective juveniles, *Nematologica*, 26, 214, 1980.

43. **Lindegren, J. E., Rij, R. E., Ross, S. R., and Fouse, D. C.**, Respiration rate of *Steinernema feltiae* infective juveniles at several constant temperatures, *J. Nematol.*, 18, 221, 1986.

44. **Kaya, H. K., and Burlando, T. M.**, Infectivity of *Steinernema feltiae* in fenamiphos-treated sand, *J. Nematol.*, 21, 434, 1989.

45. **Cook, R. J., and Baker, K. F.**, *The Nature and Practice of Biological Control of Plant Pathogens*, American Phytopathological Society, St. Paul, 1983.

46. **Mankau, R.**, Biological control of nematode pests by natural enemies, *Annu. Rev. Phytopathol.*, 18, 415, 1980.

47. **Kerry, B. R.**, Nematophagous fungi and the regulation of nematode populations in soil, *Helminthol. Abst. Ser. B*, 53, 1, 1984.

48. **Gray, N. F.**, Fungi attacking vermiform nematodes, in *Diseases of Nematodes*, Vol. 2, Poinar, G. O., Jr., and Jansson, H-B., Eds., CRC Press, Boca Raton, FL, 1988, 3.

49. **Stirling, G. R.**, Biological control of plant-parasitic nematodes, in *Diseases of Nematodes*, Vol. II, Poinar, G. O., Jr., and Jansson, H-B., Eds., CRC Press, Boca Raton, FL, 1988, 93.

50. **Ishibashi, N., and Kondo, E.**, *Steinernema feltiae* (DD-136) and *S. glaseri*: persistence in soil and bark compost and their influence on native nematodes, *J. Nematol.*, 18, 310, 1986.

51. **Kaya, H. K., Mannion, C., Burlando, T. M., and Nelsen, C. E.**, Escape of *Steinernema feltiae* from alginate capsules containing tomato seeds, *J. Nematol.*, 19, 287, 1987.

52. **Timper, P., and Kaya, H. K.**, Role of the second-stage cuticle of entomogenous nematodes in preventing infection by nematophagous fungi, *J. Invertebr. Pathol.*, 54, 314, 1989.

53. **Poinar, G. O., Jr., and Jansson, H-B.**, Infection of *Neoaplectana* spp. and *Heterorhabditis heliothidis* to the endoparasitic fungus *Drechmeria coniospora*, *J. Nematol.*, 18, 225, 1986.

54. **Poinar, G. O., Jr.**, A microsporidian parasite of *Neoaplectana glaseri* (Steinernematidae: Rhabditida), *Rev. Nematol.*, 11, 359, 1988.

55. **Poinar, G. O., Jr., and Jansson, H-B.**, Infection of *Neoaplectana* and *Heterohabditis* (Rhabditida: Nematoda) with the predatory fungi, *Monacrosporium ellipsosporum* and *Arthrobotrys oligospora* (Moniliales: Deuteromycetes), *Rev. Nematol.*, 9, 241, 1986.

56. **Ishibashi, N., Young, F. Z., Nakashima, M., Abiru, C., and Haraguchi, N.**, Effects of application of DD-136 on silkworm, *Bombyx mori*, predatory insect, *Agriosphodorus dohrni*, parasitoid, *Trichomalus apanteloctenus*, soil mites, and other non-target soil arthropods, with brief notes on feeding behavior and predatory pressure of soil mites, tardigrades, and predatory nematodes on DD-136 nematodes, in *Recent Advances in Biological Control of Insect Pests by Entomogenous Nematodes in Japan*, Ishibashi, N., Ed., Ministry of Education, Japan, Grant No. 59860005, 1987, 158.

57. **McInnis, T. M., and Jaffee, B. A.**, An assay for *Hirsutella rhossiliensis* spores and the importance of phialides for nematode inoculation, *J. Nematol.*, 21, 229, 1989.

58. **Jaffee, B. A., Gaspard, J. T., and Ferris, H.**, Density-dependent parasitism of the soil-borne nematode *Criconemella xenoplax* by the nematophagous fungus *Hirsutella rhossiliensis*, *Microbiol. Ecol.*, 17, 193, 1989.

59. **Alatorre-Rosas, R., and Kaya, H. K.**, Interspecific competition between entomopathogenic nematodes in the genera *Heterorhabditis* and *Steinernema* for an insect host in sand, *J. Invertebr. Pathol.*, 55, 179, 1990.

60. **Beavers, J. B., McCoy, C. W., and Kaplan, D. T.**, Natural enemies of subterranean *Diaprepes abbreviatus* (Coleoptera: Curculionidae) larvae in Florida, *Environ. Entomol.*, 12, 840, 1983.

61. **Akhurst, R. J., and Bedding, R. A.**, Natural occurrence of insect pathogenic nematodes (Steinernematidae and Heterorhabditidae) in soil in Australia, *J. Aust. Entomol. Soc.*, 25, 241, 1986.

62. **Hominick, W. M., and Briscoe, B. R.**, Occurrence of entomopathogenic nematodes (Rhabditida: Steinernematidae and Heterorhabditidae) in British soils, *Parasitology*, 1990, in press.

63. **Kaya, H. K., and Burlando, T. M.**, Development of *Steinernema feltiae* (Rhabditida: Steinernematidae) in diseased insect hosts, *J. Invertebr. Pathol.*, 53, 164, 1989.

64. **Barbercheck, M. E., and Kaya, H. K.**, Interactions between *Beauveria bassiana* and the entomogenous nematodes, *Steinernema feltiae* and *Heterorhabditis heliothidis*, *J. Invertebr. Pathol.*, 55, 225, 1990.

65. **Bird, A. F., and Bird, J.**, Observations on the use of insectparasitic nematodes as a means of biological control of root-knot nematodes, *Int. J. Parasitol.*, 16, 511, 1986.

66. **Molyneux, A. S., and Bedding, R. A.**, Influence of soil texture and moisture on the infectivity of *Heterorhabditis* sp. D1 and *Steinernema glaseri* for larvae of the sheep blowfly, *Lucilia cuprina, Nematologica*, 30, 358, 1984.

67. **Choo, H. Y.**, unpublished data, 1987.

68. **Geden, C. J., Axtell, R. C., and Brooks, W. M.**, Susceptibility of the lesser mealworm, *Alphitobrus diaperinus* (Coleoptera: Tenebrionidae) to the entomogenous nematodes *Steinernema feltiae, S. glaseri* (Steinernematidae) and *Heterorhabditis heliothidis* (Heterorhabditidae), *J. Entomol. Sci.*, 20, 331, 1985.

69. **Akhurst, R. J.**, Controlling insects in soil with entomopathogenic nematodes, in *Fundamental and Applied Aspects of Invertebrate Pathology*, Samson, R. A., Vlak, J. M., and Peters, D., Eds., Proc. 4th Int. Coll. Invertebr. Pathol., Wageningen, 1986, 265.

70. **Jackson, J. J., and Brooks, M. A.**, Susceptibility and immune response of western corn rootworm larvae (Coleoptera: Chrysomelidae) to the entomogenous nematode, *Steinernema feltiae* (Rhabditida: Steinernematidae), *J. Econ. Entomol.*, 82, 1073, 1989.

71. **Dunphy, G. B., and Thurston, G. S.**, Entomopathogenic nematodes and insect immunity, in *Entomopathogenic Nematodes in Biological Control*, Gaugler, R., and Kaya, H. K., Eds., CRC Press, Boca Raton, FL, 1990, chap. 16.

72. **Fowler, H. G., and Garcia, C. R.**, Parasite-dependent protocooperation, *Naturwissenschaften*, 76, 26, 1989.

73. **Kaya, H. K.**, unpublished data, 1988.

74. **Kaya, H. K.**, Entomopathogenic nematodes in biological control of insects, in *New Directions in Biological Control: Alternatives for Suppression of Agricultural Pests and Diseases*, Baker, R., and Dunn, P. E., Eds., Alan R. Liss, New York, 1990, 189.

75. **Bedding, R. A., and Akhurst, R. J.**, A simple technique for the detection of insect parasitic rhabditid nematodes in soil, *Nematologica*, 21, 109, 1975.

76. **Kaya, H. K.**, unpublished data, 1989.

77. **Poinar, G. O., Jr.**, Recognition of *Neoaplectana* species (Steinernematidae: Rhabditida), *Proc. Helminthol. Soc. Wash.*, 53, 121, 1986.

78. **Blackshaw, R. P.**, A survey of insect parasitic nematodes in Northern Ireland, *Ann. Appl. Biol.*, 113, 561, 1988.

79. **Mrácek, Z.**, Horizontal distribution in soil, and seasonal dynamics of the nematode *Steinernema kraussei*, a parasite of *Cephalcia abietis*, *Z. Ang. Entomol.*, 94, 110, 1982.

80. **Mrácek, Z.**, The use of "*Galleria* traps" for obtaining nematode parasites of insects in Czechoslovakia (Lepidoptera: Nematoda, Steinernematidae), *Acta Entomol. Bohemoslov.*, 77, 378, 1980.

81. **Akhurst, R. J., and Brooks, W. M.**, The distribution of entomophilic nematodes (Heterorhabditidae and Steinernematidae) in North Carolina, *J. Invertebr. Pathol.*, 44, 140, 1984.

82. **Mrácek, Z., and Jenser, G.**, First report of entomogenous nematodes of the families Steinernematidae and Heterorhabditidae from Hungary, *Acta Phytopathol. Entomol. Hungarica*, 23, 153, 1988.

83. **Deseö, K. V., Fantoni, P., and Lazzari, G. L.**, Presenza di nematodi entomopatogeni (*Steinernema* spp., *Heterorhabditis* spp.) nei terreni agricoli in Italia, *Atti Giornate Fitopathol.*, 2, 269, 1988.

84. **Sexton, S. B., and Williams, P.**, A natural occurrence of parasitism of *Graphognathus leucoloma* (Boheman) by the nematode *Heterorhabditis* sp., *J. Aust. Entomol. Soc.*, 20, 253, 1981.

85. **Berg, G. N., William, P., Bedding, R. A., and Akhurst, R. J.**, A commercial method of application of entomopathogenic nematodes to pasture for controlling subterranean insect pests, *Pl. Prot. Quart.*, 2, 174, 1987.

6.   Dehydration Survival and Anhydrobiotic Potential

Christopher Z. Womersley

I. INTRODUCTION

Few subjects in nematological research have caused more interest, excitement, and consternation than the ability of nematodes to survive severe dehydration, a condition referred to as anhydrobiosis. The confusion that still exists over their anhydrobiotic capabilities is understandable when one considers the diversity of nematode habitats, and the ways in which these essentially aquatic pseudocoelomates have evolved in response to the dehydration stresses to which they are normally exposed. Initially, it was believed that this mode of survival was a rarity amongst nematodes.[1] Such a viewpoint was based on the well-documented abilities of a few plant-parasitic forms (specifically those parasitizing the aerial parts of plants) to survive rapid and prolonged periods of dehydration (see Evans and Perry[2] and Evans and Womersley[3] for review). However, over the last 10 years, research has shown this belief to be incorrect. For example, pure and applied studies on nematodes inhabiting more hospitable environments than the aerial parts of plants have demonstrated that there are at least two basic groups of anhydrobiotes, slow-dehydration and fast-dehydration strategists, with most species studied fitting into the former group (see Womersley[4] for review).

The above realizations have helped reduce the confusion as to why different nematode species require radically different conditions to facilitate successful induction into anhydrobiosis but have substituted other problems in their place. To the purist, the term "anhydrobiosis" applies to those organisms capable of surviving either rapid or slow dehydration up to and beyond the point at which metabolism is fully arrested, resulting in a state of suspended animation, more commonly referred to as the "cryptobiotic condition". Because cryptobiosis is not exclusively associated with dehydration stress but can also be induced by extreme osmotic (osmobiosis), temperature (cryobiosis), or oxygen (anoxybiosis) stress, organisms capable of surviving dehydration to this degree are termed "cryptobiotic anhydrobiotes".[4] However, many slow-dehydration strategists are rarely, if ever, exposed to such rigors of dehydration. Although they are more than competent at surviving the dehydration stresses to which they would normally be subjected and the substantial water losses that occur, they may not enter a fully cryptobiotic state.[4-7] Rather, a quiescent phase, involving a reduction in metabolic activity and behavioral, morphological, and biochemical adaptations identical to or reminiscent of those exhibited by cryptobiotic anhydrobiotes, is the norm. Thus, even though their ability to survive dehydration in absolute terms is less than that shown by cryptobiotic anhydrobiotes, their anhydrobiotic potential is maximized within their natural habitat.

We are presently unaware of where entomopathogenic nematodes fit into this picture. Those species studied to date are exclusively slow-dehydration strategists and evidence I shall present suggests that they may not be capable of attaining a fully cryptobiotic state. In the remainder of this chapter, I will focus on what is known about the anhydrobiotic potential of some of the species concerned, their ability to withstand dehydration, the adaptations they employ, and how these factors may relate to solving problems of storage and field application that have arisen in the applied arena.

II. ECOLOGICAL CONSIDERATIONS FOR EVALUATING ANHYDROBIOTIC POTENTIAL

To determine the anhydrobiotic potential of any organism, the habitats in which they are normally found must be assessed with respect to the rate and level of dehydration stress that may occur. It is an impossible task to cover all habitats occupied by entomopathogenic nematodes; thus, I have restricted the scope to those illustrated in Figure 1. The five habitats presented are divided on the basis of rate of evaporative water loss and level of dehydration stress encountered. The reasons for such restrictions are simple. Nearly all research efforts on the use of nematodes for the biological control of insects have involved representatives of the families Steinernematidae, Heterorhabditidae, and Mermithidae. The infective-juvenile stage of steinernematid and heterorhabditid nematodes is predominantly associated with the upper soil profile[7] (Figure 1.2) with implications involving the soil-air interface (Figure 1.3). Those mermithids studied are usually found in similar environments or can be exclusively aquatic (i.e., *Romanomermis culicivorax*[8,9]) and thus, may be exposed to the dehydration regimes associated with the pond sediment or the pond sediment-air interface (Figures 1.4 and 1.5, respectively).

Even though pond sediments (Figure 1.4) have allowed the evolution of some of the more successful slow-dehydration cryptobiotic anhydrobiotes,[4,10,11] the level of dehydration stress that can be tolerated by *R. culicivorax* infective juveniles or eggs would appear to be negligible.[12] Similarly, although steinernematid and heterorhabditid nematodes infect a plethora of insect hosts and present the most impressive potential for biological control,[7,13] their apparent lack of anhydrobiotic potential has caused problems with their successful commercial storage or field persistence. Only the infective juvenile, complete with mutualistic bacteria (genus *Xenorhabdus*), is free-living and occurs outside the host, predominantly in the upper soil profile (Figure 1.2). This is the most stable environment in which anhydrobiotes occur.[4,10] Organisms inhabiting this region are usually only subjected to extremely slow rates of evaporative water loss and seem to have developed little resistance to control this loss.

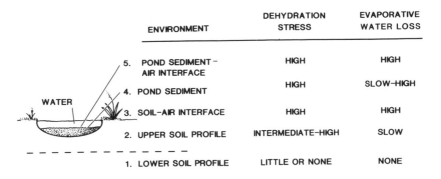

ENVIRONMENT	DEHYDRATION STRESS	EVAPORATIVE WATER LOSS
5. POND SEDIMENT – AIR INTERFACE	HIGH	HIGH
4. POND SEDIMENT	HIGH	SLOW–HIGH
3. SOIL–AIR INTERFACE	HIGH	HIGH
2. UPPER SOIL PROFILE	INTERMEDIATE–HIGH	SLOW
1. LOWER SOIL PROFILE	LITTLE OR NONE	NONE

Figure 1. Classification (based on rate of evaporative water loss and possible level of dehydration stress) of the anhydrobiotic potential of terrestrial and temporary aquatic habitats in which entomopathogenic nematodes are found.

For the most part this scenario appears to be true for steinernematids and heterorhabditids. This does not mean, however, that they are unable to withstand high dehydration stress, only that the rate of water loss is critical. For example, in certain cases successful induction into anhydrobiosis can only be achieved if the target organism is dried at a rate that simulates the drying of its natural environment.[10] Clearly, the actual rate of drying which affords entry into anhydrobiosis is critically important for all steinernematids and heterorhabitids and will be greatly influenced by the drying characteristics of their natural environment.

We can therefore presume that the infective juveniles are highly unlikely to have evolved strategies for tolerating the rigors of rapid dehydration stress imposed at the soil-air and plant-air interfaces. Thus, the limited success of foliar and soil surface applications of steinernematid and heterorhabditid infective juveniles is not surprising (see Gaugler[7] and Kaya[14] for reviews). In addition, solar radiation limits the viability of the nematodes placed in these exposed environments.[15] Only with inclusion of ultraviolet protectants[16] or antidesiccants[17-19] can the detrimental effects of rapid dehydration and exposure to ultraviolet light be reduced. Collectively, these results underline the fragile balance that exists between an organism and its natural environment and suggest that steinernematid and heterorhabditid nematodes may only realize their full biological control potential in the soil environment. This latter conclusion has been noted by other authors.[7,14,20]

III. PHYSICAL FACTORS AFFECTING DEHYDRATION SURVIVAL

Because slow drying appears to be a prerequisite for dehydration, and presumably the successful induction of anhydrobiosis in steinernematid and heterorhabditid nematodes, the question arises as to how slow this process must be. The answer is not so simple because many factors, including soil structure and the behavior and morphology of the organisms concerned, affect final viability. For example, Crowe and Madin[21] demonstrated that exposing mixed juvenile and adult aggregates of the free-living mycophagous nematode, *Aphelenchus avenae*, to elevated (97%) relative humidities (rh) for 3 days was sufficient to induce anhydrobiosis. An almost identical situation also exists for the mushroom nematode, *Ditylenchus myceliophagus*.[22,23] In both cases the behavioral adaptations of coiling[4,10] and clumping for mutual protection[24] are major factors involved in physically reducing the rate of evaporative water loss. However, clumping is a specialized phenomenon not generally applicable to soil-dwelling nematodes. Although reports of aggregates in close proximity to insect larvae and fecal pellets have been noted experimentally for *Steinernema carpocapsae*, (=*S. feltiae*),[25] no such observations have been made for other entomopathogenic nematode species. If we assume that most steinernematids and heterorhabditids (like many other soil-dwelling nematodes) dehydrate as individuals rather than as aggregates within the soil, then these results do not define the drying rates necessary for the successful induction of anhydrobiosis under field conditions. Field observations support this view in that a number of nematodes (e.g., *Rotylenchulus reniformis*, *D. myceliophagus*, *A. avenae*, etc.) are able to coil and survive anhydrobiotically as individuals in the soil but are unable to survive direct exposure to 97% rh.[10,26]

As individuals, the infective stages of steinernematid and heterorhabditid nematodes are exclusively associated with the microenvironment provided by the interstitial spaces of the soil (Figure 2), the water dynamics of which vary depending on moisture availability. At field capacity (Figure 2.1), nematode movement can (depending on species) be severely restricted due to the lack of surface tension forces necessary for locomotion. Under these conditions nematode viability may also be adversely affected if anaerobic conditions prevail, because the nematodes concerned are obligate aerobes.[13] As noted by Gaugler,[7] there is probably a moisture window[27] where the amount of moisture present in wet, aerated soils (Figure 2.2) provides for optimal movement and survival of the nematodes. However, as free water is removed from the upper soil layers via the evaporative sink at the soil-air interface, nematode movement will again be restricted, eventually leading to coiling and the induction of anhydrobiosis (Figure 2.3). In *R. reniformis* it has been demonstrated conclusively that coiling is an adaptation which can be used to signify successful induction into anhydrobiosis.[10] The rate at which water evaporates from the upper soil layers is largely dependent on the physical characteristics of the soil itself (i.e., soil porosity, water-holding capacity of soil aggregates, etc.) and the nature of the water table.[28,29] The cumulative effect of all these factors is that water removal an extremely gradual process with relative humidities approaching 100% being maintained in the interstitial spaces of the drying soil.[4,10]

Conceivably, the relationship between natural drying regimes for soils and anhydrobiotic survival in steinernematids and heterorhabditids may not be as distinct as that shown for *R. reniformis*.[10] For example, retention of the second-

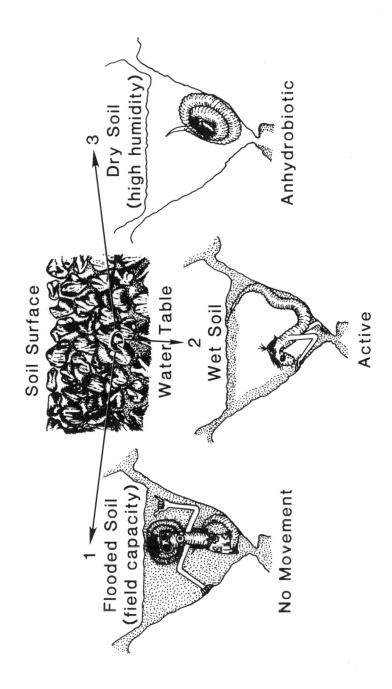

Figure 2. Microenvironmental conditions (imposed by moisture availability) experienced by soil-dwelling nematodes occupying the interstitial spaces of the soil. Rates of evaporative water loss are determined by the physical structure of the soil itself and the nature of the water table. Usually water loss is gradual and relative humidity is high (i.e., >99% in the interstitial spaces).

stage cuticle by the infective juveniles will presumably slow the rate of drying.[24,30] Nevertheless, the fact that a coiling response has not been demonstrated in most entomopathogenic nematodes suggests that they have yet to be dehydrated at relative humidities that allow for the induction of true anhydrobiosis. We have induced coiling in *S. glaseri* exposed to elevated (>97%) relative humidities[31] and in *S. carpocapsae* (All strain)[32] when dehydrated on model substrates that mimic the drying of soils after the methods of Womersley and Ching.[10] These results will be discussed in more detail in the following section.

IV. DEHYDRATION SURVIVAL OF STEINERNEMATIDS AND HETERORHABDITIDS
A. Direct Exposure to Controlled Relative Humidities

Attempts to induce anhydrobiosis in steinernematid and heterorhabditid infective juveniles by direct exposure to reduced relative humidities can be split into two distinct categories, experimental and commercial. The first has exclusively involved controlled dehydration of modest numbers (10-1000 individuals) of *S. carpocapsae* infective juveniles, either individually or in clumps, to ascertain their dehydration survival capabilities in relation to natural conditions. Simons and Poinar[20] initially induced anhydrobiosis in groups of 1000 *S. carpocapsae* (Agriotos) infective juveniles by holding them on membrane filters for 12 hr periods at 96% and then 93% rh before exposing them to rh levels of 79.5% or less. Under these conditions survival was inversely proportional to the final rh level with over 90% viability after 12 days at 79.5% rh and 80% viability after 4 days at 48.4% rh. Survival was reduced to essentially zero after 12 days at 48.4% rh and death occurred rapidly at 10% rh and below. Although promising, these results may be misleading in that, as discussed previously, the clumping ability of infective juveniles in the soil has yet to be assessed, and thus, the protective effects afforded by aggregation may not be a strategy employed in the natural environment. Ishibashi et al.[33] demonstrated that an increase in the size of the nematode clump (ranging from 20-150 mg) did enhance the survival of *S. carpocapsae* (DD-136) infective juveniles, but no nematode numbers were given. In addition, they showed that a gradual decline in rh level from 98-34% over 9 days was far more effective in facilitating survival (i.e., >30% viability) than was direct exposure to 34% rh for the same period of time which resulted in only 4.2% survival. Again, although mutual protection is an obvious factor, the latter results underline the importance of gradual vs. immediate and rigorous desiccation for enhancing the dehydration survival capabilities of this nematode.

Our experiments with *S. carpocapsae* (All) infective juveniles have concentrated more on the survival of individuals with the protective effects of aggregation being avoided by the use of low juvenile numbers (ca. 50 per experimental point). In these experiments infective juveniles were preconditioned at 97%

rh for 3 days before being placed at other rh levels, followed by gradual rehydration at 100% rh for 24 hr before the addition of bulk water. Behaviorally, many rehydrated infective juveniles assumed an inactive or quiescent state (i.e., slightly curved or straight body posture with bent tail) identical to that reported by Hara and Kaya.[34] To accurately assess survival, physical stimulation was necessary to induce movement in these infective juveniles. Continued exposure to 97% rh for a further 8 days seems to have little effect on nematode survival. However, exposure to 80% rh after preconditioning results in a gradual decrease in survival (ca. 60%) until the 5th day, after which survival decreases at a much faster rate, approaching zero after 8 days (Figure 3). Exposure to 60% rh has given little survival after 2 days and no infective juveniles appear able to survive 24 hr exposure to 0% or 40% rh.

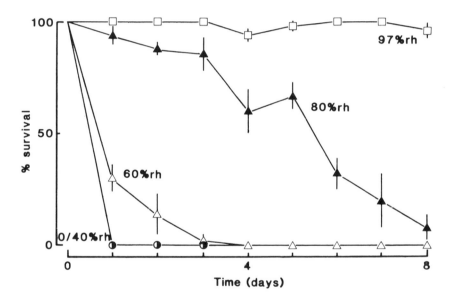

Figure 3. Survival of *Steinernema carpocapsae* (all) infective juveniles on membrane filters after 3 days preconditioning at 97% relative humidity (rh) followed by direct exposure to reduced relative humidities for up to 8 days (N=3, bars indicate + SD; approximately 50 infective juveniles per sample).[35]

The high survival of *S. carpocapsae* at 97% and even 80% rh are somewhat surprising in being far better than that recorded for *R. reniformis* second-stage juveniles,[10] or individual juveniles and adults of *A. avenae* and *D. myceliopha-gus*.[26] This enhanced survival may well be due to the effects of retention of the second-stage cuticle by the infective stage, slowing their rate of drying. Retention of the second-stage cuticle in infective juveniles of *Haemonchus contortus* has been shown to increase survival for up to 20 days at 47% rh compared with total lack of recovery in exsheathed forms after 8 hr exposure to the same rh level.[24] Preconditioning infective juveniles at 97% rh also appears to be an important factor, because Ohba[36] was only able to maintain an LT_{50} of 120 min for individual *S. carpocapsae* (DD-136) infective juveniles when exposed directly to 70% rh at 25°C.

The second category of research has involved dehydration of extremely large numbers of infective juveniles (10^6-10^9 nematodes) for commercial level dry storage of steinernematids and heterorhabditids. By preconditioning infective juveniles at 97% rh for 3 days before exposing them to lower humidities, Popiel et al.[37] maintained high viability in *S. carpocapsae*, *S. feltiae* (=*S. bibionis*), and *Heterorhabditis bacteriophora* (=*H. heliothidis*) for 3-5 months at 25°C, with optimal survival at rh levels ranging from 85-94%. Some of these results are summarized in Table 1. These improvements in viability over experimental results with individual or small groups of infective juveniles are clearly due to the slower rates of drying afforded by the large aggregates. It would be interesting to know the level of juvenile hydration within the nematode mass with respect to individuals on the outside and compare their survival rates with time stored. The same researchers also noted that, although oxygen consumption was reduced from 6 ml in control infective juveniles to less than 1 ml/min per 10^6 nematodes in dehydrated ones, oxygen was still required to maintain shelflife.[37] These data strongly suggest that the nematodes were in a quiescent anhydrobiotic state, but how closely this approached the cryptobiotic condition (i.e., no measurable metabolic activity) is unknown.

TABLE 1. Percent Viability of Infective Juveniles of Three Species of Entomopathogenic Nematodes After Storage at Various Relative Humidity (rh) Levels for 3-5 Months Following Preconditioning for 3 Days at 97% rh[37]

Species	Final rh	Time stored	% Viability
Steinernema carpocapsae	85%	5 months	>95%
Steinernema feltiae	94%	3 months	>95%
Heterorhabditis bacteriophora	97%	3 months	>95%
	94%	3 months	>90%

In a different approach Bedding[38] employed attapulgite clays, diatomaceous clays, and kieselguhr, either forming a homogenous nematode/clay suspension or a clay/nematode cream/clay sandwich in which nematodes could be stored. In the first technique *S. carpocapsae* (All), *S. feltiae*, and *H. bacteriophora* (NZ) retained viabilities of 70-80% after being stored on calcined attapulgite chips for up to 24 weeks at 4°C. However, the actual or even approximate rh level finally reached during storage is difficult to interpret from the data. The use of a clay/nematode cream/clay sandwich held at about 95% rh for approximately 3 days and then gradually reduced to an apparent rh of 60% over a period of 7 weeks resulted in a viability of about 90% after two months storage at 23°C. These results are even more difficult to interpret. To assume that after 2 weeks, an atmospheric rh level of 60% in a closed container containing the "sandwich" equates to a water activity of 0.6 throughout the nematode mass is probably incorrect. The differential drying characteristics of a 2-cm deep layer of *S. carpocapsae* infective juveniles are presently unknown, but this loss will likely be extremely gradual (i.e., far more gradual than would allow equilibration of the whole nematode mass to 60% rh during the 2-week interval). Thus, although the nematodes in the outer layers of the nematode mass may be in equilibrium with atmospheric rh levels, those in the deeper layers will not. Rather, these nematodes will, in all probability, exhibit variable water contents which increase to the highest levels of hydration at the center of the aggregate. This suggests that, as in Popiel et al.,[37] these infective juveniles are in varied stages of quiescent anhydrobiosis as opposed to cryptobiotic anhydrobiosis. The rapid reductions in nematode viability under anaerobic conditions, particularly at temperatures above 10°C, support this view.[38] Presumably then, the superior survival of desiccated infective juveniles at low temperatures (i.e., 4°C) is facilitated by reduced metabolic rates imposed by the temperature regime (thus reducing oxygen requirements) and not by a suspension of metabolic activity due to entry into a cryptobiotic condition induced by dehydration stress.

B. Dehydration Survival on Model Substrates

An obvious problem with the dehydration studies on individuals or small aggregates of *S. carpocapsae* discussed so far is that all have failed to demonstrate any coiling response of the nematode during dehydration stress. Similarly, although dehydration of large numbers of infective juveniles enhances survival, presumably because of slower rates of evaporative water loss, no observations have been made of individuals coiling on the outside of the nematode mass. Coiling would be difficult on the inside of the aggregate due to the physical constraints imposed by close packing of the individuals themselves.[4] If coiling is a behavioral response that signifies the induction of anhydrobiosis, then the lack of this response after direct exposure to reduced rh levels suggests that (1) coiling may not be an adaptation employed by *S. carpocapsae* infective juveniles, or (2) the drying rates experienced by individual infective juveniles and those on the outside of aggregates have been too severe to induce this response. If this latter scenario is the case, it seems to indicate that successful induction into anhydrobiosis has not been achieved in direct exposure experiments.[35] The poor survival rates of individual nematodes tend to support this view.

Our present attempts to increase the survival of and induce coiling in infective juveniles by subjecting them to reduced rates of water loss that mimic the drying of soil have met with mixed results. Following the experimental protocols of Womersley and Ching,[10] we have applied infective juveniles to model substrates (0.5% agar and 1.0% agarose) whose moisture loss characteristics, when exposed to 97% rh, simulate the natural drying rates of soils maintained at the same rh. Moisture release characteristics of these substrates in comparison to a wet Hawaiian soil are shown in Figure 4. Water loss from the model substrates is similar to that from wet soil for the first 48 hr, after which the soil stabilizes due to moisture bound into the soil macrofabric.[39] In contrast, moisture continues to be lost from the model substrates, eventually approaching 0% rh after 14 days. In essence, this continued water loss represents prolonged loss of pore water from the interstitial spaces of the soil,[10] and thus, nematodes dehydrated on these substrates lose water far more gradually than if exposed directly to 97% rh.

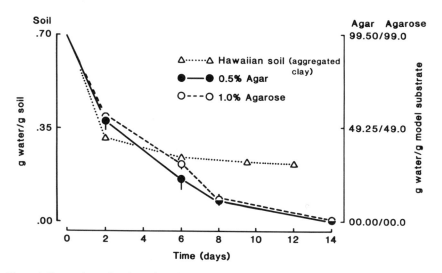

Figure 4. Comparison of moisture loss characterisitic curves for a wet low humic latosol and model substrates (modified after Womersley and Ching[10]).

Preliminary survival data of infective juveniles dried on 1% agarose at various rh levels after preconditioning at 97% rh are presented in Figure 5.[32] Survival has remained high for infective juveniles exposed to 90% rh, stabilizing at about 80% survival after 8 days. Similarly, infective juveniles placed at 80% rh survive far better in comparison to direct exposure experiments, with over 50% survival after 8 days. Even at 60% rh, more than 50% of the juveniles survive for 3 days compared with little or no survival after direct exposure to the same rh level for 2 days (Figure 3). A coiling response is also evident at all relative humidities except 0% rh, the response maximizing at 90% and 80% rh with 30-40% of all individuals coiling. The morphology of a dehydrated

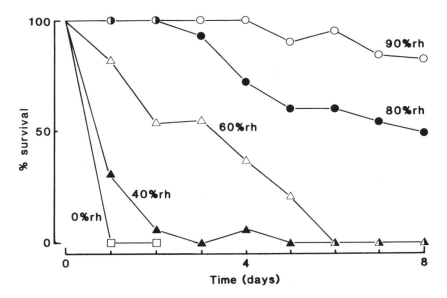

Figure 5. Survival of *Steinernema carpocapsae* (All) infective juveniles on 1% agarose after exposure to reduced relative humidities for up to 8 days following 3 days preconditioning at 97% relative humidity (rh) (approximately 50 infective juveniles per sample point).[32]

coiled infective nematode and one that has undergone rehydration for 4 hr at 100% rh are shown in Figure 6.

We are presently unable to reproduce the coiling response consistently and have yet to ascertain whether coiled individuals survive better than uncoiled ones. However, coiling seems to have little effect on survival at 40% rh and below where, like direct exposure experiments, there is little tolerance to these dehydration regimes despite the slower rates of water loss provided by the model substrate. A change in induction period from 3 days at 97% rh to 24 hr periods at 97% and 90% rh, respectively, before exposure to 80% rh and below appears to facilitate higher survival rates (ca. 50%) for *S. carpocapsae* dried on 1% agarose for 7 days at 60% rh (Figure 7).[35] Conversely, there is little difference between the survival rates afforded by direct drying and drying on 0.5% agar (Figure 7), and little or no survival occurs on exposure to 40% rh and below, irrespective of the substrate used. Interestingly, individuals dried on 0.5% agar tend to burrow into the substrate during dehydration, and thus, little or no coiling response occurs.

The above results appear to confirm the data reported for large aggregate drying experiments to some extent in that *S. carpocapsae* (if dried slowly enough) can apparently enter a quiescent anhydrobiotic condition and persist under relatively dry conditions (i.e., down to 60% rh). However, if exposed to harsher dehydration regimes, the nematodes seem unable to enter a cryptobiotic state and quickly die. The fact that *S. carpocapsae* may only be capable of surviving as a quiescent anhydrobiote rather than a cryptobiotic anhydrobi-

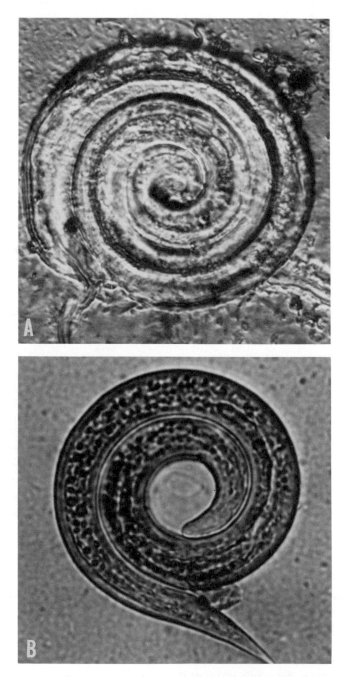

Figure 6. Pictorial presentations of the coiling response exhibited by some *Steinernema carpocap-sae* (all) infective juveniles dried slowly at 97% rh and below on 1% agarose, (A, top figure) dehydrated coiled, (B, bottom figure) coiled following rehydration at 100% rh for 4 hr.

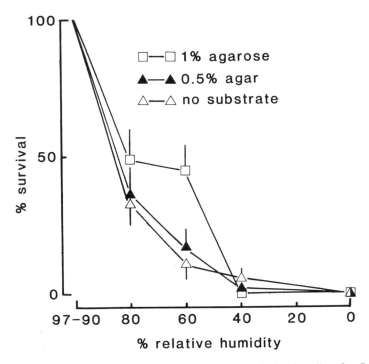

Figure 7. Survival of *Steinernema carpocapsae* (All strain) infective juveniles after 7 days' exposure to reduced relative humidities following preconditioning for 24 hr periods at 97% and 90% reltive humidity, with or without model substrates (approximately 50 infective juveniles per sample).[35] No substrate = nematodes on a glass slide.

ote is interesting in itself and supports our present data on other soil-dwelling nematodes.[4,10] The adaptive significance of many infective juveniles maintaining this quiescent state after rehydration is unknown. It may allow protracted survival by conserving energy stores during periods of host absence, the mechanical/physical stimulation required for activation being provided by the physical presence of the host and by the host's activities within the soil matrix.

V. PHYSIOLOGICAL AND BIOCHEMICAL ASPECTS OF DEHYDRATION SURVIVAL

Clearly, like other slow-dehydration strategists, steinernematid and heterorhabditid infective juveniles must be dried extremely slowly to ensure any survival. The rates of water loss required to induce anhydrobiosis remain to be fully defined, but will probably be similar to the moisture loss characteristics of the nematodes' natural habitats. Even so, this statement is presumptive in that slow dehydration does not always facilitate survival at low water activities.[4] For example, ensheathed, infective juveniles of *H. contortus* dry far more slowly than some of the aerial plant-parasitic nematode anhydrobiotes but cannot survive to the same extent.[24] Results of this nature indicate that other adaptations are involved in retaining cellular integrity as water is removed. Because water is an essential component of living cells and is important in the maintenance of the structural and functional integrity of biomembranes and biomacromolecules (see Womersley[4] for review), these additional adaptations must be intimately concerned with the molecular fabric of the cell itself.

Crowe and Madin[21] suggested that gradual water loss was a prerequisite for anhydrobiotic survival, enabling the organisms concerned to make the necessary metabolic and biochemical adaptations needed to enter cryptobiosis and thus, survive rigorous and extended periods of dehydration. Such adaptations would allow for the stabilization of an essentially dry biological system and preclude the irreversibly damaging effects normally associated with the removal of bulk and structural water.[4,40] Subsequently, the same workers showed that, in *A. avenae*, induction into anhydrobiosis was associated with an accumulation of glycerol and trehalose (a disaccharide of glucose) at the expense of glycogen and lipid reserves.[41] These results correlate well with analyses made on encysted embryos of the brine shrimp, *Artemia salina*.[42] Only over the last two years have similar analyses been conducted on *S. carpocapsae*, the results of which are encouraging.

Ishibashi et al.[33] first demonstrated that *S. carpocapsae* (DD-136) decreased insoluble saccharides (presumably glycogen) and increased soluble saccharides (tentatively identified as trehalose with a minor glucose component) during dehydration for 6 days at 98% rh. However, it is difficult to compare the amount of trehalose present with that found in anhydrobiotic *A. avenae* or *A. salina* because results were expressed as glucose equivalents (μM/mg dry wt) rather than on a % dry wt basis. Our preliminary experiments with small samples (<1 mg) of *S. carpocapsae* (All) infective juveniles (provided by R. Gaugler) have involved both carbohydrate and lipid analyses.[43] Fresh, unstressed samples have an initial lipid content of ca. 30% dry wt and a glycogen content of ca. 4% dry wt, but these levels are reduced to ca. 24% dry wt and essentially zero, respectively, after preconditioning for 3 days at 97% rh. The decrease in lipid represents a 20% reduction in total lipid reserves, which is much less than that recorded for either *D. myceliophagus* or *A. avenae*.[22,23,41] Conversely, trehalose levels start at less than 0.2% dry wt in fully hydrated samples and increase to just below 7.0% dry wt after dehydration. Glucose and glycerol levels have yet to be quantified accurately. Although the elevated trehalose levels we have recorded in *S. carpocapsae* are not as high as those reported for *A. avenae*, they do compare favorably with the levels found in other nematode anhydrobiotes.[22,23,44] Thus, it would appear that *S. carpocapsae* infective juveniles do possess the biochemical adaptations that have been associated with anhydrobiotic survival in other nematodes.[4]

The preferential storage of trehalose by anhydrobiotes has been linked to its use in the replacement of bulk and structural water and in the inhibition of oxidative damage (see Womersley[4] for review). For example, research on model membrane systems has shown that the presence of trehalose in physiologically relevant amounts (0.3 g trehalose/g membrane) can preserve their structural and functional integrity by preventing phospholipid phase transitions and bilayer deformations as structural water is removed.[45-49] Other research has demonstrated that trehalose can prevent dehydration-induced fusion between phospholipid bilayers in synthetic membrane systems and, by inference, may have the potential to protect anhydrobiotic systems from similar irreversible structural damage.[50] However, the ways in which these model membrane studies relate to the actual anhydrobiotes themselves is presently unknown. Our research on carbohydrate levels in *D. myceliophagus*[22,23] and on membrane compositions in a number of nematode anhydrobiotes[51] has caused us to begin to reevaluate the protective role played by trehalose in whole organisms during anhydrobiosis.

If an increase in trehalose is indeed indicative of successful entry into cryptobiotic anhydrobiosis the question remains, why are *S. carpocapsae* infective juveniles unable to achieve cryptobiosis? One would predict that once the required amount of trehalose had been synthesized, stabilization of cellular structure and maintenance of functional integrity would become automatic and would no longer be dependent on maintaining a slow rate of evaporative water loss. This has been reported to be the case for *A. avenae*.[41] However, our research has shown that this is not the case for juveniles and adults of *D. myceliophagus*.[11,22,23] Nematodes cultured on the mycelium of *Agaricus bisporus* or *Rhizoctonia cerealis* and then dehydrated for up to 5 days at 97% rh show similar declines in lipid and glycogen stores, but trehalose levels are very different (i.e., ca. 3% dry wt vs. over 8% dry wt, respectively) (Figure 8). Moreover, if natural dehydration regimes are substituted for enforced dehydration at 97% rh by allowing nematodes to swarm from fungal cultures and dehydrate in coiled aggregates over a period of 6-9 weeks,[22,23] then trehalose levels are elevated above 15% dry wt (Figure 9). We do not understand the rise in lipid content seen during the first 48 hr of swarming and need to confirm these levels.

When all these differently treated samples are placed at reduced rh levels following induction, they exhibit the same survival characteristics, regardless of the amount of trehalose present. That is, all survive equally well down to 40% rh but are unable to maintain survival at 10% rh and below, with mortality reaching 100% after 24 hr at 0% rh. If, however, these nematodes are dried slowly through successive rh levels down to 0% rh, then survival is maintained above 50%.[51] We can infer from these results that, although elevated trehalose levels may be indicative of induction into quiescent anhydrobiosis, other adaptations are involved, either separately or in conjunction with trehalose that allow entry into the cryptobiotic state. Ultimately, the correct implementation of these adaptations would appear to rely on maintaining a gradual rate of water loss rather than the actual amount of trehalose present after induction.

Because of these results our attention has turned to membrane associated lipid constituents of nematode anhydrobiotes to determine whether these

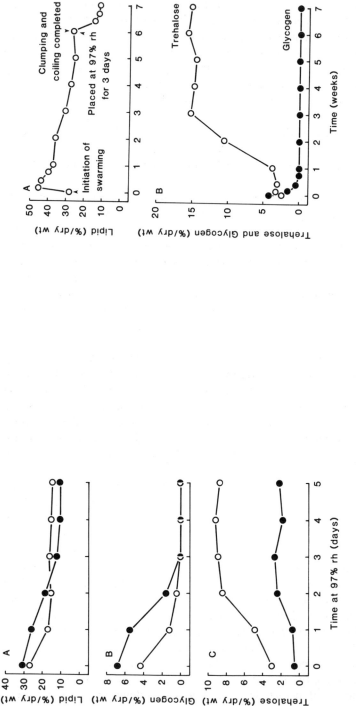

Figure 8 (Left). Comparison of changes in lipid, glycogen, and trehalose contents occurring in mixed juvenile and adult aggregates of *Ditylenchus myceliophagus*, cultured on *Agaricus bisporus* (●) and *Rhizoctonia cerealis* (○), during the induction of anhydrobiosis by enforced exposure to 97% relative humidity (rh).[22] Figure 9 (Right). Changes in lipid, glycogen, and trehalose contents occurring in mixed juvenile and adult samples of *Ditylenchus myceliophagus* at the initiation of swarming, during swarming, and after the completion of coiling and clumping under natural dehydration regimes.[22]

constituents remain stable in nematodes containing elevated levels of trehalose, or if their composition is modified during dehydration. In all nematode species we have analyzed, changes are evident in the nature of the molecular speciation of specific phospholipids either during the induction of or on rehydration from anhydrobiosis. Our studies are not so complete for *S. carpocapsae* in that actual analyses of phospholipid molecular speciation have yet to be undertaken. However, our analyses of the major lipid constituents show some interesting trends (Table 2).[52] Phospholipids and cholesterol (both of which are important membrane components) remain relatively constant and, as expected, triglycerides are gradually depleted during the dehydration period. The increase in free-fatty acids observed initially is not maintained and stabilizes at low levels after 8 hr exposure to 97% rh. Concomitant with this is a rapid increase in another lipid component (tentatively identified as cholesterol esters) over the first 8 hr of dehydration, after which this component stabilizes and is apparently unaffected by continued drying. Again, we can only speculate on the adaptive significance of this, but interestingly, esterification of free fatty acids to cholesterol esters may provide a mechanism for their immobilization during dehydration and reduce the potential for oxidative damage. Further, the preferential storage of cholesterol esters could allow for modification of membrane fluidity through modulation of free cholesterol. However, since free cholesterol levels appear to change little during dehydration, this suggests that any effects on membrane fluidity may only be apparent during rehydration.

TABLE 2. Changes in Lipid Constituents of *Steinernema carpocapsae* (All Strain) Infective Juveniles during Dehydration at Controlled Relative Humidities (rh) and the Presumptive Induction of Anhydrobiosis[52]

	Constituent lipids/% total lipids				
	Phospho-lipids	Choles-terol	Free fatty acids	Trigly-cerides	Choles-terol esters
0 min[a]	23.5	7.2	13.2	53.5	2.6
30 min/97% rh	21.6	6.5	15.8	52.9	3.2
60 min/97% rh	20.9	9.0	9.3	53.4	7.4
8 hr/97% rh	20.6	5.5	6.3	51.3	16.3
24 hr/97% rh	23.2	7.0	5.4	49.5	14.9
72 hr/97% rh	25.5	8.8	6.0	43.2	16.5
72 hr/97% rh + 6 days at 90% rh	25.2	9.3	4.4	43.7	17.4

[a] Fresh nematodes from an aqueous suspension.

The interpretations drawn from the biochemical data presented here are not applicable to all nematode anhydrobiotes. Differences in biochemical adaptations and their adaptive significance have already been shown to exist between slow- and fast-dehydration strategist nematode anhydrobiotes.[4,44] Similarly, any results concerning biochemical adaptations occurring in slow-dehydration strategists are themselves open to interpretation because little data are available for any given anhydrobiote. Moreover, the observed biochemical changes may be a product of the experimental regime rather than that which might occur under more natural conditions. Trehalose does appear to be an integral part of the adaptive process at the cellular level in many anhydrobiotes, but its definitive role in stabilizing cellular structure may not be so distinct as previously thought.[45-49] There are other metabolic profiles besides carbohydrates (i.e., organic acids, amino acids, etc.) that need to be analyzed and which may also show significant changes in character during dehydration. In addition, examination of the morphology and molecular organization of intracellular components (i.e., membrane systems) has yet to be fully realized, and the ways in which changes in molecular structure may interface with changes in metabolic profiles remain unknown. Nevertheless, results obtained for *S. carpocapsae* clearly show that although we are unable to interpret fully the reasons why certain biochemical adaptations occur, they are present in this nematode and are comparable to those found in proven nematode anhydrobiotes.[44]

VI. CONCLUSIONS

In comparison to other nematode anhydrobiotes that are slow-dehydration strategists, we obviously know very little about the dehydration survival capabilities of entomopathogenic nematodes. This lack of knowledge is not surprising when one considers that, with the exception of a few isolated studies, research concerning their ability to survive dehydration has been directed towards commercial storage applications, rather than identifying the anhydrobiotic potential of the organisms within the context of their natural environment. Consequently, their anhydrobiotic potential is difficult to assess at the present time. The available evidence suggests that the steinernematids and heterorhabditids studied so far are exclusively quiescent anhydrobiotes at best. If so, this realization alone is an important one. Lack of knowledge concerning the anhydrobiotic potential of these nematodes has already led to them being placed in environments whose moisture characteristics have been partially or totally unsuitable for their survival, resulting in a high degree of failure in the field.

In addition to defining their environmental limitations for field applications, the probability that steinernematids and heterorhabditids are quiescent anhydrobiotes reduces the prospect of their eventual successful storage in the dry state. Again, this would indicate that the partially hydrated and aerobic conditions reported to give high viability may already define limitations for their commercial storage and fall short of the attainment of a cryptobiotic state.

Anomalies do exist, however, in assessing these nematodes as quiescent anhydrobiotes. Preliminary studies on small groups of *S. carpocapsae* have shown that infective juveniles are capable of making the metabolic adjustments considered necessary for the successful induction of cryptobiotic anhydrobiosis. This fact does not support the contention that *S. carpocapsae* is a quiescent anhydrobiote. On the contrary, these results suggest that this and similar nematode species have yet to be dried at rates that are conducive to the induction of the cryptobiotic state. This could be the case for large aggregates of infective juveniles that, because natural dehydration regimes are unknown, have been dried following the techniques used for much smaller numbers of other nematode anhydrobiotes. In these instances, we are uncertain whether the apparent quiescent cryptobiotic condition attained is a consequence of the differential drying characteristics of the nematode mass, or a true representation of the anhydrobiotic potential of the nematodes studied.

In contrast, problems of differential drying do not exist in dehydration experiments involving individual infective juveniles, where viability remains high at elevated humidities, but cannot be maintained below 60% rh, even with the use of model substrates. These experiments provide the strongest evidence in support of the quiescent condition. If metabolic adjustments are occurring here (and there is no reason to suspect they are not), this would suggest that the biochemical adaptations normally associated with cryptobiotic anhydrobiosis are merely a prelude to the initiation of further cellular adaptations that, either independently or in concert with initial biochemical adaptations (i.e., trehalose storage), support this ametabolic state. That metabolic adaptations may be apparent in quiescent anhydrobiotes emphasizes our lack of knowledge as to the ways in which dry biological systems are stabilized at the cellular level. Moreover, if proven, the use of metabolic adaptations by quiescent anhydrobiotes will demand a reorganization of our thinking on the biochemical adaptations necessary for the successful induction of anhydrobiosis at both the quiescent and the cryptobiotic levels.

REFERENCES

1. **Van Gundy, S.D.,** Factors in survival of nematodes, *Annu. Rev. Phytopathol.*, 3, 43, 1965.
2. **Evans, A. A. F., and Perry, R. N.,** Survival strategies in nematodes, in *The Organization of Nematodes*, Croll, N. A., Ed., Academic Press, New York, 1976, 383.
3. **Evans, A. A. F., and Womersley, C.,** Longevity and survival in nematodes: models and mechanisms, in *Nematodes as Biological Models,* Vol. 2, Zuckerman, B. M., Ed., Academic Press, New York, 1980, 193.
4. **Womersley, C.,** A reevaluation of strategies employed by nematode anhydrobiotes in relation to their natural environment, in *Vistas on Nematology,* Veech, J., and Dickson, D. W., Eds., Society of Nematologists, Hyattsville, 1987, 165.
5. **Demeure, Y., and Freckman, D. W.,** Recent advances in the study of anhydrobiotic nematodes, in *Plant Parasitic Nematodes*, Vol. 3, Zuckerman, B. M., and Rohde, R. A., Eds., Academic Press, New York, 1981, 205.

6. **Freckman, D. W., and Womersley, C.,** Physiological adaptations of nematodes in Chihuahuan desert soils, in *New Trends in Soil Biology*, Proc. VIII Int. Coll. Soil Zool., 1983, 395.

7. **Gaugler, R.,** Ecological considerations in the biological control of soil-inhabiting insects with entomopathogenic nematodes, *Agric. Ecosys. Environ.*, 24, 351, 1988.

8. **Webster, J. M.,** Biocontrol: the potential of entomophilic nematodes in insect management, *J. Nematol.*, 12, 270, 1980.

9. **Platzer, E. G.,** Biological control of mosquitoes with mermithids, *J. Nematol.*, 13, 257, 1981.

10. **Womersley, C., and Ching, C.,** Natural dehydration regimes as a prerequisite for the successful induction of anhydrobiosis, *J. Exp. Biol.*, 143, 359, 1989.

11. **Womersley, C.,** Biochemical changes associated with anhydrobiotic survival in the mushroom nematode, *Proc. 2nd Int. Cong. Comp. Physiol. Biochem.*, 1988, 610.

12. **Poinar, G. O., Jr.,** *The Natural History of Nematodes*, Prentice-Hall, Englewood Cliffs, NJ, 1983, 189.

13. **Woodring, J. L., and Kaya, H. K.,** Steinernematid and heterorhabditid nematodes: a handbook of techniques, *South. Coop. Ser. Bull.*, 331, 1988.

14. **Kaya, H. K.,** Entomogenous nematodes for insect control in IPM systems, in *Biological Control in Agricultural IPM Systems*, Hoy, M. A., and Herzog, D. C., Eds., Academic Press, New York, 1985, 283.

15. **Gaugler, R., and Boush, G. M.,** Effects of ultraviolet radiation and sunlight on the entomogenous nematode *Neoaplectana carpocapsae, J. Invertebr. Pathol.*, 32, 291, 1978.

16. **Gaugler, R., and Boush, G. M.,** Laboratory tests on ultraviolet protectants of an entomogenous nematode, *Neoaplectana carpocapsae, Environ. Entomol.*, 8, 810, 1979.

17. **MacVean, C. M., Brewer, J. M., and Capinera, J. L.,** Field tests of antidesiccants to extend the infection period of an entomogenous nematode, *Neoaplectana carpocapsae*, against the Colorado potato beetle, *J. Econ. Entomol.*, 75, 97, 1982.

18. **Nash, R. F., and Fox, R. C.,** Field control of the Nantucket pine tip moth by the nematode DD-136, *J. Econ. Entomol.*, 62, 660, 1969.

19. **Webster, J. M., and Bronskill, J. F.,** Use of Gelgard M and an evaporation retardant to facilitate control of larch sawfly by a nematode-bacterium complex, *J. Econ. Entomol.*, 61, 1370, 1968.

20. **Simons, W. R., and Poinar, G. O., Jr.,** The ability of *Neoaplectana carpocapsae* (Steinernematidae: Nematodea) to survive extended periods of desiccation, *J. Invertebr. Pathol.*, 22, 228, 1973.

21. **Crowe, J. H., and Madin, K. A. C.,** Anhydrobiosis in nematodes: evaporative water loss and survival, *J. Exp. Zool.*, 193, 323, 1975.

22. **Womersley, C.,** Morphological and biochemical adaptations to anhydrobiosis in artificially and naturally dehydrated populations of *Ditylenchus myceliophagus* (Nematoda), *Am. Zool.*, 28(4), 76, 1988.

23. **Womersley, C.,** unpublished data, 1989.

24. **Ellenby, C.,** Dormancy and survival in nematodes, *Symp. Soc. Exp. Biol.*, 23, 572, 1969.

25. **Gaugler, R.,** Biological control potential of neoaplectanid nematodes, *J. Nematol.*, 13, 241, 1981.

26. **Womersley, C.,** unpublished data, 1988.

27. **Wallace, H. R.,** The influence of pore size and moisture content of the soil on the migration of larvae of the beet eelworm, *Heterodera schachtii* Schmidt, *Ann. Appl. Biol.*, 46, 74, 1958.

28. **Ekern, P. C.,** Evaporation from bare low humic latosol in Hawaii, *J. Appl. Meteorol.*, 5, 431, 1966.

29. **Simons, W. R.,** Nematode survival in relation to soil moisture, *Meded. Landbouwhogesch. Wageningen*, 73, 1, 1973.

30. **Timper, P., and Kaya, H. K.,** Role of the second-stage cuticle of entomogenous nematodes in preventing infection by nematophagous fungi, *J. Invertebr. Pathol.*, 54, 314, 1989.

31. **Womersley, C.,** unpublished data, 1984.
32. **Womersley, C., and Higa, L. M.,** unpublished data, 1989.
33. **Ishibashi, N., Tojo, S., and Hatate, H.,** Desiccation survival of *Steinernema feltiae* str. DD-l36 and possible desiccation protectants for foliage application of nematodes, in *Recent Advances in Biological Control of Insect Pests by Entomogenous Nematodes in Japan,* Ishibashi, N., Ed., Ministry of Education, Japan, Grant No. 59860005, 1987, 139.
34. **Hara, A. H., and Kaya, H. K.,** Toxicity of selected organophosphate and carbonate pesticides to infective juveniles of the entomogenous nematode *Neoaplectana carpocapsae* (Rhabditida:Steinernematidae), *Environ. Entomol.,* 12, 496, 1983.
35. **Womersley, C., and Foltz, N.,** unpublished data, 1989.
36. **Ohba, K.,** Desiccation survival of *Steinernema feltiae* (str. DD-l36), in *Recent Advances in Biological Control of Insect Pests by Entomogenous Nematodes in Japan,* Ishibashi, N., Ed., Ministry of Education, Japan, Grant No. 59860005, 1987, 145.
37. **Popiel, I., Holtemann, K. D., Glaser, I., and Womersley, C.,** Commercial storage and shipment of entomogenous nematodes, U.S. Patent Appl. PCT/0587/02043, 1987.
38. **Bedding, R. A.,** Storage of entomopathogenic nematodes, Int. Patent WO88/08668, 1988.
39. **Sharma, M. L., and Uehara, G.,** Influence of soil structure on water relations in low humic latosols: water retention, *Proc. Soil Sci. Soc. Am.,* 32, 765, 1968.
40. **Womersley, C.,** Biochemical and physiological aspects of anhydrobiosis, *Comp. Biochem. Physiol.,* 70B, 669, 1981.
41. **Madin, K. A. C., and J. H. Crowe,** Anhydrobiosis in nematodes: carbohydrate and lipid metabolism during rehydration, *J. Exp. Zool.,* 193, 335, 1975.
42. **Clegg, J. S.,** Metabolic studies of cryptobiosis in encysted embryos of *Artemia salina,* *Comp. Biochem. Physiol.,* 20, 801, 1967.
43. **Womersley, C., Higa, L. M., and Ching, C.,** unpublished data, 1988.
44. **Womersley, C., and Smith, L.,** Anhydrobiosis in nematodes. I. The role of glycerol, myo-inositol and trehalose during desiccation, *Comp. Biochem. Physiol.,* 70B, 579, 1981.
45. **Crowe, J. H., Crowe, L. M., and Jackson, S. A.,** Preservation of structural and functional activity in lyophilized sarcoplasmic reticulum, *Arch. Biochem. Biophys.,* 220, 477, 1983.
46. **Crowe, J. H., Crowe, L. M., and Chapman, D.,** Preservation of membranes in anhydrobiotic organisms: the role of trehalose, *Science,* 223, 701, 1984.
47. **Crowe, J. H., Crowe, L. M., and Chapman, D.,** Infrared spectroscopic studies on interactions of water and carbohydrates with a biological membrane, *Arch. Biochem. Biophys.,* 232, 400, 1984.
48. **Crowe, J. H., and Crowe, L. M.,** Stabilization of membranes in anhydrobiotic organisms, in *Membranes, Metabolism, and Dry Organisms,* Leopold, C. A., Ed., Cornell University Press, Ithaca, NY, 1985, 188.
49. **Crowe, L. M., Mouradian, R., Crowe, J. H., and Womersley, C.,** Effects of carbohydrates on membrane stability at low water activities, *Biochim. Biophys. Acta,* 769, 141, 1984.
50. **Womersley, C., Uster, P. S., Rudolph, A. S., and Crowe, J. H.,** Inhibition of dehydration-induced fusion between liposomal membranes by carbohydrates as measured by fluorescence energy transfer, *Cryobiology,* 23, 245, 1986.
51. **Womersley, C.,** unpublished data, 1989.
52. **Womersley, C. and Ayau, E.,** unpublished data, 1989.

7.   Behavior of Infective Juveniles

N. Ishibashi and E. Kondo

I. INTRODUCTION

Soil nematodes possess a rich variety of intriguing behaviors, including chemotaxis, thermotaxis, osmotic avoidance, and dauer formation and recovery.[1-6] Mutants defective in each of these behaviors have been isolated for the free-living nematode *Caenorhabditis elegans*,[6] permitting detailed behavioral analyses. Using *C. elegans* as a model, the basic functions behind many nematode behaviors may become resolved, and we should be able to apply this information for controlling injurious nematodes or for enhancing beneficial nematodes.

Infective juveniles of entomopathogenic nematodes in the families Steinernematidae and Heterorhabditidae are known to exhibit many of the same specific and complicated behavioral patterns of *C. elegans* dauer juveniles. Yet, interpreting the behavior of either of these nematodes remains speculative. Although some investigations and interpretations of entomopathogenic nematode behavior are referred to in the present text, this research is still in its infancy. Our main concern at this time is how to exploit the behavior of infective juveniles of entomopathogenic nematodes for the control of insect pests.

II. INFECTIVE VS. NORMAL DEVELOPMENTAL STAGE JUVENILES

Infective juveniles of entomopathogenic nematodes are more motile than the respective third-stage juveniles of the normal developmental cycle.[7] This attribute is an adaptation for a free-living existence in which orientation and movement for host-finding can be critical. In contrast, the third-stage developmental juveniles are parasitic, and so there is little advantage or adaptation for high motility. For example, the movement of *Steinernema carpocapsae* (DD-136 strain) infective juveniles, believed to be among the least motile of entomopathogenic nematodes, is still nearly twice that of developmental juveniles. Although there are no significant differences in wavelength between the two forms (44 μm for developmental vs. 49 μm for infective juveniles) when placed on 2% agar, the wave amplitudes were 27 μm for developmental and 17 μm for infective juveniles (Figure 1). Consequently, infective juveniles can migrate 245 μm/min on agar compared with 132 μm/min for developmental juveniles.

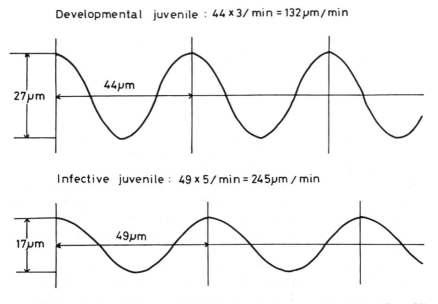

Figure 1. Schematized undulatory movement of *Steinernema carpocapsae* from tracks left on a 2% water-agar film.

III. SURVIVAL STRATEGIES

Infective juvenile survival or persistence remains a key obstacle to the effective use of entomopathogenic nematodes in insect control. Applications against foliage-feeding insects have generally been discouraged because infective juveniles are not adapted to the environmental extremes characteristic of exposed surfaces. The soil is the natural reservoir for entomopathogenic nematodes, and these organisms have evolved at least three major behavioral strategies for persistence in this buffered environment: aggregation, inactivity, and anhydrobiosis. The latter topic has been extensively reviewed by Womersley.[8]

A. Aggregation

Aggregation commonly occurs in many nematode species. This behavior may be driven by mechanosensitivity or tactile sensitivity, both of which are fundamental properties of nematodes.[4] Aggregation has survival value in protecting nematodes from desiccation and sunlight.[9] Nematodes at the periphery of aggregations usually die to form a barrier against unfavorable environmental stresses, so that survival within a clump is greater than that of isolated individuals on the same substrate. Aggregation is also a frequently observed laboratory phenomenon with entomopathogenic nematodes (Figure 2), with the aggregating nematodes assuming a straight posture like a bundle of wire.

Figure 2. Aggregation of *Steinernema carpocapsae* infective juveniles on an insect cadaver.

B. Inactivity

Active movement has advantages to entomopathogenic nematodes, because it increases chances for encountering a susceptible host and for survival by permitting escape from unfavorable habitats. However, some entomopathogenic nematodes may employ inactivity as a strategy to enhance their survival: an inactive nematode reduces energy usage, attractiveness to predators (e.g., mesostigmatid mites do not attack motionless infective juveniles),[10] and chances for encountering nematode natural enemies.[11-13]

Inactivity in soil has been well documented with *S. carpocapsae*, which tends to remain at the point of application.[14,15] Inactivity may be responsible for the sharp decline in infective juvenile recovery from soil using the Baermann funnel method, which depends on nematode movement, compared with centrifugal flotation.[16] Further evidence supporting inactivity is that infective juveniles stored in water often assume a straight nonmotile posture.

Coiling is another form of inactivity, a posture related to anhydrobiosis in some nematode species.[8] Entomopathogenic nematodes do not coil as readily as aphelenchid nematodes undergoing anhydrobiosis.[8] *S. carpocapsae* infective juveniles coil loosely, whereas *S. glaseri* coils more frequently and somewhat tighter. In both nematodes, however, true coiling is rare. Consequently, the infective juveniles of these nematodes are generally not as resistant to desiccation nor is their inactivity as fixed as the tightly coiling aphelenchids.

IV. DISPERSAL

Dispersal is a behavioral mechanism entomopathogenic nematodes use to locate habitats for survival and infection. The free-living infective juvenile is the dispersal stage for these nematodes. Dispersal can be active or passive.[17] Active dispersal under the nematode's own power is short-range, and may include movement away from the host cadaver, to sheltered microenvironments, or to preferred soil depths. Active dispersal is constrained by soil particle size, pore diameter, water content, temperature, and the relative activity of the nematode species. Passive dispersal is longer range and is affected through the action of other agents, including water, infected insects,[18] phoretic hosts,[19] or soil (e.g., farm machinery, nursery plants). Dispersal plays a particularly important role in entomopathogenic nematode colonization of new localities, as demonstrated by their cosmopolitan geographical distribution.[17]

V. NICTATION

Nictation, the behavior of lifting all but the posterior portion of the nematode's body from the substrate and waving the extended body from side to side, occurs in infective juveniles of entomopathogenic nematodes (Figure 3), although there are differences among species. *Steinernema carpocapsae* actively nictates, alternatively waving and pausing in a straight extended state; *S. feltiae* (=*bibionis*) nictates but only for a short time and without pausing in the straight form; *S. glaseri* nictates rarely and only when moving to neighboring particles.[20] This behavior is unique to the infective stage and is not present in the developmental stages. Although this behavior is not fully understood, it may be related to "bridging" and "leaping" (i.e., active dispersal), or host-finding (i.e., orientation).

A. Nictation Substrate

An infective juvenile's ability to nictate is dependent upon substrate. It seldom occurs on a water-agar film with less than 2% agar, although some nematodes nictate on the surface of dog-food medium containing 2% agar. Nictation is most frequently observed on rough surfaces, where substrate particle size greatly influences the behavior.[21]

Nictation is more often observed in covered than open Petri dishes, suggesting that high relative humidity is preferred for this behavior. Nictation can occur over a wide range of substrate moistures, but saturated surfaces show a lower rate than somewhat dry substrates. Moreover, infective juveniles can elevate almost their entire bodies, perching on their tails, on drier surfaces with less surface tension. They cannot nictate on water-saturated substrates, and those that attempt standing only initiate the process before resuming their

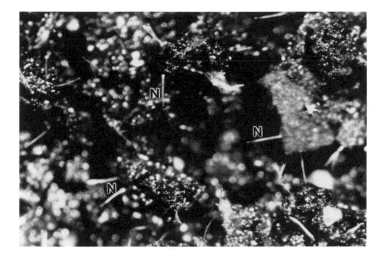

Figure 3. Infective juveniles of *Steinernema carpocapsae* nictating on bark compost. N indicates nictating nematodes.

gliding behavior on the substrate surface. The water content of bark compost as a substrate for nictation was best at 60% on a weight basis in open vessels, but when the vessel was covered no significant difference in nictation occurred over a moisture content range of 30 to 60%.[7]

B. Behavioral Sequences

Croll and Matthews[4] suggested parasitic nematode activity is likely fixed as a behavioral sequence, which once initiated results in a series of activities not interrupted until the sequence is completed. Nictation of *S. carpocapsae* infective juveniles seems to follow a series of behaviors as described for other nematodes.[7]

Nictation starts when infective juveniles elevate their anterior end and wave their heads from side to side. This behavior includes large waving, which can last for several hours, followed by small waving, pendulum waving, riding, and/or straight standing (Figure 4). Initially, they make large waves of the head, with some individuals forming bridges with their bodies to adjacent surface projections, as if the large waving is intended for attachment. Pendulum waving and straight standing nematodes occasionally provide a surface for other infective juveniles to start nictating; nematodes will crawl up these infective juveniles to become riders. When many nematodes participate in the "riding" formation, the aggregated clump looks like one moving organism. Small waving and pendulum waving infective juveniles can also shift to "leaping", which may have evolved from these behaviors. When they leap, the small waving individuals suddenly make a large wave and leap with a sudden reaction following the loop formation described by Reed and Wallace.[22] Leap-

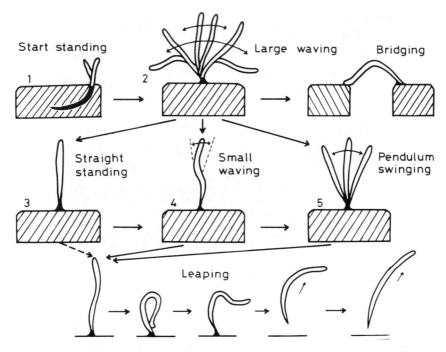

Figure 4. A sequential behavior from standing to leaping, *Steinernema carpocapsae* infective juveniles.

ing can be induced by a sudden change of milieu like flashing a light or lifting the lid of the culture vessel.[7] Neither nictating nor leaping is fully understood to be associated with infection, although soil moistures suitable for nictation also provide high insect mortality.[20,23] Nictating infective juveniles can often be seen on setae or the body surface of host insects.

VI. INFECTION

There are two major behavioral components to infection: host-finding and penetration. Infection is unlikely to be achieved without host-finding mediated by host kairomones, followed by penetration into the host.

A. Host-Finding

Nematode attraction to chemical stimuli has often been attributed to klinotactic orientation.[3] The side-to-side movement or headwaving of nematodes has been interpreted as sampling the stimuli;[24,25] presumably the distance between the amphids is insufficient to permit comparison of the intensity of stimulation without head movement. Infective juveniles of *S. carpocapsae* exhibit klinotactic orientation in responding positively to CO_2 and aggregating at the source.[25] Although CO_2 is believed to be important in host-finding, it appears to be a short-range attractant and may play its most critical role in host penetration through the spiracles. CO_2 may also play a role in observed aggregations of infective juveniles of *S. glaseri* at plant roots,[26] a response which places them in the feeding zone of root-feeding insect hosts.

Host fecal components including uric acid, xanthine, allantoin, ammonia, and arginic acid have also been reported to cause aggregations of infective juveniles of *S. carpocapsae*,[27,28] although it is unclear whether the nematodes were attracted to or arrested by these compounds. In our experiments, fecal pellets of common cutworm, *Spodoptera litura*, repelled *S. carpocapsae* juveniles.[29]

Infective juveniles of *S. carpocapsae* are also reported to accumulate around the nematode's symbiont bacterium, *Xenorhabdus nematophilus*.[30] We have confirmed this observation, although it is again unclear whether the bacteria acted as attractants or locomotion arrestants. Interestingly, we found few juveniles accumulated near heat-killed symbiont bacteria. The significance of attraction to the symbiont is unclear.

Temperature has been suggested as an additional cue by which entomopathogenic nematodes locate host insects.[31] However, these invertebrate parasites seem unlikely to respond as positively to the insect body temperature as do nematode parasites of warm-blooded hosts (e.g., hookworms). Burman and Pye[32] found that infective juveniles of *S. carpocapsae* tend to aggregate at or near their culture temperature. This tendency is similar to the acclimatization of *Ditylenchus dipsaci* and other soil nematodes.[33,34] Conversely, starved and dauer juveniles of *C. elegans* appeared to disperse from the acclimatized temperature.[35] The significance of entomopathogenic nematode temperature-related behavior is uncertain, and more investigations are needed.

Although many stimuli attract entomopathogenic nematodes or cause aggregations, it seems improbable that any one stimulus is predominant. Host-finding is likely the result of a sequence of host stimuli which integrate searching behavior.[25]

B. Penetration

The portal of entry for steinernematid and heterorhabditid infective juveniles is through natural openings (i.e., mouth, anus, and spiracles) and, much less commonly, wounds that permit direct access to the host's hemocoel. In addition, heterorhabditids possess a large terminal tooth that may be used to penetrate soft intersegmental areas of the cuticle.[36] This type of penetration is usually preceded by the nematode probing cuticular folds, and abrading the cuticle with the tooth until an opening occurs, allowing nematode entry.

The portal of entry can be restricted depending on the developmental stage of the host. For example, the only portal of entry for nematodes entering pupae is the spiracles. In larval stages, the spiracles of scarabs are protected with sieve plates which may have evolved as a means of protection against pathogens, especially entomopathogenic nematodes.[37] The oral route is the only penetration portal for black fly larvae, but buccal size determines whether infective juveniles are able to enter without being ruptured by the mouthparts. Similarly, the oral route of invasion does not occur in insects with sclerotized proventricular valves such as in cockroaches.[38]

Infective juveniles of *S. carpocapsae* enter and aggregate in the ventral eversible vesicle beneath the esophagus of *Spodoptera litura* (Figure 5), which may serve as a means of entry into the hemocoel,[21] although further information is needed about this structure.

The ability of infective juveniles to penetrate into the host is an essential step in the entomopathogenic nematode's life cycle. Nevertheless, the probability of any individual juvenile successfully penetrating and establishing in even highly susceptible hosts is low.[39] The inability of most infective juveniles to penetrate a host is cause for concern in using these nematodes as biological control agents, and may reflect another reason why so many nematodes are needed in field efficacy trials.

VII. ACTIVATION

The infective juveniles of steinernematid nematodes tend to become motionless in the absence of a host insect and take on a straight posture (Figure 6A). Sinusoidal activity can be restored by placing the juveniles near a host or agitating them. Various chemicals can also induce movement. For example, dilute oxamyl (an insecticide/nematicide) (Figure 6B), or juice from kale or aloe plants can stimulate activity.[40] The response of infective juveniles to attractants or repellents on agar film is amplified by using "activated" infective juveniles. For example, the difference in attraction between living and heat-killed symbiont bacteria is more clearly shown by infective juveniles activated in kale juice.[21] Chemically activated nematodes do not always display normal movements. The movements are of several types depending on the properties of the chemicals, showing typical large sinusoidal undulations, as well as kinks, twitches, or convulsions, and other uncoordinated movements.[41]

Chemical activating agents have been used in field trials to achieve higher insect mortality rates.[40,42] Oxamyl at 10 to 50 ppm stimulates *S. carpocapsae*'s normal movement for more than 2 days at room temperature. When oxamyl granules (3.5 kg AI/ha) were mixed with *S. carpocapsae* infective juveniles ($2.5 \times 10^5/m^2$), 70-99% field mortality of the turnip moth larvae, *Agrotis segetum*, was achieved. By contrast, infective juveniles alone resulted in 22-78% mortality, and oxamyl alone had virtually no effect.[42]

Not all chemicals activate infective juveniles to attack insects; infective juveniles treated with some organophosphates and carbamates can show partial paralysis with a curled or coiled posture.[43] On the other hand, oxamyl, and some insecticides and herbicides, are nontoxic to *S. carpocapsae* at the concentrations of practical field use.[44,45] Most chemicals in our laboratory experiments, except dilute oxamyl and acephate, also imparted abnormal movement to infective juveniles of *S. carpocapsae*, despite giving more effective results in mixed field applications. It is uncertain whether the chemicals acted on the

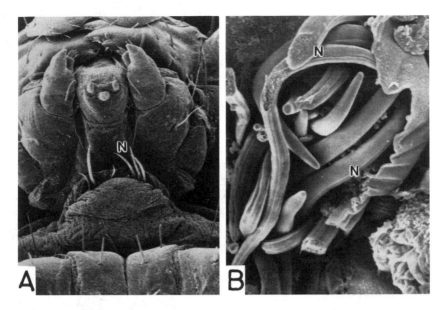

Figure 5. Aggregation of *Steinernema carpocapsae* infective juveniles (N) in ventral eversible vesicle of common cutworm, *Spodoptera litura*. (A) Ventral view of a larva with nematodes in partly everted vesicle. (B) Nematode aggregation in vesicle.

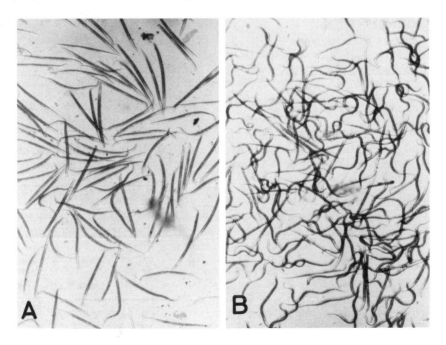

Figure 6. Oxamyl activation of infective juveniles of *Steinernema carpocapsae*. (A) Juveniles in distilled water. (B) "Activated" juveniles in an oxamyl solution (30 ppm) 2 days after treatment.

insect host to increase nematode invasion or whether the nematodes were activated to find and infect the insects.[46] Substances which stimulate infective juvenile activity are worthy of further investigation to enhance nematode efficacy against insect pests.

VIII. CONCLUSIONS

Our understanding of entomopathogenic nematode infective juvenile behavior is clearly deficient, particularly with regard to heterorhabditid species. What little is understood about steinernematid behavior is based largely on one nematode species, *S. carpocapsae*, studied on artificial substrates. Studies of infective stage behavior in the natural habitat, soil, are formidable, restricting present research to consequences rather than processes. It is achievable, however, to assess the generality of *S. carpocapsae* as a behavioral model for other entomopathogenic nematode strains and species. Finally, efforts to integrate behavior with function should be encouraged, as such studies will provide fundamental knowledge useful in increasing our ability to use them effectively in biological control.

REFERENCES

1. **Croll, N. A.**, *The Behaviour of Nematodes*, Edward Arnold, London, 1970.
2. **Wharton, D. A.**, Movement and co-ordination, in *A Functional Biology of Nematodes*, Croom Helm, London, 1986, chap. 3.
3. **Dusenbery, D. B.**, Behavior of free-living nematodes, in *Nematodes as Biological Models*, Vol. 1, Zuckerman, B. M., Ed., Academic Press, New York, 1980, chap. 3.
4. **Croll, N. A., and Matthews, B. E.**, *Biology of Nematodes*, Blackie, Glasgow, 1977, chap. 4.
5. **Croll, N. A.**, Behavioral activities of nematodes, *Helminthol. Abstr.*, A41, 359, 1972.
6. **Chalfie, M., and White, J.**, The nervous system, in *The Nematode Caenorhabditis elegans*, Wood, W. B., Ed., Cold Spring Harbor Laboratory, Cold Spring Harbor, NY, 1988, chap. 11.
7. **Ishibashi, N., Tabata, T., and Kondo, E.**, Movement of *Steinernema feltiae* infective juveniles with emphasis on nictating behavior, *J. Nematol.*, 19, 531, 1987.
8. **Womersley, C. Z.**, Dehydration survival and anhydrobiotic potential, in *Entomopathogenic Nematodes in Biological Control*, Gaugler, R., and Kaya, H. K., Eds., CRC Press, Boca Raton, FL, 1990, chap. 6.
9. **Wallace, H. R., and Doncaster, C. C.**, A comparative study of the movement of some microphagous, plant parasitic and animal parasitic nematodes, *Parasitology*, 54, 313, 1964.
10. **Ishibashi, N., Young, F. L., Nakashima, M., Abiru, C., and Haraguchi, N.**, Effects of application of DD-136 on silkworm, *Bombyx mori*, predatory insect, *Agriochpororus dohrni*, parasitoid, *Trichomalus apanteloctenus*, soil mites, and other non-target soil arthropods, with brief notes on feeding behavior and predatory pressure on soil mites, tardigrades, and predatory nematodes on DD-136 nematodes, in *Recent Advances in Biological Control of Insect Pests by Entomogenous Nematodes in Japan*, Ishibashi, N., Ed., Ministry of Education, Japan, Grant No. 59860005, 1987, 158.

11. **Ishibashi, N., and Kondo, E.**, *Steinernema feltiae* (DD-136) and *S. glaseri*: persistence in soil and bark compost and their influence on native nematodes, *J. Nematol.*, 18, 310, 1986.

12. **Ishibashi, N., and Kondo, E.**, Dynamics of the entomogenous nematode, *Steinernema feltiae*, applied to soil with and without nematicide treatment, *J. Nematol.*, 19, 404, 1987.

13. **Timper, P., and Kaya, H. K.**, Role of the second-stage cuticle of entomogenous nematodes in preventing infection by nematophagous fungi, *J. Invertebr. Pathol.*, 54, 314, 1989.

14. **Moyle, P. L., and Kaya, H. K.**, Dispersal and infectivity of the entomogenous nematode, *Neoaplectana carpocapsae* Weiser (Rhabditida: Steinernematidae), in sand, *J. Nematol.*, 13, 295, 1981.

15. **Georgis, R., and Poinar, G. O., Jr.**, Effect of soil texture on the distribution and infectivity of *Neoaplectana carpocapsae* (Nematoda: Steinernematidae), *J. Nematol.*, 15, 308, 1983.

16. **Ishibashi, N., and Kondo, E.**, A possible quiescence of the applied entomogenous nematode, *Steinernema feltiae*, in soil, *Jpn. J. Nematol.*, 6, 66, 1986.

17. **Kaya, H. K.**, Soil ecology, in *Entomopathogenic Nematodes in Biological Control*, Gaugler, R., and Kaya, H. K., Eds., CRC Press, Boca Raton, FL, 1990, chap. 5.

18. **Timper, P., Kaya, H. K., and Gaugler, R.**, Dispersal of the entomogenous nematode *Steinernema feltiae* (Rhabditida: Steinernematidae) by infected adult insects, *Environ. Entomol.*, 17, 546, 1988.

19. **Epsky, N. D., Walter, D. E., and Capinera, J. L.**, Potential role of nematophagous microarthropods as biotic mortality factors of entomogenous nematodes (Rhabditida: Steinernematidae, Heterorhabditidae), *J. Econ. Entomol.*, 81, 821, 1988.

20. **Kondo, E., and Ishibashi, N.**, Nictating behavior and infectivity of the entomogenous nematode, *Steinernema* spp., to the larvae of common cutworm, *Spodoptera litura* (Lepidoptera: Noctuidae), on the soil surface, *Appl. Entomol. Zool.*, 21, 553, 1986.

21. **Ishibashi, N., and Kondo, E.**, unpublished data, 1989.

22. **Reed, E. M., and Wallace, H. R.**, Leaping locomotion by an insect parasitic nematode, *Nature*, 206, 210, 1965.

23. **Kondo, E., and Ishibashi, N.**, Effects of soil moisture on the survival and infectivity of the entomogenous nematode, *Steinernema feltiae* (DD-136), *Proc. Assoc. Pl. Prot. (Kyushu)*, 31, 186, 1985.

24. **Green, C. D.**, Nematode sex attractants, *Helminthol. Abstr.*, B49, 81, 1980.

25. **Gaugler, R., LeBeck, L., Nakagaki, B., and Boush, G. M.**, Orientation of the entomogenous nematode, *Neoaplectana carpocapsae*, to carbon dioxide, *Environ. Entomol.*, 8, 658, 1980.

26. **Bird, A. F., and Bird, J.**, Observations on the use of insect parasitic nematodes as a means of biological control of root-knot nematodes, *Int. J. Parasitol.*, 16, 511, 1986.

27. **Schmidt, J., and All, J. N.**, Chemical attraction of *Neoaplectana carpocapsae* (Nematoda: Steinernematidae) to insect larvae, *Environ. Entomol.*, 7, 605, 1978.

28. **Schmidt, J., and All, J. N.**, Attraction of *Neoaplectana carpocapsae* (Nematoda: Steinernematidae) to common excretory products of insects, *Environ. Entomol.*, 8, 55, 1979.

29. **Ishibashi, N.**, unpublished data, 1987.

30. **Pye, A. E., and Burman, M.**, *Neoaplectana carpocapsae*: nematode accumulations on chemical and bacterial gradients, *Exp. Parasitol.*, 51, 13, 1981.

31. **Byers, J. A., and Poinar, G. O., Jr.**, Location of insect hosts by the nematode, *Neoaplectana carpocapsae*, in response to temperature, *Behaviour*, 79, 1, 1982.

32. **Burman, M., and Pye, A. E.**, *Neoaplectana carpocapsae*: movements of nematode populations on a thermal gradient, *Exp. Parasitol.*, 49, 258, 1980.

33. **Wallace, H. R.**, The orientation of *Ditylenchus dipsaci* to physical stimuli, *Nematologica*, 6, 222, 1961.

34. **Croll, N. A.**, Acclimatization in the critical thermal response of *Ditylenchus dipsaci*, *Nematologica*, 13, 385, 1967.

35. **Hedgecock, E. M., and Russel, R. L.**, Normal and mutant thermotaxis in the nematode, *Caenorhabditis elegans, Proc. Natl. Acad. Sci.*, 72, 4061, 1975.
36. **Bedding, R. A., and Molyneux, A. S.**, Penetration of insect cuticle by infective juveniles of *Heterorhabditis* spp., *Nematologica*, 28, 354, 1982.
37. **Gaugler, R.**, Ecological considerations in the biological control of soil-inhabiting insect pests with entomopathogenic nematodes, *Agric. Ecosystems Environ.*, 24, 351, 1988.
38. **Georgis, R.**, unpublished data, 1987.
39. **Gaugler, R., Campbell, J. F., and McGuire, T. M.**, 1990, Fitness of a genetically improved entomopathogenic nematode, *J. Invertebr. Pathol.*, 1990, in press.
40. **Ishibashi, N.**, Integrated control of insects/nematodes by mixing application of steinerne-matid nematodes with chemicals, in *Recent Advances in Biological Control of Insect Pests by Entomogenous Nematodes in Japan*, Ishibashi, N., Ed., Ministry of Education, Japan, Grant No. 59860005, 1987, 165.
41. **Ishibashi, N.**, unpublished data, 1987.
42. **Ishibashi, N., Choi, D-R., and Kondo, E.**, Integrated control of insects/nematodes by mixing application of steinernematid nematodes and chemicals, *J. Nematol.*, 19, 531, 1987.
43. **Hara, A. H., and Kaya, H. K.**, Toxicity of selected organophosphate and carbamate pesticides on infective juveniles of the entomogenous nematode *Neoaplectana carpocapsae* (Rhabditida: Steinernematidae), *Environ. Entomol.*, 12, 496, 1982.
44. **Fedorko, A., Kamionek, M., Kozlowska, J., and Mianowska, E.**, The effects of some insecticides on nematodes of different ecological groups, *Pol. Ecol. Stud.*, 3, 79, 1977.
45. **Fedorko, A., Kamionek, M., Kozlowska, J., and Mianowska, E.**, The effects of vydate-oxamyl on nematodes of different ecological groups, *Pol. Ecol. Stud.*, 3, 89, 1977.
46. **Hatsukade, M., Katayama, H., and Yamanaka, S.**, Pathogenicity of the entomogenous nematode, *Steinernema* sp., to scarabaeid larvae injurious to turfgrass, *J. Jpn. Soc. Turfgrass Sci.*, 17, 53, 1988.

Commercialization and Application Technology

8.   Commercial Production and Development

Milton J. Friedman

I. STRATEGIES OF COMMERCIAL DEVELOPMENT

Development of production technology, distribution, and sales of ento-mopathogenic nematode-containing products is occurring now at a rapid pace. The strategies of commercial development which are selected will have a significant effect on how production technology develops and on which technologies are eventually implemented. Three factors distinguish these strategies of development: the source of investment capital, the level of investment compared to sales, and the degree of patent or other competitive protection which can be obtained.

The source of investment influences the rate of technology development by determining the level of funding available. Higher levels of investment and more rapid rates of development will lead to improved chances of developing proprietary, patent-protected technology. Protection of technology from competition in turn assures a significant return on investment. Whether such proprietary technology can be developed and whether sales will follow to justify such investment are the unknown factors that promote more conservative approaches.

The simplest form of investment and the most conservative is owner, or "sweat", equity, in which the business revenues provide both the owner's salary and development funding. Several small nematode-producing concerns have been of this general type. In the owner equity model, investment capital is largely limited to the level of sales. As a result, the rate of technology development is comparatively slow and rarely results in the sustained development of proprietary, patented improvements. Nevertheless, if the technology is labor (rather than capital) intensive, if there is no effective patent barrier to using the technology, and if there is no price competition with more efficient companies, this model can be and has been successful. The greatest vulnerability of this business strategy is an inability to respond to competition from capital intensive, patent-protected, and efficient production companies.

More rapid technology development requires a higher level of investment than what is available from direct sales, and two types of investment can be used for this purpose. One is social investment, in which, through grants, guaranteed loans, or other social subsidies, a government or nonprofit organization supports technology development directly or by allowing the commercial use of public-sponsored research under limited license. In the social investment model, the amounts of funding available are highly variable and responsive to noncommercial factors. Examples of these factors include socially recognized needs for technology development that can ameliorate suffering, needs for strengthening national economies through diversification or self-sufficiency, and needs to develop jobs through industry relocation. Normally, the development of technology is performed separately from the commercialization of the technology in the social investment model. As a result, social investment projects build gradually, based on long-term plans of ten years or more. Projects on large-scale nematode production in New Jersey during the 1930s and at the CSIRO in Australia during the 1980s are examples of social investment. Working over a relatively long period of time, technologies were developed (described below) which provided effective production of entomopathogenic nematodes for widespread use in both the public and private sectors. In the absence of more efficient competition, the social investment model can lead to significant patent-protected technologies. Limitations on technology licenses imposed by social investment, such as requirements on the site of manufacture or the unavailability of exclusive licenses to use the technology, can be disadvantages. Companies which are based on technology development in the nonprofit sector are most vulnerable to competition from more rapidly developing companies, and the inefficiencies described above further detract from their ability to compete.

Venture, or risk, capital investment can provide levels of investment that greatly exceed sales during the development phase, but this type of investment requires a shorter development time than the model above and the development of strategic barriers to competition to provide an attractive return on investment. Internal corporate investment is similar. The most common competitive barriers sought are patents. Overall, the venture model can be successful when proprietary technologies can be developed in a relatively short time and when market development is possible in an equivalent time period. The greatest vulnerability of a venture investment strategy is to errors in estimating the time required to develop a proprietary, efficient production process and to inefficiencies in management of the development effort.

The best model of development is not any one of the above, but a combination. For example, Biosys was founded by a small group of private individuals. Through a combination of venture funding for in-house proprietary technology development and out-of-house cooperation with the USDA and university researchers, Biosys established basic feasibility data and a patent-protected technology base. Subsequent development is being accomplished by additional venture funding for product and market development and government support in Alberta, Canada for establishment of a production facility.

In the earliest stage of commercial development, owner investment provides the greatest incentives to the initiation and invention processes and provides a nucleus of feasibility which can attract both social and venture investment. Of these two, venture investment should be used for the development of proprietary technology and patents. In this way, the patent rights are clearly controlled by the company itself without the constraints or inefficiencies of the social investment model. Social investment can be used effectively for nonproprietary developments such as field demonstrations, production plant financing, and market introductions.

Similarly, Bioenterprises, Ltd. of Australia is a private concern with access to technology developed under the government-sponsored CSIRO, and Agricultural Genetics of the U.K. benefits from government-sponsored research for its commercial efforts. These individual efforts represent varied combinations of the models described above. In general, the Biosys technology is more capital intensive and more dependent on patent protection from competition. The efficiencies of different processes and the efficacy of different products will ultimately determine which combination of models will prove most successful.

II. STRATEGIES OF TECHNOLOGY DEVELOPMENT

Existing technologies for the production of entomopathogenic nematodes have developed over a period of 60 years. In this time, entomopathogenic nematodes have been produced by a variety of means — by insect infection or on artificial media — by axenic or monoxenic culture, in solid and liquid phase fermentation. The means chosen have depended on the amounts of product required and the time, resources, and knowledge available. The methods discussed here have application to both the steinernematid and heterorhabditid entomopathogenic nematodes. In practice, modifications to the general methods of production are required for each nematode, and the reader is referred to the citations provided for more information. In the case of axenic culture, other members of the superfamily Rhabditoidea are referred to in the text or in citations and are assumed here to be similar in culture characteristics to the entomopathogenic nematodes. While many aspects of large-scale production currently used by industry are proprietary, the general conditions and constraints of production can be and are discussed.

The large-scale production of entomopathogenic nematode-containing products must be planned in a context composed of the production plant, distribution system, and application environment if the production system is to be profitable and self-sustaining. In conventional commercial operation, plant is large scale, distribution is through third parties, and application is unsupervised. Until recently, virtually all applications of nematode products have been limited to the nonprofit sector, in which production has been small scale, distribution has been direct and rapid, and application has been closely supervised by experts. The only commercial applications of nematodes have been closer to the nonprofit model than to conventional commercial operation and as a result have provided little profit stability. More recently, several efforts are being made to close the gap between what is practiced and what is profitable.

The economy of scale for a production process is a key factor in commercial development. As distribution area, market penetration, and competition increase, the fraction of the product's price which is available for production shrinks. Other costs such as transportation, inventory, and sales and marketing activities comprise larger and larger shares of the total cost as products develop. If, in parallel, the cost of production decreases, profits will be stable or increase. If production costs remain constant as scale increases, profits will decline. The sensitivity of the production cost to scale is relatively easy to determine by computer modeling of the production process and its costs. Such a model is based on the flow of materials through a production plant. Included in the model are algorithms to calculate equipment sizing and cost, utilities loads, and labor and material requirements. As the total capacity of the plant is changed, the model responds by recalculating the cost of unit production. A graph of the unit cost vs. capacity function reveals the economy of scale. When the model includes transportation costs and alternate modes of production (i.e., capitalized plant vs. contract operations, centralized vs. distributed operations, etc.), the optimal size for each of several distributed plants can also be determined. Such models are useful for guiding research and development priorities even in the earliest stages of research. Objectives such as shorter cycle times, increased production yields, and lower cost materials can be compared for their effects on production cost at different scales. When the model is applied diligently, production processes can be developed in a way that is consistent with long-term product development and marketing.

The desire for nematode products to be storable and easy to use derives from constraints in the distribution and marketing systems already in place for pest control products. Such systems have developed over time for products that can be exposed to more extreme conditions of temperature than are normally withstood by living materials. Exposures to freezing temperatures on loading docks and to very high temperatures in unventilated warehouses represent only the most extreme of these conditions. In addition, many products might remain in the distribution system for up to three years with only rudimentary provisions for inventory control of shelf life or replacement of expired stock. While the pest control industry is certainly the most logical conduit for marketing nematode products, it is not clear today whether nematodes will ever be compatible with existing distribution systems. Where better controlled distribution systems can be used, however, there still remain compelling reasons to develop products with long shelf lives. Plant load scheduling, inventory planning, transportation scheduling, and distribution are all more efficient with products having shelf lives of at least 12 months.

Applications of pest control products are usually performed by the users themselves with only the packaged instructions as a guide. In such a setting, special handling requirements are often ignored. In the design of nematode-containing pest control products, there are few options to developing products that can be treated much like commodity chemicals. Where this cannot be achieved, safeguards must be provided to assure that the quality of application equals the quality of production.

Consideration of these three issues, scalability of production, product stability in distribution, and ease and quality of application, provides the environment in which nematode-based products must be developed to succeed in the commercial setting. These issues are largely to partially technical and each is dependent on the others in the development process.

III. METHODS OF NEMATODE PRODUCTION

A. Economies of Scale

Economies of scale are cost decreases which result from increasing scale of operation and derive from three sources: capital, labor, and materials. Capital economy can be obtained when the production yield of the plant increases at a higher rate than the capital cost of equipment and materials. Processes performed in tanks, for example, provide yields that increase as the cube of tank diameter while capital costs for tanks increase roughly as the square. In this case, cost reductions from capital expenditures are obtained in direct proportion to the extent of increased scale. Labor economy can be obtained from two sources: increases in equipment scale or increases in automation. An increase in tank scale provides decreased costs when the labor requirement is determined more by the number of tanks than by their size. Automation, while decreasing labor, always has a cost, and the economy of automation can only be determined by comparing the decreased labor costs with any increased capital cost. Raw material economy can be obtained when the materials used are those which can be produced in large-scale commercial processes. Specialized materials may not produce any economy of scale, and, in the extreme case, the use of specialized materials can negate capital economies.

An additional economy of scale can be obtained as an economy of quality, where the consistency of product quality is increased with scale. Although difficult to predict on a theoretical basis, an economy of quality significantly affects the proportion of total useable product, a proportion that varies with different biological processes. The alternative modes of nematode production can be compared within this framework, and the processes which provide the best economies of scale can be identified.

B. *In Vivo* vs. *In Vitro*

As early as 1931, Glaser recognized the value of developing culture methods for entomopathogenic nematodes,[1] and he was able to produce the entire life cycle of *Steinernema glaseri* on artificial media (the first such success with a parasitic nematode of any kind). In the following 10 years, Glaser made several important discoveries. First, he found that microbes contaminating the cultures caused a high degree of variability in yields.[2] He developed the practice of allowing either yeast or "the natural flora from the nematodes" (now known to be the symbiont, *Xenorhabdus*) to cover the surface of his agar-based media prior to inoculation with nematodes as a means for suppressing contaminant growth.[3] Using a dextrose-veal infusion medium, his cultures were productive for only 7-10 subcultures. Glaser[3] hypothesized that an essential nutrient was lost during this time, and he found that only bovine ovarian substance as an additive could restore the vigor of the cultures. While living yeast could provide some of the essential nutrient, the "natural flora" could not. Nevertheless, Glaser et al.[4] established the first large-scale production process for nematodes using a variety of media preseeded with living yeast and supplemented with antimicrobials for additional protection against contaminants. The costs for the process, however, were high, and the yields inconsistent. The problems of identifying an easily produced essential nutrient and of controlling contamination remained. Glaser never identified the essential nutrient, but the problem of contaminating organisms was solved by producing nematodes in an insect host (*in vivo*) and developing axenic culture.

Entomopathogenic nematodes can be produced by *in vivo* processes, in which insects serve as small biological reactors, as long as a source of insects is available. Most investigators today use the wax moth larva, *Galleria mellonella*, for this purpose. The inoculum and the product of the *in vivo* production process is the third-stage infective juvenile, which in nature survives outside the insect host and actively seeks new hosts. Poinar and Himsworth[5] describe the process by which an infective juvenile of *Steinernema carpocapsae* infects a *Galleria* larva by penetrating the midgut wall of the larva and invading its hemocoel. The infective juvenile carries in its digestive tract cells of the symbiotic bacterium, *Xenorhabdus nematophilus*, and within 1 hr of entering the hemocoel, the nematode begins pharyngeal pumping and releases the bacteria. These bacteria multiply and, within an additional 7 hr, the tissues of the insect larva begin to disintegrate. Death of the host is closely associated with these events. Bovien[6] observed first that nematode development awaits the death of the host; in retrospect, an indication that the reproduction of the bacterial symbiont is a necessary precursor to nematode growth.

Stoll[7] reminds us that the parasitic infection of insects by *Steinernema* is unusual in that the nematode can complete an entire reproductive cycle within a single host, as well as completing development to a third-stage infective juvenile. Growth and development of the nematode are continuous, but punctuated by juvenile molts. Within 3 days of invasion, and providing that the sexes are both represented, eggs are produced. The size of the females and the number of eggs produced by each female vary depending on nematode density with higher reproductive rates at lower female densities.[8] Juvenile development proceeds along one of two possible paths. One path leads, as with most other parasitic nematodes, to the infective stage. The other path allows direct development of second- to third- to fourth-stage nematodes and the formation of second generation adults. Control of this developmental bifurcation has been studied by Popiel et al.,[9] who found that infective juvenile formation was promoted by increased population levels and decreased nutrient.

The developmental events of steinernematid infection of an insect combine to allow an efficient conversion of insect tissue to infective juvenile mass. First, the symbiont is released and proceeds to liquify the insect tissues, making a high proportion of the tissues available to the nematode. Second, the reproductive rate of the nematode is adjusted in response to nematode density, apparently to maximize reproduction up to the capacity of the host tissue. Third, a second round of reproduction is allowed to further maximize reproduction. The eventual outcome of these processes act, in a production setting, to provide a consistently high yield of infective juveniles. The life cycle of the heterorhabditids[10,11] is similar in many respects and characterized by the same efficiency of production.

Methods of rearing the insect host, of nematode infection and incubation, and of harvesting infective juveniles have been described.[12,13] The harvest usually depends on the tendency of infective juveniles to migrate away from the insect cadaver and into a water trap.[14] Where the water source and the cadavers are well separated, the nematodes leave behind debris and contaminants to produce a relatively clean suspension.

The cost of producing nematodes by insect infection is high. Estimates of cost exceed one dollar U.S. per million infective juveniles. Moreover, the process lacks any economy of scale. The equipment used is simple — trays and shelves — and is dependent for capacity on surface area. Therefore, a doubling of capacity requires a doubling of area and a doubling of the capital cost. Without automation, labor also increases as a linear function of capacity since the process is highly labor intensive. In addition, one of the required materials, of course, is insect larvae, the cost for which similarly will not demonstrate economies of scale. Perhaps more important than these considerations is the lack of an economy of quality in increasing scale. To the contrary, just as disease outbreaks are characteristic of high intensity animal husbandry practices, *in vivo* production of nematodes is increasingly sensitive to biological variations and catastrophes as scale increases. The process will be strongly sensitive to the health, physiological state, and pathogen load of the insect host used in the process, and these factors will be difficult to control without stringent practices.

The above considerations suggest several features that a desirable production process should have. One is freedom from reliance on a living host. The benefits from use of an artificial medium come primarily from an economy of quality. As scale of production increases, uniformity of raw materials must be both attainable and ascertainable.

C. Axenic vs. Monoxenic Culture

The second means developed for avoiding the growth of contaminating organisms on artificial media was to grow the nematode axenically. Glaser was the first to develop an axenic culture process for *S. glaseri*[2] and later *S. carpocapsae* (=*chresima*).[15] The axenic inoculum was obtained by a strenuous process of cleaning and disinfection applied to infective juveniles. This inoculum was introduced into a simple agar medium containing 1 g of sterile rabbit kidney on 10 ml of 1% agar. Glaser was able to obtain over 14 transfers or 50 generations of *S. glaseri* with this medium. Moreover, he demonstrated that the nematodes could live equally well in liquid media containing kidney extract. For the first time, the need to increase surface area to achieve increased capacity was weakened, because the feeding, development, mating, and reproduction of the nematode did not require a solid substrate. Successful exploitation of this discovery awaited a better defined medium and a better definition of the conditions required for optimal reproduction.

Stoll[7] continued Glaser's work in liquid medium and found that raw liver extract (RLE), never heated in its preparation, added to veal infusions, gave consistently good yields of axenic *S. glaseri* when grown in test tubes shaken 100 times per minute. Stoll found that this rate of shaking was important, since rates less than 100 per minute gave lower yields,[16] while higher rates were also deleterious.[7] The contribution of RLE, apparently a replacement for Glaser's ovarian substance, remained unexplained, though its activity seemed associated with its degree of precipitation and was variable.[16] Clearly, the objective of finding an artificial medium free of the variability and the uncertainty characteristic of living materials had not yet been achieved.

During the next decade, the requirement for an unidentified essential nutrient in nematode axenic liquid culture remained, and similarly, the culture medium of the free-living nematode, *Caenorhabditis briggsae*,[17] while highly defined and complex, was also found to require "a single crude preparation in small amounts". Continuing efforts to define the requirements of *S. glaseri* were successful at achieving reproduction in a completely defined medium,[18,19] but did not achieve high reproductive levels. A bovine plasma fraction was found to partially replace RLE for high yield cultures, and the requirement for aeration was restated. Meanwhile, axenic culture techniques were extended to *S. carpocapsae* (DD-136).[20]

The benefit of good aeration of liquid axenic cultures and its hazards were clearly shown by Hansen and Cryan[21] when they achieved up to 5-fold increase in yield by growing nematodes in a thin liquid film. In attempting to culture large numbers of nematodes, the authors found that shaking, rolling, and gassing tubes of liquid medium failed to produce high yields. Instead, the authors filled a separatory funnel with glass wool and allowed a relatively small volume of liquid to adsorb on the surface. The configuration of their culture vessel provided a combination of high gas exchange, no liquid agitation or shear, and a reticulated surface which allowed an increase in the dimensionality of the culture without sacrificing the possible benefits of nematode-surface contact. The concept of a reticulated matrix was to prove valuable in later attempts to culture the nematodes, but another means for providing aeration was to prove equally effective.

In 1971, Buecher and Hansen[22] reported that the simple bubbling of liquid medium in bottles provided yields equal to those in the glass wool, and, for the first time, presented a means for producing large quantities of axenic nematodes, including *S. carpocapsae*, in an artificial medium. The discovery that bubbling was an acceptable means for providing aeration to liquid nematode cultures put the process on solid, well-travelled ground.

Bubbled cultures are easy to scale-up,[23] and the technique of Buecher and Hansen can be used as the basis of a plant design to predict the economy of scale obtained by axenic liquid culture of entomopathogenic nematodes (Figure 1). Unlike the *in vivo* process, the axenic liquid process shows appreciable capital and labor economies of scale. Economies of quality are also expected because the nematodes would be produced in a homogeneous and reproducible environment. The overall manufacturing cost, however, does not show an economy of scale after 10×10^{12} nematodes per month, because it is largely determined by the cost of the medium. As previously discussed, specialized raw materials can eliminate economies of scale that would otherwise characterize a process.

The medium used in the axenic liquid process received much more attention in subsequent years. Consistently, researchers focused on identifying the essential nutrient, or "growth factor", required by nematodes in culture. The studies of Buecher et al.[24,25] on *C. briggsae* centered on the role of protein and heme precipitates that enhanced growth. Rothstein,[26] building on earlier work by Hieb and co-workers,[27,28] reported a beneficial effect of hemoglobin and cholesterol in a yeast and peptone-based medium, a medium that, significantly, could be autoclaved. Vanfleteren[29] working from the results of Rothstein[26], developed an inexpensive medium for *Caenorhabditis elegans* and, for the first time, reported results of high yields of nematodes from 10 l fermenters. Aeration was by bubbling, and suspension was assisted by slow stirring. Pace et al.[30] has reported that a similar medium and agitation system supports the growth of *S. carpocapsae*. With the Rothstein-Vanfleteren medium, we can expect a production process which shows improvements over Buecher and Hansen (Figure 2). The process gives an estimated cost of production less than one-third that of the Buecher process. At plant capacities greater than 20×10^{12} nematodes per month, however, further improvements develop slowly. Examination of the model shows that additional improvements can be anticipated from less expensive media and improved yields per generation.

The theme of precipitated factors, originating in the studies of Stoll,[7,16] was recapitulated by Vanfleteren,[31] who proposed the necessity of heme-containing particles for efficient reproduction. He also wrote that some studies had reported benefits from living bacterial cells.

D. Monoxenic Culture-Solid vs. Liquid Phase

The finding that particulates benefited the culture of entomopathogenic nematodes followed by several years the practice of growing these nematodes together with their bacterial associate. By 1964, Dutky et al.[12] were referring to the "nematode-bacterium complex" which they maintained on peptone-glucose agar and pork kidney. In 1965, House et al.[32] reported a culture system that

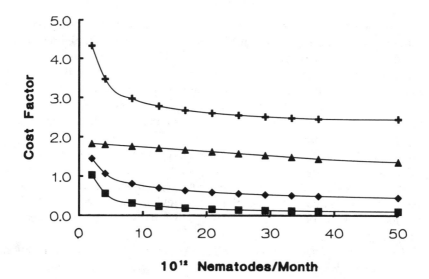

Figure 1. Sensitivity of nematode production costs to plant capacity based on the axenic liquid process described by Buecher and Hansen.[22] The mathematical model used to generate this figure estimates equipment size, utility loads, personnel loads and costs, materials costs, and capital costs which include engineering and construction. The cost elements, capital (◆), material (▲), labor (■), and the total (✛) are in arbitrary units. The assumption was made that the process would be as efficient in 100,000 l bubble columns as it was in the small scale.

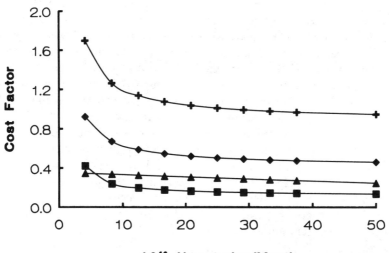

Figure 2. Sensitivity of nematode production costs to plant capacity based on the axenic liquid process described by Vanfleteren.[29] The same assumption of scalability was made here as described in the Figure 1 legend and the same model was used. The cost elements, capital (◆), material (▲), labor (■), and the total (✛) are in arbitrary units.

relied on the use of dog food and provided a medium that was inexpensive, widely available, and autoclavable. Their observation that "the bacteria are probably necessary for the success of the culture" is now realized to be an understatement. By 1967, Dutky et al.[33] had found that a medium of peptone-glucose, with the bacterial associate, and with cholesterol, could support the reproductioin of DD-136, a strain of *S. carpocapsae*. These reports demonstrated that the symbiotic bacterium, *Xenorhabdus*, provided significant nutritional benefits in a production sstem and allowed the use of inexpensive medium materials. Additional benefits were described by Hansen et al.[34] who showed that incorporation of *Xenorhabdus* into a defind liquied medium markedly stimulated exsheathment of infective juveniles. From the literature alone, it is difficult to appreciate how the concept of monoxenic culture developed. It is clear, however, that Glaser,[1-3] Dutky et al.[12] and finally Poinar and Thomas[20] (who describe the monoxenic infection of *Galleria*) benefited from working directly wit h the insect infection, observing first hand how these two organisms cooperate.

The solid phase culture systems of House et al.[32] and the liquied phase monoxenic system suggested by Hansen et al.[34] are ameable to scale-up. Both are more complex than axenic systems in that they involve two organisms. Of critical note, however, are the observations that (1) bacteria other than *Xenorhabdus* serve less well in monoxenic culture,[20] (2) the phase one variant of the symbiont is preferred by the nematode,[35] and (3) nematode species have evolved close and specific interactions with their symbionts and grow best with the symbionts with which they have evolved.[36] Important to monoxenic culture are the facts that (1) the bacteria produce abundant amounts of protease,[37] (2) the nematodes, as well as feeding on bacteria, absorb nutrients avidly and directly from the medium,[38] and (3) the bacteria respire[39] and therefore require oxygen.

The production of protease by the symbiont explains in large measure the ability of almost any high protein medium to support monoxenic culture.[40] The ability of the nematode to absorb nutrients directly suggests that various additives may significantly benefit production efficiency. The need of the bacteria for oxygen increases the overall requirement for oxygen in a monoxenic culture. The scale-up of monoxenic cultures becomes a problem largely of scaling up the delivery of oxygen, and this problem can meet with different solutions in solid or liquid phase culture.

Hara et al.[41] described a scaled-up version of the House et al.[32] method in which Petri dishes containing dog food medium are inoculated, incubated for 20-30 days, and then harvested over a 3-week period by nematode migration into sterile water. As oxygenation is accomplished by diffusion through the atmosphere to an agar surface, the yields of the process related to the total surface area of agar used. By this process, a single technician can produce 100 million nematodes per week. This was confirmed by Poinar[39] who reported on the practice of a commerical operation. The costs of large-scale production based on a dog food-agar process can be compared with an *in vivo* production process. The material costs are considerably less, and an economy of quality would be obtained due to the consistency and availability of the medium and to the ability to limit contamination. The limitations on achieving economies of scale in both capital and labor, however, like the *in vivo* process, limit the commercial potential of this method.

Bedding[42,43] addressed both of these limitations by increasing the dimensionality of the process. Using a variety of reticulated matrices and, finally, shredded plastic foam, Bedding increased the yields of the solid phase monoxenic process from 1 million per Petri dish to 40 million per flask. Moreover, he introduced a semi-automated harvesting process capable of handling large quantities of foam.[43] A wide variety of materials were used as media, with preference to chicken offal. Using a pig kidney-beef fat medium, production of *S. carpocapsae*, *S. feltiae* (=*bibionis*), and *S. glaseri* and four heterorhabditid species was obtained. Wouts[44] subsequently developed a medium containing nutrient broth, yeast extract, soy flour, and corn oil for heterorhabditid production using this solid phase method.

The Bedding process has been adopted by a number of commercial operations during the developmental stage and is being used for commercial production. The economy of the process is highly dependent upon the consistency of production, and the process is sensitive to contamination. If the process described by Bedding[43] is modeled (Figure 3), economies of scale are obtained up to a production level of approximately 10×10^{12} per month. Beyond that point, labor costs remain constant and significant, suggesting that more advanced technologies are needed to support larger scale production of entomopathogenic nematodes in industrial countries with high labor costs. There is little doubt, however, that the Bedding process has application in the developmental stage of the industry and in decentralized societies such as China.

Monoxenic, liquid culture has only recently been described in the scientific literature beyond the observations of Hansen et al.[34] Buecher and Popiel[45] have described the complete development of *S. carpocapsae* in a nutrient broth monoxenic culture. In addition, two international patent applications have been published which describe some of the first attempts at monoxenic culture in fermenters.[30,46] Experience in solid phase, monoxenic culture indicates that a wide variety of media could support high yields. Pace et al.[30] report that ox kidney homogenate-yeast extract or various homogenized offals can serve as media in liquid monoxenic processes. In kidney-yeast medium they report yields as high as 90,000 nematodes per milliliter in fermenters in approximately 3 weeks and yields of 190,000 per milliliter in shake flasks. The conditions necessary to achieve high proportions of infective juveniles, however, are not described. In one example, the authors describe transferring the nematodes to water at 15°C and incubating until 60-90% infective juveniles are obtained. The final count of infective juveniles is not reported. Friedman et al.[46] report yields up to 110,000 infective juveniles per milliliter in only 8 days using a medium containing soy flour, yeast extract, corn oil, and egg yolk. No special

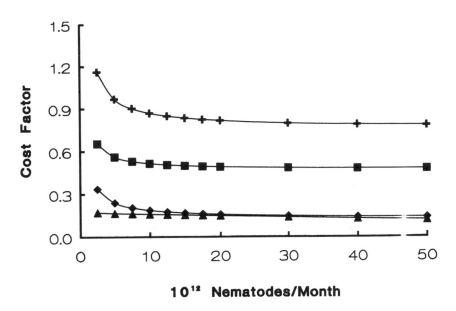

Figure 3. Sensitivity of nematode production costs to plant capacity based on the solid phase process described by Bedding.[43] The assumption was made that the process could be performed in stainless steel trays of approximately 0.5 m^2 which would each produce 10^9 nematodes. Equipment used was scaled to the most efficient size for a given capacity. The model included all of same factors as those used to generate Figures 1 and 2. The cost elements, capital (◆), material (▲), labor (■), and the total (✛) are in arbitrary units.

conditions are described as necessary to achieve a high proportion of infective juveniles.

Both Pace et al.[30] and Friedman et al.[46] found that aeration was the most difficult requirement to meet and that shear sensitivity was the most significant limitation to production efficiency. The approaches taken to resolve these problems differ. Pace et al.[30] limit agitation from stirring to the minimum needed to maintain the nematodes in suspension. Aeration is provided by bubbling. In the case where they use a downward pointing sparger, they are able to forego stirring completely. Friedman et al.,[46] on the other hand, use stirring to increase the rate of gas transfer, having discovered that juvenile nematodes are much less shear sensitive than adults. Friedman et al.[46] also report that air lift fermenters can be used to produce nematodes, in contrast to the experience of Pace et al.[30]

In the scale-up model, liquid monoxenic culture provides the best economies of scale among all of the production methods available (Figure 4). Beginning at a capacity of 1×10^{12} nematodes per month, production costs are less than other methods, and these costs decrease more rapidly up to a capacity of 50×10^{12} nematodes per month. Over this size, multiple plants may be preferred.

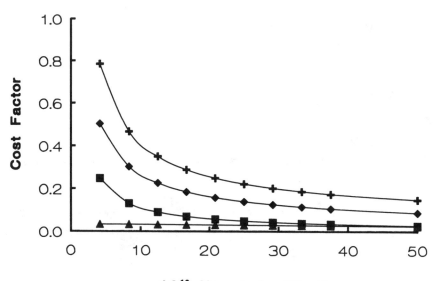

Figure 4. Sensitivity of nematode production costs to plant capacity based on the liquid monoxenic process described by Friedman et al.[46] The model used was identical to that used to generate Figure 1 except for the materials, yields, and cycle time. The cost elements, capital (◆), material (▲), labor (■), and the total (✦) are in arbitrary units.

E. Engineering of Liquid, Monoxenic Culture

In the small scale, the requirements and constraints of a process can be determined under a variety of conditions. Achieving the optimal combination of requirements during scale-up, without exceeding the constraints, is an engineering problem. Choices between alternative means to solving this problem are made based on economic considerations, just as was the determination of the process itself.

A description of the engineering correlates and design considerations useful in scaling-up nematode production is beyond the scope of this chapter. Instead, it may be valuable to note some of the qualitative factors important to success. In small volume static cultures and shaking tubes or flasks, sufficient oxygen transfer can be obtained so that nematode development is limited by other factors, such as the nutrient content of the medium. At the large scale, however, oxygen transfer can become limiting.

In the macroscale, oxygen transfer in liquid reactors is accomplished by the mixing of oxygenated liquid and bubbles into regions of low oxygen. In the microscale, diffusion of oxygen through unstirred layers mediates oxygen transfer. Since oxygen is only sparingly soluble in aqueous liquid, it diffuses slowly. Therefore, oxygen transfer can be limited by the thickness of any unstirred layer of liquid surrounding an organism. The thickness of these layers will in turn depend on disruptive fluid forces impinging on the unstirred volume and on the viscosity and viscoelastic properties of the liquid. Fluid forces which are of a sufficient energy to disrupt the unstirred layer enhance oxygen transfer to the organism, but these forces must not be of sufficient energy to disrupt the organism. Minimizing oxygen demand will increase the probability that sufficient oxygen can be delivered. Figure 5 shows the growth of *Xenorhabdus nematophilus* on five different media. Bacterial growth on Medium 1 requires the least oxygen per 10^7 bacteria, although Medium 1 still supports high bacterial densities. Such a medium would be expected to support higher yields of nematodes in fermenters than the other media shown.

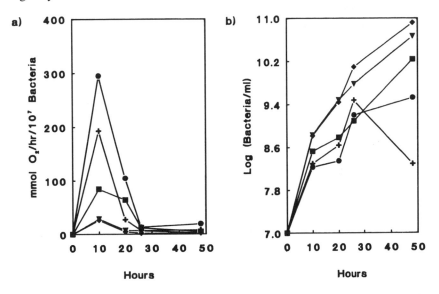

Figure 5. Growth of *Xenorhabdus nematophilus* on a family of related media, M1 through M5 (♦, ▼, ■, ✤, ●). Graph A shows the oxygen demand rate of a bacterial culture grown in a shake flask at 25°C. Graph B shows the bacterial densities in the same cultures. Oxygen demand was determined in a closed, stirred microcell with an oxygen electrode. Demand is expressed in units of mmol O_2 used per hour by 10^7 bacterial cells.

The unstirred layer surrounding the air bubbles themselves can also limit oxygen transfer, especially when surface-active agents accumulate on the surface of air bubbles and impede oxygen transfer and surface mobility. These surface active agents include materials used to control foam. Therefore, in any particular system, it is important to characterize the oxygen transfer process and identify the elements, such as viscosity or antifoam agents, which are limiting oxygen transfer.

Understanding the factors involved in controlling shear forces is more difficult than understanding those which determine oxygen transfer. Not only are fluid forces in aerated and mixed tanks largely uncharacterized, but the forces capable of disrupting a given organism are usually unknown. The characterization of these forces must be done with the understanding that a biological reactor presents transient rather than constant forces and allows recovery periods of variable duration. Only now are attempts being made to address each of these problems. The simplest approach is to adapt that of Buecher and Hansen[22] and use bubbling, which nematodes tolerate to some extent. As an alternative, Friedman et al.[46] attempted to measure the shear sensitivity of nematodes suspended in the 2 mm gap between a bob and a concentric cup. They found that shear sensitivity varied throughout the nematode life cycle, with adults disrupted at 1800 rpm and first through third-stage juveniles disrupted between 2800 and 3000 rpm. Based on these results, the authors developed a variable agitation regime in stirred fermenters that provided optimal reproduction (Figure 6).

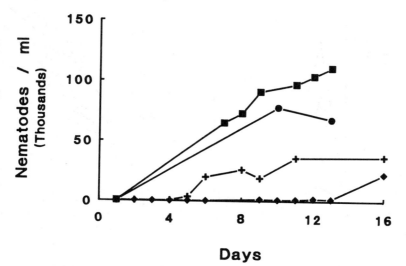

Figure 6. Effects of controlled agitation rates on cultures of *S. carpocapsae.*[46] Cultures of *X. nematophilus* and *S. carpocapsae* in 16 liters of medium were grown in 20 liter fermenters. Stirring rates were either 50 rpm (✤), 125 rpm (●), 500 rpm (◆), or variable rpm (■). In the case of variable rpm, the fermenter was stirred at 250 rpm until juvenile reproduction began. The rpm was then increased in response to the increasing oxygen demand of the culture.

During the past 3 years, Biosys has scaled-up the liquid monoxenic culture of entomopathogenic nematodes to the 15,000 l level and heterorhabditid nematodes to the 7500 l level. The Biosys effort has been characterized by a close collaboration between biologists and engineers. Developmental biology, nutrition, reactor design, and microbiology have all contributed to the overall success of the process.

IV. STORAGE OF NEMATODES

The ability to store nematodes in product form is critical to commercial success, but little basic work has been done on the factors that limit survival. Dutky et al.[12] described storing *S. carpocapsae* in insulated jugs held at 7.1°C.

A 3.8-l jug was filled with one liter of water containing 50 million nematodes. If we assume that the diameter of the jug equalled its height, the nematode layer must have been about 1 mm thick. With oxygenation of the jug's atmosphere every few weeks, the nematodes were reported to survive up to 5 years. This early work established that, at refrigerated temperatures, thin layers of nematodes up to 1 mm thick can be stored for long periods if the layer is in an oxygenated environment. Such thin layers have been achieved by applying a thick paste of nematodes to thin sheets of plastic foam[47] and storing these foam pieces in the cold. In addition, Poinar[39] mentions that materials such as charcoal or wood chips can be used as a supporting matrix for shipment as long as the matrix is kept moist and aerated. Products containing *S. carpocapsae* which are designed in this manner typically provide up to 6 months of refrigerated shelf life and up to 1 month of shelf life at temperatures up to 25°C with losses of less than 15%. A notable example is the Biosys consumer product, Biosafe, which contains 10 million nematodes trapped in a thin gel supported by a mesh screen.

The above methods emphasize thin layers of nematodes in moist, aerated environments. In contrast, Burman and Pye[48] report that they were able to maintain 100% survival of *S. carpocapsae* at 20°C for a period exceeding 43 days by sparging with nitrogen down to an oxygen tension of 0.75 mmHg. This level of oxygen is below that normally required by aerobic organisms for survival, and suggests that nematodes have some ability to survive extended periods in microaerobic conditions. This method has not been adopted by industry.

The storage of entomopathogenic nematodes in an anhydrobiotic or desiccated form has been attempted by many investigators. Popiel et al.[49] describe a method for the induction of a desiccated state requiring carefully controlled conditions of evaporation. A similar induction period is described by Bedding[50] with the nematodes mixed in ground clays. The latter method has been used by Bioenterprises in a commercial product.

Other patents have been applied for that claim various means of storing nematodes for extended periods, but, with the above exceptions, none have been used commercially. Further advances in storage technology will be required before the full commercial potential of entomopathogenic nematodes is achieved

V. CONCLUSIONS

A principal characteristic desired in a commercial process is economy of scale. Therefore, methods of producing entomopathogenic nematodes have been reviewed and analyzed from the perspective of scale-related costs. It is well known that production methods have steadily improved in reliability and efficiency over the last 60 years. It is less well known that currently used processes differ substantially in their response to scaling. In models described here, liquid monoxenic culture proves to be the most robust of the current production methods for development to large scale. Successful implementation of large-scale liquid monoxenic culture will depend on achieving consistent product, and benefits will be obtained from creating stable inoculum and material sources, developing fault tolerant processes, and constructing low-fault equipment.

Technologies for formulation and packaging have not been developed which are as easy to scale-up as nematode production technology. While current technologies appear sufficient to support preliminary commercial introductions, progress in these areas is necessary before entomopathogenic nematodes can be used to their full potential for the control of insect pests. Nevertheless, we can anticipate that such progress will be forthcoming. The level of sales which can be supported by existing technology can support existing and higher levels of investment. Investment in the further development of storage technology is easily justified by the manyfold increase in sales that are expected to follow increases in storage stability.

REFERENCES

1. **Glaser, R. W.,** The cultivation of a nematode parasite of an insect, *Science*, 73, 614, 1931.
2. **Glaser, R. W.,** The bacteria-free culture of a nematode parasite, *Proc. Soc. Exp. Biol. Med.*, 43, 512, 1940.
3. **Glaser, R. W.,** Continued culture of a nematode parasitic in the Japanese beetle, *J. Exp. Zool.*, 84, 1, 1940.
4. **Glaser, R. W., McCoy, E. E., and Girth, H. B.,** The biology and economic importance of a nematode parasitic in insects, *J. Parasitol.*, 26, 479, 1940.
5. **Poinar, G. O., Jr., and Himsworth, P. T.,** *Neoaplectana* parasitism of larvae of the greater wax moth, *Galleria mellonella*, *J. Invertebr. Pathol.*, 9, 241, 1967.
6. **Bovien, P.,** Some types of association between nematodes and insects, *Vidensk. Medd. Dansk Naturh. Foren.*, 101, 1, 1937.
7. **Stoll, N.,** Axenic cultivation of the parasitic nematode, *Neoaplectana glaseri*, in a fluid medium containing raw liver extract, *J. Parasitol.*, 39, 422, 1953.
8. **Popiel, I., and Lanier, W.,** unpublished data, 1986.
9. **Popiel, I., Grove, D. L., and Friedman, M. J.,** Infective juvenile formation in the insect parasitic nematode *Steinernema feltiae*, *Parasitology*, 99, 77, 1989.
10. **Khan, A., Brooks, W. M., and Hirschmann, H.,** *Chromonema heliothidis* n. gen., n. sp. (Steinernematidae, Nematoda), a parasite of *Heliothis zea* (Noctuidae, Lepidoptera), and other insects, *J. Nematol.*, 8, 159, 1976.
11. **Wouts, W. M.,** The biology and life cycle of a New Zealand population of *Heterorhabditis heliothidis* (Heterorhabditidae), *Nematologica*, 25, 191, 1979.
12. **Dutky, S. R., Thompson, J. V., and Cantwell, G. E.,** A technique for mass propagation of the DD-136 nematode, *J. Insect Pathol.*, 6, 417, 1964.
13. **Woodring, J. L., and Kaya, H. K.,** Steinernematid and Heterorhabditid Nematodes: A Handbook of Biology and Techniques, South. Coop. Ser. Bull. 331, Arkansas Agric. Exp. Stat., 1988.
14. **White, G. F.,** A method for obtaining infective nematode larvae from cultures, *Science*, 66, 302, 1927.

15. **Glaser, R. W., McCoy, E. E., and Girth, H. B.,** The biology and culture of *Neoaplectana chresima*, a new nematode parasitic in insects, *J. Parasitol.*, 28, 123, 1942.

16. **Stoll, N. R.,** Favored RLE for axenic culture of *Neoaplectana glaseri, J. Helminthol, R. T. Leiper Suppl.,* 169, 1961.

17. **Dougherty, E. C., Hansen, E. L., Nicholas, W. L., Mollett, J. A., and Yarwood, E. A.,** Axenic cultivation of *Caenorhabditis briggsae* (*Nematoda: Rhabditidae*) with unsupplemented and supplemented chemically defined media, *Ann. N. Y. Acad. Sci.*, 77, 176, 1959.

18. **Jackson, G. J.,** The parasitic nematode, *Neoaplectana glaseri*, in axenic culture II. Initial results with defined media, *Exp. Parasitol.*, 12, 25, 1962.

19. **Jackson, G. J.,** *Neoaplectana glaseri*: essential amino acids, *Exp. Parasitol.*, 34, 111, 1973.

20. **Poinar, G. O., Jr., and Thomas, G. M.,** Significance of *Achromobacter nematophilus*, Poinar and Thomas (Achromobacteraceae: Eubacteriales) in the development of the nematode, DD-136 (Neoaplectana sp. Steinernematidae), *Parasitology*, 56, 385, 1966.

21. **Hansen, E. L., and Cryan, W. S.,** Continuous axenic culture of free-living nematodes, *Nematologica*, 12, 138, 1966.

22. **Buecher, E. J., and Hansen, E. L.,** Mass culture of axenic nematodes using continuous aeration, *J. Nematol.*, 3, 199, 1977.

23. **Jackson, M. L., and Shen, C-C.,** Aeration and mixing in deep tank fermentation systems, *AICHE*, 24, 63, 1978.

24. **Buecher, E. J., Perez-Mendez, G., and Hansen, E. L.,** The role of precipitation during activation treatments of growth factor for *Caenorhabditis briggsae, Proc. Exp. Biol. Med.*, 132, 724, 1969.

25. **Buecher, E. J., Hansen, E. L., and Yarwood, E. A.,** Growth of nematodes in defined medium containing hemin and supplemented with commercially available proteins, *Nematologica*, 16, 403, 1970.

26. **Rothstein, M.,** Practical methods for the axenic culture of the free-living nematodes *Turbatrix aceti* and *Caenorhabditis briggsae, Comp. Biochem. Physiol.*, 49B, 669, 1974.

27. **Hieb, W. F., and Rothstein, M.,** Sterol requirement for reproduction of a free-living nematode, *Science*, 160, 778, 1968.

28. **Hieb, W. F., Stokstad, E. L. R., and Rothstein, M.,** Heme requirement for reproduction of a free-living nematode, *Science*, 168, 143, 1970.

29. **Vanfleteren, J. R.,** Large scale cultivation of a free-living nematode (*Caenorhabditis elegans*), *Experientia*, 32, 1087, 1976.

30. **Pace, G. W., Grote, W., Pitt, D. E., and Pitt, J. M.,** Liquid culture of nematodes, Int. Patent WO 86/01074, 1986.

31. **Vanfleteren, J. R.,** Axenic culture of free-living, plant-parasitic and insect-parasitic nematodes, *Annu. Rev. Phytopathol.*, 16, 131, 1978.

32. **House, H. L., Welch, H. E., and Cleugh, T. R.,** Food medium of prepared dog biscuit for the mass production of the nematode DD-136 (Nematoda: Steinernematidae), *Nature*, 206, 847, 1965.

33. **Dutky, S. R., Robbins, W. E., and Thompson, V. V.,** The demonstration of sterols as requirements for the growth, development, and reproduction of the DD-136 nematode, *Nematologica*, 13, 140, 1967.

34. **Hansen, E. L., Yarwood, E. A., Jackson, G. L., and Poinar, G. O., Jr.,** Axenic culture of *Neoaplectana carpocapsae* in liquid media, *J. Parasitol.*, 54, 1236, 1968.

35. **Akhurst, R. J.,** Morphological and functional dimorphism in *Xenorhabdus* spp. bacteria symbiotically associated with the insect pathogenic nematodes *Neoaplectana* and *Heterorhabditis, J. Gen. Microbiol.*, 121, 303, 1980.

36. **Dunphy, G. G., Rutherford, T. A., and Webster, J. M.,** Growth and virulence of *Steinernema glaseri* influenced by different subspecies of *Xenorhabdus nematophilus, J. Nematol.*, 17, 476, 1985.

37. **Schmidt, T. M., Bleakley, B., and Nealson, K. H.,** Characterization of an extracellular protease from the insect pathogen, *Xenorhabdus luminescens, Appl. Environ. Microbiol.,* 54, 2793, 1988.

38. **Devidas, P., Jean-Baptiste, B., Eliane, B., and Christian, L.,** A method to employ tritium-labelled compounds in studies on host-parasite relations between *Galleria mellonella* and *Neoaplectana carpocapsae, Indian J. Nematol.,* 12, 370, 1982.

39. **Poinar, G. O., Jr.,** *Nematodes for Biological Control of Insects,* CRC Press, Boca Raton, 1979.

40. **Bedding, R. A.,** New methods increase the feasibility of using *Neoaplectana* spp. (Nematoda) for the control of insect pests, *Proc. 1st. Int. Coll. Invertebr. Pathol.,* Kingston, 1976, 250.

41. **Hara, A. H., Lindegren, J. E., and Kaya, H. K.,** Monoxenic mass production of the entomogenous nematode, *Neoaplectana carpocapsae* Weiser, on dog food/agar medium, USDA/SEA, AAT-W-16, 1981.

42. **Bedding, R. A.,** Low cost in vitro mass production of *Neoaplectana* and *Heterorhabditis* species (Nematoda) for field control of insect pests, *Nematologica,* 27, 109, 1981.

43. **Bedding, R. A.,** Large scale production, storage and transport of the insect parasitic nematodes, *Neoaplectana* spp. and *Heterorhabditis* spp., *Ann. Appl. Biol.,* 104, 17, 1984.

44. **Wouts, W. M.,** Mass production of the entomogenous nematode *Heterorhabditis heliothidis* (Nematoda: Heterorhabditidae) on artificial media, *J. Nematol.,* 13, 467, 1981.

45. **Buecher, E., and Popiel, I.,** Growth of *Steinernema feltiae* in liquid culture, *J. Nematol.,* 21, 500, 1989.

46. **Friedman, M. J., Langston, S. L., Pollitt, S.,** Mass production in liquid culture of insect-killing nematodes, Int. Patent WO 89/04602, 1989.

47. **Finney, J. R.,** Shipping and storage package for nematodes and their eggs with container and light foam filling pieces around storage container, U.S. Patent 4 417 545, 1983.

48. **Burman, M., and Pye, A. E.,** *Neoaplectana carpocapsae:* respiration of infective juveniles, *Nematologica,* 26, 214, 1980.

49. **Popiel, I., Holtemann, K. D., Glaser, I., and Womersley, C.,** Commercial storage and shipment of entomogenous nematodes, Int. Patent WO 87/78515, 1987.

9.   Formulation and Application Technology

Ramon Georgis

I. INTRODUCTION

Proper application is one of the keys to success for any biological pesticide treatment. Simply stated, the application process delivers the infective stage of steinernematid or heterorhabditid nematodes to the target insect. This process usually involves using a carrier in a liquid, capsule, or bait to transport the nematodes to the intended surface or target.[1]

Many factors affect our ability to place quantities of nematodes on or in close proximity to the target host to produce optimal results at the lowest possible cost. Application method and timing are crucial to success; however, unless proper consideration is given to a reliable and stable formulation, successful application of entomopathogenic nematodes is not very probable. Certainly, as with chemical pesticides, selection and use of equipment are of utmost importance and deserve major emphasis when considering nematode application.

II. FORMULATION

Formulation science is a broad field because it deals with formulation development, production, and storage as well as the interaction of the ingredients with plants, insects, other invertebrates, mammals, soil, air, and water.[2] Pesticide formulations include aqueous solutions or suspensions, emulsifiable concentrates, wettable powders, and granular and controlled release substances. Some formulation ingredients include biologically active agents, clays, solvent diluents, surfactants, and polymers.[2] In general, formulation is required to stabilize a product and to improve its efficacy in the field.

The formulation and storage of entomopathogenic nematodes present unique problems not encountered with a chemical pesticide. Prior to formulation, nematodes are commonly stored under refrigeration in aerated aqueous suspension or on sponges (Table 1). However, the temperature conditions for nematode storage tend to be species specific. Thus, the maintenance of nematode virulence for steinernematids is generally between 5 and 10°C, and for heterorhabditids is between 10 and 15°C. Additionally, the oxygen and moisture parameters required by each species must be assessed. Once the requirements for the nematodes are defined, selection of the formulation type, ingredients, packaging size, and storage conditions during each step of product development and distribution can be undertaken (Table 1).

Components used to formulate some commercially available nematode products, even though termed "inert", may possess properties that enhance

TABLE 1. Storage and Shipping Requirements for Current Formulations[a] of Steinernematid and Heterorhabditid Nematodes

Steps in Product Development and Distribution	Storage parameters		
	Method	Condition	Period[b]
Postharvest	Aqueous suspension[c] or on sponge	Refrigerated	6-12 months
Formulation	Formulated product	Refrigerated	6-12 months
Shipping	Formulated product	Refrigerated	1-12 days
		Nonrefrigerated	1-7 days
Distributor and user	Formulated product	Refrigerated	1-6 months
		Room temperature	1-3 months

[a] Immobilized or partially desiccated nematodes (e.g., alginate, clay, and polyacrylamide gel).
[b] Depends on nematode species, concentration, formulation, and package design.
[c] Aerated.

handling, application, persistence, and storage. Among these are alginate, clays, activated charcoals, and polyacrylamide gels. These carriers, which immobilize the nematodes or partially desiccate them, reduce their metabolism and improve their tolerance to temperature extremes. Nematodes are also commercially available on a moist substrate such as sponge, vermiculite, and peat, although extended storage at room temperature is generally poor because nematode viability declines rapidly. Nematode metabolism is temperature driven, and a warm (20-30°C) environment increases metabolic activity, thus reducing nematode pathogenicity and viability. Anhydrobiotic nematodes, which have an arrested metabolism, have the potential of an extended shelf life, and future development of this area warrants investigation.

A. Alginate

Alginate is a water-soluble polysaccharide with excellent gel-forming properties. Entomopathogenic nematodes can easily be entrapped in a calcium alginate matrix by a fast, gentle, aqueous, room temperature process. The gelatinous matrix produced by this process is biodegradable in the environment.

In order to perform their function as biological control agents, encapsulated nematodes must escape or be liberated from the capsule after application in order to seek out hosts (Tables 2 and 3). When infective juveniles of *Stein-*

ernema or *Heterorhabditis* were encapsulated at a concentration of about 300 nematodes per gel bead (capsule) and fed to larvae of the beet armyworm, *Spodoptera exigua*, the nematodes were released when the insect larvae fed on the gel beads.[3] The mortality rate was excellent provided adequate moisture was present to prevent death of the nematodes by desiccation. This formulated *S. carpocapsae* maintained its infectivity when stored for 8 months in closed containers at 4°C. When capsules were placed in several habitats with adequate moisture, most nematodes migrated out of the capsules within 1 week.[4] When bacteria were abundant in the surrounding environment, decomposition of the capsule matrix accelerated nematode dispersal. Kaya and Nelsen[3] suggested that capsules could be incorporated into baits to attract insects such as cutworms or grasshoppers. This concept was tested against grasshoppers (*Melanoplus* spp.) and the black cutworm (*Agrotis ipsilon*) with moderate success.[5,6]

Another potential use of the capsules is to incorporate nematodes with seeds for prophylactic control of insects such as the cabbage root maggot (*Delia*

TABLE 2. Formulations and Form of Application of Steinernematid and Heterorhabditid Nematodes

Formulation	Form of Application	Ref.
Activated charcoal	Aqueous spray of nematode and charcoal	11
Alginate capsules	Spreading capsules or capsules in aqueous suspension	5,7
Alginate sheet	Aqueous spray after dissolution of the alginate	8
Baits	Spreading baits	5,6,27
Clay	Aqueous spray of nematodes and clay	10
Evaporetardants	Aqueous spray	15-17
Ultraviolet protectants	Aqueous spray	18
Gel-forming polyacrylamides	Aqueous spray of extracted nematodes	8
Absorbent pad	Baited traps	68
Peat	Aqueous spray of nematodes and peat	83
Polyether-polyurethane sponge	Aqueous spray of extracted nematodes	14
Vermiculite	Aqueous spray of extracted nematodes	84

TABLE 3. Application Methodology and Target Insects of Steinernematid and Heterorhabditid Nematodes

Form of Application	Carrier	Target Insects[a]	Application Site	Ref.
Capsules	Alginate[b]	Grasshoppers, black cutworm	Soil surface	5,6
		Western corn rootworm	In soil at planting	69
Capsules	Alginate[c]	Red imported fire ant	Laboratory	70
		German cockroach	Laboratory	69
Liquid baits	Sucrose	Western yellow jacket	Laboratory	24
		Lepidopterous larvae	Laboratory	23,25
Pellet baits	Bran[d]	Grasshoppers, black cutworm	Soil surface	5,6
		Black cutworm, tawny mole cricket	Laboratory	27
Spray	Water[e]	Soil insects	Soil surface, injection	1,14,28
		Foliar insects	Leaves, trunks	1,14,28
		Borer insects	Fruits, trunks	1,14,28
Traps	Pads[d]	House fly	Laboratory	68
		German cockroach	Laboratory	69

[a]   Scientific names in text except for german cockroach, *Blatella germanica*, and house fly *Musca domestica*.
[b]   With or without bran.
[c]   With liquid bait.
[d]   With attractants.
[e]   With or without additives.

radicum), onion maggot (*D. antiqua*), or corn rootworms (*Diabrotica* spp.).[3] This approach was tested by placing a quantity of *S. carpocapsae* and a single tomato seed into each alginate gel bead.[7] Nematodes were released from the beads during seed germination and were in position to protect seedlings from insect attack. Successful bioassays with these gel bead formulations were conducted in soil against larvae of the greater wax moth, *Galleria mellonella*. Larvae were killed even when seeds were absent in the soil-applied capsules. This indicated that some nematodes were able to escape on their own from the capsules. However, the practical use of encapsulated nematodes as a slow-release mechanism for prophylactic control of insects may be limited by the rapid biodegradable nature of the capsules in the environment.

A new commercial product uses a sheet of calcium alginate as a substrate to maintain nematode viability and to provide a convenient approach to application.[8] The nematodes are entrapped in a thin layer of the alginate spread on a plastic mesh screen. Before application sodium citrate is used to dissolve the gel matrix so that the nematodes can be applied as an aqueous suspension to the target insect (Tables 2 and 3).

B. Clay

Clay compounds are common diluents in agrichemical formulations. Their structural differences arise from anion-cation arrangement, composition, and crystalline structure. Montmorillonite and attapulgite clays exhibit a greater degree of congruity with microbial organisms than kaolinite clays.[9] The presence of a compatible diluent with a biological insecticide will increase shelf life by decreasing viability during storage. Clay diluents function by buffering pH shifts, absorbing suppressive metabolites (including antibiotics), and physically sheltering organisms against environmental stresses.

Recently, Bedding[10] reported a formulation for nematodes comprised of a homogenous mixture of an aqueous cream of infective nematodes with clay. Nematodes were placed in contact with an adsorbent clay resulting in gradual absorption of water from the nematodes (i.e., partial desiccation) and enabling them to withstand room temperatures better than unformulated nematodes. A commercial product utilizing clay as a carrier is currently available (Table 2). An aqueous nematode and clay suspension can be applied with conventional sprayers, although some difficulty occurs if the nozzles become clogged by the clay particles.

C. Activated Charcoal

Nematodes have also been formulated by mixing an adsorbent material such as activated charcoal with a thick paste of nematodes and an antimicrobial agent.[11] Storage conditions are initially aerobic but will tend to have less oxygen as it is used by the nematodes. Prolonged storage of nematodes in this formulation is possible because infective nematodes do not require much oxygen (aeration) or high surface to volume ratios during cold storage. A sealable container has been designed for storage and transport of this formulation. Although an aqueous suspension of nematodes and charcoal can be applied with conventional sprayers, this formulation is no longer available, probably due to the development of better formulations that are easier to mix and apply.

D. Gel-Forming Polyacrylamides

Gel-forming, cross-linked polyacrylamides, available commercially, are used as conditioners where short-term moisture deficits or persistent drought inhibits plant growth.[12] These commercial products absorb water many times their own weight, improving the water retention properties of porous soils and delaying the onset of permanent wilting by plants. Addition of such a polymer to an aqueous nematode suspension significantly enhanced the field persistence of *S. carpocapsae* applied for control of the sugarcane root stalk borer (*Diaprepes abbreviatus*) in citrus.[13]

Steinernematids and heterorhabditidis are compatible with many gel-forming polyacrylamides. Currently, one commercial product utilizes a polyacrylamide gel as a carrier to ease mixing of suspensions and to maintain nematode viability.[8] Nematodes are placed on the gel and packed inside a mesh bag from which they are easily extracted in water (Table 2).

E. Polyether-Polyurethane Sponge

For experimental usage and sales in small market segments, steinernematids and heterorhabditids can be placed onto clean, polyether-polyurethane sponges at rates of 500-1000 nematodes per cm^2 surface area. An aqueous suspension is prepared by hand-squeezing the sponge in a small volume of water to extract nematodes from the sponge. For large commercial applications using, for example, a rate of 2.5 billion nematodes per hectare, squeezing 50 sponges (50 million nematodes per sponge) to extract the nematodes is time-consuming and physically demanding, making sponge formulation impractical.

F. Vermiculite and Peat

Vermiculite and peat are used as moistened carriers for transporting and storing nematodes. The nematode-vermiculite or nematode-peat mixtures are commonly applied as a mulch for insect control. The mixtures can only be applied through a conventional sprayer if the sprayer nozzle is removed to avoid clogging.

G. Evaporetardants and Ultraviolet Protectants

Environmental factors such as temperatures above 30°C, relative humidities below 90%, wind, and ultraviolet light were listed as reasons for the discouraging results of nematodes against foliar insects.[14] Evaporetardants[15-17] and ultraviolet protectants[18] improved the survival of steinernematid nematodes but not sufficiently to enable their use against foliage-feeding insects. Finding new, effective protective additives along with genetic improvements may enable nematodes to be used effectively in certain foliar environments as part of a pest management system.[19]

H. Anhydrobiotic Nematodes

The term anhydrobiotic applies to organisms capable of surviving either rapid or slow dehydration up to and beyond the point at which metabolism is fully arrested.[20] Significant progress has been made in inducing infective juveniles of *S. carpocapsae* into anhydrobiosis.[21] Desiccated preparations could form the basis for an improved product formulation because they would be lightweight, stable, and easy to handle, store, and apply.

Desiccated nematodes have enhanced tolerance for extremes of abiotic factors as compared with nondesiccated nematodes.[22] Therefore, such preparations could be utilized in the form of baits, capsules, or sprays to control insects that are presently difficult or impractical to control with nondesiccated nematodes. Such approaches were tested utilizing partially desiccated nematodes against a wide spectrum of insect pests with encouraging results.[23-25] In these experiments, partially desiccated nematodes mixed with baits rehydrated inside the insect's gut, penetrated into the body cavity, and infected the host. The capability and technology necessary to desiccate and rehydrate steinernematids at industrial scales have been developed,[22] but their shelf life falls short of agrichemical industry standards due to unexplained degradation of the nematodes.

I. Bait

Bait formulations generally consist of a mixture of a carrier (e.g., corncob grits, wheat bran, and peanut hulls), a feeding stimulant (e.g., sucrose, malt extracts, glucose, and molasses), and a toxicant. In some situations, baits are the only feasible formulation because irrigation and soil incorporation are not practical. Baits often offer the most cost effective and ecologically sound use of available insecticides.[26] This concept was recently validated under laboratory conditions in tests conducted on black cutworms (*A. ipsilon*) and tawny mole crickets (*Scapteriscus vicinus*)[27] by adding hydroxyethyl cellulose and glycerine to nematodes to minimize their migration from the bait. Under field conditions, the bait formulation did not outperform the aqueous nematode suspension when applied against black cutworms, although it produced a significant pest reduction compared with controls.[6]

III. APPLICATION TECHNOLOGY

As with chemical pesticides, optimal application strategies are needed to maximize field effectiveness of entomopathogenic nematodes. Substantial efforts are now being directed toward developing new technologies that may close the efficacy gap between nematodes and chemicals.

A. Method of Application

Nematodes can be applied with common agrichemical equipment including small pressurized sprayers, mist blowers, electrostatic sprayers, fan sprayers, and helicopters (Table 4). Entomopathogenic nematodes can withstand application pressures of 300 lb/in.^{2} and can be delivered with all common nozzle type sprayers (e.g., "01" nozzles) with openings as small as 50 microns in diameter. However, some types of pumping equipment produce a considerable amount of heat, and should the temperature in the sprayer plumbing rise above 32°C, the nematodes could be adversely affected.[28] Fortunately, this rarely occurs and is usually only a problem when the spray tank is almost empty and little water is available as a heat sink.

Other methods of delivering nematodes to the target are injection, baits, and alginate capsules (Table 3). Practical implementation of these techniques will require further investigation. It is expected that these application strategies

TABLE 4. Examples of Equipment Systems Used to Apply Steinernematid and Heterorhabditid Nematodes

System	Target Insects[a]	Crop	Ref.
Conventional hand or ground sprayers	Various	Various	14
Helicopter and aircraft	Navel orangeworm	Almond	71
	Black vine weevil	Cranberries	56,72,73
	Cranberry girdler	Cranberries	56,72
Overhead irrigation	Black vine weevil	Cranberries	56,72,73
	Cranberry girdler	Cranberries	56,73
Microjet irrigation	Sugarcane root stalk borer	Citrus	75
Center pivot irrigation	Corn rootworms	Corn	76
Drip irrigation	Striped cucumber beetle	Cucumber	77
	Mole crickets	Turf	78
	Black vine weevil	Strawberry	74
Electrostatic sprayer	Navel orangeworm	Almond	56
	Artichoke plume moth	Artichoke	79
Mist sprayer	Navel orangeworm	Almond	80
Photoelectric intermittent sprayer	Artichoke plume moth	Artichoke	79
Soil injecting sprayer	Redheaded pasture cockchafer	Pasture	41
	Corn rootworm	Corn	69
	White grubs	Turf	42

[a]   Scientific names in text except for cranberry girdler *Chrysoteuchia topiaria,* striped cucumber beetle *Acalymma vittatum,* and redheaded pasture cockchafer, *Adoryphorus couloni.*

requires fewer nematodes or a lower spray volume for effective control because they will provide better protection for nematodes from environmental extremes. Nematodes may also be applied by drip and sprinkler irrigation systems (Table 4). Reasons for the increased interest in applying nematodes through irrigation systems include minimal labor requirements, automatic provision of sufficient moisture for the nematodes, availability of permanently installed equipment, and flexibility in the timing of applications. Against wood-boring insects, nematodes have been delivered into galleries with a syringe, cotton swab plug, oil can, and backpack sprayer.[14] Application of nematodes to nylon pack cloth bands lined with Pellon fleece or terry cloth and placed around tree trunks is a control tactic documented for control of late stage gypsy moth larvae (*Lymantria dispar*),[29] or overwintering codling moth prepupae (*Cydia pomonella*).[30] These studies provided some degree of control but may be more effective if nematodes are integrated with other control measures.

B. Timing of Application

Abiotic factors can influence the optimal time of nematode application. Early morning and preferably early evening or night application of nematodes is recommended to avoid solar radiation[31] and high temperatures.[32] On the other hand, low soil temperature (10-16°C) has been attributed to the inconsistent results of *Heterorhabditis bacteriophora* HP88 applied against the spring generation of white grubs (Scarabaeidae) infesting turf during 1984-1988 (24 trials). Tests (n=36) against the fall generation at temperatures between 18 and 26°C produced more consistent results.[33]

As with chemical insecticides, surface spraying of nematodes is the application method most often used. In some situations nematodes were found to have potential against the cabbage maggot, *D. radicum*, and the large pine weevil, *Hylobius abietis*, when applied at planting time.[34,35] When nematode persistence was required for four to six weeks in the soil before the appearance of the target insect stage (i.e., corn rootworms, *Diabrotica* spp., and crucifer flea beetle, *Phyllotreta cruciferae*), insect damage was not prevented.[36,37] Inadequate nematode persistence and nonsusceptible early larval stages could explain these results. Therefore, to optimize the interaction between nematode and insect host, applications should be conducted when susceptible target stages are present in the environment.

C. Irrigation and Spray Volume

Pre- and postapplication irrigation and continuing moderate soil moisture are essential for nematode movement, persistence, and pathogenicity. According to Simons and Poinar,[38] most nematodes applied to dry soil perished; however, if they were applied to damp soil which later became dry, many survived for some period because of their ability to withstand gradual desiccation. Field applications of nematodes to dry turf with 0.64 cm pretreatment irrigation followed by 0.64 cm posttreatment irrigation with normal rainfall resulted in Japanese beetle grub (*Popillia japonica*) control comparable with a standard insecticide.[39] In nonirrigated turf, less insect control by nematodes was observed. Similarly, field trials with *H. bacteriophora* HP88 (32 tests) in turf receiving frequent irrigation yielded more consistent results than trials (16 tests) in nonirrigated turf.[33] However, if the soil contains too much moisture, nematodes are less effective.[40]

Spray volume depends upon characteristics of the turfgrass or soil being treated. In order to penetrate deep thatch (thatch is a tightly bound layer of dead and living roots and stems that accumulate between the soil surface and green vegetation in turfgrass), a high spray volume may be needed for nematodes to reach the depth occupied by the target insect. Under such conditions, 1400-1870 l/ha (150-200 gal/acre) followed immediately with irrigation is usually acceptable. Recent attempts to inject the nematodes into soil appear to be an efficient method for delivering the nematodes when water is limited,[41,42] but more research is needed to determine the proper spray volume in relation to soil type and insect activity.

Foliar application against insects such as the artichoke plume moth (*Platyptilia carduidactyla*) in artichokes requires adequate coverage and placement of nematodes. This can only be achieved at volumes of ca. 2800 l/ha (300 gal/acre). In these situations, nematode application was less practical (expensive) compared with registered insecticides which only required 935 l/ha (100 gal/acre). Better formulations and delivery systems are needed to increase the cost-effectiveness of nematodes.[43]

D. Pest Population and Behavior

If one or a few individuals of a pest species are sufficient to cause tangible economic damage, then the economic threshold of the crop is low. In this situation, an insecticide with excellent persistence and effectiveness against all larval stages is required to provide adequate protection. Lacking these features, biological insecticides generally cannot provide adequate protection compared with insecticides against pests infesting crops with a low economic threshold. Thus, such crop-pest systems are not ones for which nematodes currently have the highest potential.

Similarly, frequent applications or integration of the nematodes with other control measures would be needed for pests that are characterized by the presence of feeding stages throughout the year, such as mole crickets (*Scapteriscus* spp.) and artichoke plume moth (Table 5). In contrast, a narrow "window of opportunity" exists for nematodes applied against insects with susceptible stages occurring for only a few days in soil, such as the cabbage root maggot and the western corn rootworm (*Diabrotica virgifera virgifera*).

In some cases, successful control of an insect pest can be achieved with a single application. For example, Japanese beetle populations were controlled in such a fashion because of the susceptibility of the larval stage and the ability of the nematode to recycle and persist in the field.[33]

Little information is available on insect responses to nematode application, although host behavior may greatly influence efficacy of the nematodes. The strong tendency of red imported fire ant workers (*Solenopsis invicta*) to relocate colonies to satellite mounds after treatment currently limits nematode efficacy for fire ant control.[44] Moreover, termites respond to nematodes by walling off infected individuals or by withdrawing from the infested areas, resulting in unsatisfactory control.[45] Some scarabaeid species may avoid

TABLE 5. Successful Attempts to Utilize *Steinernema carpocapsae* and *Heterorhabditis* sp. with Other Insect Control Strategies

Target Insect[a]	System	Site	Ref.
Red imported fire ant	*S. carpocapsae* and insecticide (carbaryl, diazinon, or acephate)	Field mounds	61
Artichoke plume moth	Insecticide treatment (methidathion), followed by *S. carpocapsae*[b]	Artichoke field	81
Black vine weevil	*Heterorhabditis* sp. and insecticide (propoxur or methomyl)	Greenhouse	62
Leaf miner	*S. carpocapsae* for larvae and insecticide for adults	Greenhouse	82
Cutworm	*S. carpocapsae* and insecticide (oxamyl or DDVP)	Field plots	60
Artichoke plume moth	*S. carpocapsae* and sex pheromone	Artichoke field	81
Black vine weevil	Insecticide (acephate) for adults and *S. carpocapsae* for larvae[c]	Cranberry field	56

[a] Scientific names in text except for leaf miner *Liriomyza trifolii*, and cutworm *Agrotis segutum*.
[b] Two treatments of *S. carpocapsae*.
[c] *S. carpocapsae* 3-4 weeks after insecticide treatment.

nematode infection because of their low carbon dioxide output, sieve plates that deny nematodes access to the spiracles, highly anaerobic hindgut and high defecation rate, and their ability to sense and remove nematodes in the vicinity of the mouth.[46]

E. Field Dosage and Optimum Species

Field concentrations exceeding 2.5 billion nematodes/ha (1 billion/acre) are usually applied to ensure that a sufficient nematode population will come in contact with the target insects to provide control. A high concentration is needed to overcome the negative impacts of the abiotic and biotic soil environment. High nematode concentrations are also needed against certain insects which only remain in soil for a few days before tunneling into roots, such as cabbage maggots (*D. radicum*), and against insects which are not highly susceptible to nematode infection because of their small size and active movement. For example, high nematode concentrations have been employed against early immature stages of mole crickets (*Scapteriscus* spp.), root maggots (*Delia* spp.), and corn rootworms (*Diabrotica* spp.).

Large differences in field efficacy exist among the species and strains of nematodes. Thus, heterorhabditids are highly effective against the Japanese beetle and the bluegrass billbug (*Sphenophorus parzulus*), but provide unsatisfactory control against mole crickets. In contrast, steinernematids and, in particular, *S. scapterisci* (Uruguay strain)[47] are effective against mole crickets. As Gaugler[48] has pointed out, differences in field efficacy may, in part, be related to the vertical distribution of the target insect in the soil. Steinernematids show a preference for soil near the surface[49,50] and, therefore, may be best adapted to attack insects like mole crickets and cutworms which feed at the soil litter interface or on the soil surface. Heterorhabditids have a greater tendency to move downwards[51] and have superior host-seeking abilities relative to steinernematids,[52,53] making them efficacious against Japanese beetles, billbugs, and root weevils (*Otiorhynchus* spp.), which are present at the root zones of host plants.

F. Crop Morphology and Phenology

The feeding behavior and location of a particular insect pest may vary according to the crop, cultivar, and age. These factors may directly affect the effectiveness of nematodes against the insect host.

The ability of *H. bacteriophora* to find and infect wax moth larvae was not impaired by the presence of sparse corn roots (0.33 g dry weight). However, at high root density (1.55 g dry weight), insect mortality was significantly lower than in treatments with no roots.[52] *S. glaseri* and *S. carpocapsae* are known to orient toward plant roots,[54,55] which may impair their host-finding ability, especially if the target insect occurs away from the roots.

Field trials against the navel orangeworm larvae, *Amyelois transitella*, infesting Le Grande cultivar almonds were more encouraging than trials conducted on other cultivars. Larvae tend to feed on the inner side of the hull before feeding on the nut meat in the Le Grande variety. This behavior provides sufficient time for nematodes to infect and kill the larvae before they damage the nut.[56] Additionally, because the Le Grande variety has a relatively synchronous hull split, one properly timed application of nematodes will be sufficient to infect most larvae. As a result of limited field persistence in almond orchards,[57] nematodes provide inconsistent protection for other cultivars with a longer hull split period.

The amount of chemical insecticide reaching white grubs infesting turfgrass is frequently less than 5% of that applied to the surface. This is especially true in turfgrass having 2 cm or greater thickness of thatch. This thatch may also act as a barrier to nematode entry to the feeding site of the grubs and may explain some unsuccessful field trials.[33,58]

G. Integrating Nematodes with Other Control Strategies

Integrated pest management is a systematic approach to compatible use of the available forms of pest control, including mechanical, biological, chemical, and natural control methods.[59] Because more than 90% of insect species spend at least part of their life cycle in soil, most can be considered candidates for suppression by steinernematid and heterorhabditid nematodes.[46] Consequently, these nematodes can fit into control programs for a broad spectrum of insects (Table 5).

The potential of combining nematodes with other control strategies must be assessed broadly in terms of whether their incorporation results in an additive or synergistic effect on pest mortality or damage reduction. Additionally, consideration must be given to whether the cost/benefit ratios are improved by the use of nematodes. Increased costs may be incurred in the integrated use of nematodes with other control tactics. However, the potential benefits of a reduction in chemical insecticide use are great. These benefits include decreased probability of insecticide resistance in the target insect, less environmental contamination by chemical pesticides, and less likelihood of secondary pest outbreaks.

If these nematodes are to be incorporated in an integrated pest management program in agriculture, they must be used with other tactics to control insects, plant-parasitic nematodes, weeds, and plant pathogens. Entomopathogenic nematodes are compatible with some chlorinated hydrocarbon, organophosphate, and carbamate insecticides in water mixtures.[14] Certain fungicides, herbicides, miticides, and nematicides have little or no adverse effect on the infective stage. However, more research is needed to ascertain the compatibility of nematodes with new agrichemical products at concentrations specific to different agricultural practices. Pesticides should not be acutely toxic to nematodes at field rates. It may be possible to reduce the pesticide rate to a level compatible with the use of nematodes,[60] or nematodes could be applied before the pesticide or vice versa, the latter allowing time for the pesticide to become absorbed or degraded to a nontoxic level to the nematode. To date, several successful attempts have been made to further increase nematode efficacy when employed in conjunction with chemical[60-62] and microbial agents.[63-65]

IV. STANDARDIZATION AND QUALITY CONTROL

Infective-juvenile nematodes and their associated bacteria must possess a high and stable level of virulence or pathogenicity throughout all stages of product formulation and application. Maintaining high quality of the final product is the most important reason behind the development of standards by industry. Presently, the most useful method for nematode quality control is the insect bioassay.[66] The usual parameters measured are LC_{50}, LD_{50}, or LT_{50} (i.e., the concentration, dose, or time needed to kill 50% of test insects). These techniques confine numerous nematodes in a high pathogen to host ratio (e.g., 4:1-

10:1), with more than one insect host in a small arena such as a petri dish. Molyneux et al.[66] emphasized the importance of using one insect host per arena for assay of nematodes. Recently, a reliable bioassay method was developed.[67] The method utilizes a culture cell well (Falcon, NJ, U.S.) which confines one infective *Steinernema* with one insect host (*Galleria mellonella* larva) and results in ca. 50% infection within 48 hr. Similar mortality curves have been developed for some species of *Heterorhabditis*. Sealed cells are affected only slightly by ambient relative humidity of the holding chamber. Additionally, errors in counting and diluting nematodes are avoided in this cell technique. Finally, this technique assesses pathogenicity in a portion of the nematode population which would normally go unassessed, the less pathogenic juveniles. In multiple nematode bioassays, these individuals are lost in the larger population effects of multiple nematode invasion into a host. Pathogenicity assessment alone, however, will not provide meaningful results in the field. Therefore, attention should be devoted to development of other performance indicators such as biochemical or serological methods. In the meantime, bioassay will remain the primary method for determining quality control of steinernematid and heterorhabditid nematodes.

Although great potential exists for large-scale use of steinernematid and heterorhabditid nematodes, the environmental factors affecting dispersal, survival, and infectivity must be elucidated so that their performance under field conditions may be enhanced. In addition, artificial selection for, or natural isolation of more virulent strains and development of stable, reliable formulations must be performed. A reliable method for measuring the virulence of each nematode species or strain against the insect host is also a necessity. Once this method has been proven, standard preparations with designated potency can be produced for each nematode considered for commercial use. Materials made by government and industrial laboratories can then be compared against these designated standards.

IV. CONCLUSIONS

Due to recent progress in mass production, steinernematid and heterorhabditid nematodes are commercially available. Reliable and stable formulations for storage, shipping, and application are the principal hurdles limiting the market size for nematode-based products. Substantial improvement in formulation stability has been obtained by immobilizing or partially desiccating the nematodes on specific carriers. Apparently, these carriers reduce nematode metabolism, thus improving their tolerance to temperature extremes. The storage period of such formulations is related to the oxygen and moisture requirements of the nematodes. For example, to achieve 30 days storage at room temperature and 6 months under refrigeration conditions, current products utilize approximately 10 million nematodes in 0.5 l containers. For large scale applications where the recommended dose is high, refrigeration (1-3 months) is suggested for products that utilize one billion nematodes per 2.3 kg net weight of gel-forming polyacrylamides. In spite of these limitations, nematode products have been introduced successfully in markets where safety and the use of restricted insecticides are an issue. Anhydrobiotic nematodes may ultimately be a solution to developing a stable formulation for maximum market distribution. The effective host range of nematodes in the field is limited to those insects found in soil and cryptic environments that display moisture and temperature parameters conducive for these nematode species, or to foliage-feeding insects in very humid environments. Unfortunately, attempts to control insects in exposed environments have been discouraging due to rapid nematode desiccation. In some situations the cost of employing nematodes is excessive compared with agrichemicals. These difficulties may be overcome by utilization of desiccated or partially desiccated nematodes in bait, capsule, or spray formulations that enhance their biological control potential. In the presence of adequate moisture (soil, insect digestive system, etc.) partially desiccated nematodes can rehydrate, penetrate into the insect, and cause infection. Currently, the shelf life of desiccated nematodes falls short of industry's standard of at least 1 year. Research is in progress to characterize and prevent the degradation reactions in desiccated nematodes.

In recent years, efforts made to narrow the "efficacy gap"[48] between chemical pesticides and nematodes have been successful in various market segments. Research efforts toward adopting a quality control procedure, selecting a suitable target environment and target insects for nematodes, identifying an effective strain and dosage, timing applications properly, and choosing means of application or delivery that increase the probability of successful contact between insect and nematode, have been instrumental in commercializing these entomopathogenic nematodes.

REFERENCES

1. **Georgis, R., and Poinar, G. O., Jr.,** Field effectiveness of entomophilic nematodes *Neoaplectana* and *Heterorhabditis*, in *Integrated Pest Management in Turf and Ornamentals*, Leslie, A. R., and Metcalf, R. L., Eds., U. S. Environmental Protection Agency, Washington, D.C., 1989, 213.
2. **Scher, H. B.,** Innovation and developments in pesticide formulations. an overview, in *Pesticide Formulations Innovation and Developments*, Cross, B., and Scher, H. B., Eds., American Chemical Society, Washington, D.C., 1988, 1.
3. **Kaya, H. K., and Nelsen, C. E.,** Encapsulation of steinernematid and heterorhabditid nematodes with calcium alginate: a new approach for insect control and other applications, *Environ. Entomol.*, 14, 572, 1985.
4. **Poinar, G. O., Jr., Thomas, G. M., Lin, K. C., and Mookerjee, K.,** Feasibility of embedding parasitic nematodes in hydrogels for insect control, *IRCS Med. Sci.*, 13, 754, 1985
5. **Capinera, J., and Hibbard, B. E.,** Bait formulations of chemical and microbial insecticides for suppression of crop-feeding grasshoppers, *J. Agric. Entomol.*, 4, 337, 1987.

6. **Capinera, J., Pelissier, D., Menout, G. S., and Epsky, N. D.,** Control of black cutworm, *Agrotis ipsilon* (Lepidoptera: Noctuidae), with entomogenous nematodes (Nematoda: Steinernematidae, Heterorhabditidae), *J. Invertebr. Pathol.*, 52, 427, 1988.

7. **Kaya, H. K., Mannion, C. M., Burlando, T. M., and Nelsen, C. E.,** Escape of *Steinernema feltiae* from alginate capsules containing tomato seeds, *J. Nematol.*, 19, 287, 1988.

8. **Pruitt, P.,** unpublished data, 1988.

9. **Ward, M. G.,** Formulation of biological insecticides: surface and diluent selection, in *Advances in Pesticide Formulation Technology*, Scher, H. B., Ed., American Chemical Society, Washington, D.C., 1984, 175.

10. **Bedding, R. A.,** Storage of entomogenous nematodes, Int. Patent WO88/08668, 1988.

11. **Biotechnology Australia,** Nematode cream contg. adsorbent and antimicrobial agent stable on prolonged storage and transport under anaerobic conditions, Int. Patent 85/03412, 1985.

12. **Hamilton, J. L., and Lowe, R. H.,** Use of a water absorbent polymer in tobacco seedling production and transplanting, *Tob. Sci.*, 26, 17, 1982.

13. **Schroeder, W. J.,** Super absorbent polymer as a survival aid to entomogenous nematodes for control of *Diaprepes abbreviatus* (Coleoptera: Curculionidae) in citrus, *J. Econ. Entomol.*, 1990, in press.

14. **Kaya, H. K.,** Entomogenous nematodes for insect control in IPM systems, in *Biological Control in Agricultural IPM Systems*, Hoy, M. A., and Herzog, D. C., Eds., Academic Press, New York, 1985, 283.

15. **Kaya, H. K., and Reardon, R. C.,** Evaluation of *Neoaplectana carpocapsae* for biological control of the western spruce budworm, *Choristoneura occidentalis*: ineffectiveness and persistence of tank mixes, *J. Nematol.*, 14, 595, 1982.

16. **MacVean, C. M., Brewer, J. W., and Capinera, J. L.,** Field tests of antidesiccants to extend the infection period of an entomogenous nematode, *Neoaplectana carpocapsae*, against the Colorado potato beetle, *J. Econ. Entomol.*, 75, 97, 1982.

17. **Shapiro, M., McLane, W., and Bell, R.,** Laboratory evaluation of selected chemicals as antidesiccants for the protection of the entomogenous nematode, *Steinernema feltiae* (Rhabditidae: Steinernematidae), against *Lymantria dispar* (Lepidoptera: Lymantriidae), *J. Econ. Entomol.*, 78, 1437, 1985.

18. **Gaugler, R., and Boush, G. M.,** Laboratory tests on ultraviolet protectants of an entomogenous nematode, *Environ. Entomol.*, 8, 810, 1979.

19. **Gaugler, R.,** Entomogenous nematodes and their prospects for genetic improvement, in *Biotechnology in Invertebrate Pathology and Cell Culture*, Maramorosch, K., Ed., Academic Press, New York, 1988, 457.

20. **Womersley, C.,** A reevaluation of strategies employed by nematode anhydrobiotes in relation to their natural environment, in *Vistas on Nematology*, Veech, J. A., and Dickson, D. W., Eds., Society of Nematologists, Hyattsville, 1987, 165.

21. **Popiel, R.,** unpublished data, 1987.

22. **Friedman, M.,** unpublished data, 1989.

23. **Wojcik, W. F., and Georgis, R.,** unpublished data, 1987.

24. **Wojcik, W. F., and Georgis, R.,** Infection of adult western yellow jackets with desiccated *Steinernema feltiae* (Nematoda), *J. Invertebr. Pathol.*, 52, 183, 1988.

25. **Georgis, R.,** Nematodes for biological control of urban insects, American Chemical Society, Division of Environmental Chemistry, 194th National Meeting, New Orleans, 27, 816, 1987.

26. **Kepner, R. L., and Yu, S. J.,** Development of a toxic bait for control of mole crickets, Orthoptera: Gryllotalpidae, *J. Econ. Entomol.*, 80, 659, 1987.

27. **Georgis, R., Wojcik, W. F., and Shetlar, D. J.,** Use of *Steinernema feltiae* in a bait for the control of black cutworm (*Agrotis ipsilon*) and tawny mole crickets (*Scapteriscus vicinus*), *Florida Entomol.*, 72, 203, 1989.

28. **Poinar, G. O., Jr.,** Entomophagous nematodes, in *Biological Plant and Health Protection,* Franz, J.M., Ed., G. Fischler Verlag, Stuttgart, 1986, 95.

29. **Reardon, R. C., Kaya, H. K., Fusco, R. A., and Lewis, F. B.,** Evaluation of *Steinernema feltiae* and *S. bibionis* (Rhabditida: Steinernematidae) for suppression of *Lymantria dispar* (Lepidoptera: Lymantriidae) in Pennsylvania, U.S.A., *Agric. Ecosys. Environ.,* 15, 1, 1986.

30. **Kaya, H. K., Joos, L. A., Falcon, L. A., and Berlowitz, A.,** Suppression of the codling moth (Lepidoptera: Olethreutidae) with the entomogenous nematode, *Steinernema feltiae* (Rhabditida: Steinernematidae), *J. Econ. Entomol.,* 77, 1240, 1984.

31. **Gaugler, R., and Boush, G. M.,** Effects of ultraviolet radiation and sunlight on the entomogenous nematode, *Neoaplectana carpocapsae, J. Invertebr. Pathol.,* 32, 291, 1978.

32. **Schmiege, D. C.,** The feasibility of using a neoaplectanid nematode for control of some forest insect pests, *J. Econ. Entomol.,* 56, 427, 1963.

33. **Georgis, R., and Gaugler, R.,** unpublished data, 1989.

34. **Georgis, R., Poinar, G. O., Jr., and Wilson, A. P.,** Practical control of the cabbage root maggot, *Hylemia brassicae,* (Diptera: Anthomyiidae) by entomogenous nematodes, *ICRS Med. Sci.,* 11, 322, 1983.

35. **Pye, A. E., and Pye, N. L.,** Different applications of the insect parasitic nematode *Neoaplectana carpocapsae* to control the large pine weevil, *Hylobius abietis, Nematologica,* 31, 109, 1986.

36. **Munson, J. D., and Helms, T. J.,** Field evaluation of a nematode DD-136 for control of corn rootworm larvae, *Proc. No. Central Branch Entomol. Soc. Am.,* 25, 97, 1970.

37. **Morris, O. N.,** Evaluation of the nematode, *Steinernema feltiae* Filipjev, for the control of the flea beetle, *Phyllotreta cruciferae* (Goeze) (Coleoptera: Chrysomelidae), *Can. Entomol.,* 119, 95, 1987.

38. **Simons, W. R., and Poinar, G. O., Jr.,** The ability of *Neoaplectana carpocapsae,* (Steinernematidae: Nematodea) to survive extended periods of desiccation, *J. Invertebr. Pathol.,* 22, 228, 1973.

39. **Shetlar, D. J., Suleman, P. E., and Georgis, R.,** Irrigation and use of entomogenous nematodes, *Neoaplectana* spp. and *Heterorhabditis heliothidis* (Rhabditida: Steinernematidae and Heterorhabditidae), for control of Japanese beetle (Coleoptera: Scarabaeidae) grubs in turfgrass, *J. Econ. Entomol.,* 81, 1318, 1988.

40. **Molyneux, A. S., and Bedding, R. A.,** Influence of soil texture and moisture on the infectivity of *Heterorhabditis* sp. D1 and *Steinernema glaseri* for larvae of the sheep blowfly, *Lucilia cuprina, Nematologica,* 30, 358, 1984.

41. **Berg, G. N., William, P., Bedding, R. A., and Akhurst, R. J.,** A commercial method of application of entomopathogenic nematodes to pasture for controlling subterranean insect pests, *Pl. Prot. Quart.,* 2, 174, 1987.

42. **Niemczyk, H.,** personal communication, 1989.

43. **Kaya, H. K.,** *Steinernema feltiae*: use against foliage feeding insects and effect on non-target insects, in *Fundamental and Applied Aspects of Invertebrate Pathology,* Samson, R. A., Vlak, J. M., and Peters, D., Eds., *Proc. 4th Int. Coll. Invertebr. Pathol.,* Wageningen, 1986, 268.

44. **Miller, R. W.,** unpublished data, 1988.

45. **Fujii, J. K.,** Effects of an Entomogenous Nematode, *Neoaplectana carpocapsae* Weiser, on the Formosan Subterranean Termite, *Coptotermes formosanus* Shiraki, with Ecological and Biological Studies on *C. formosanus,* Ph.D. thesis, University of Hawaii, Honolulu, 1975.

46. **Akhurst, R. J.,** Controlling insects with entomopathogenic nematodes, in *Fundamental and Applied Aspects of Invertebrate Pathology,* Samson, R. A., Vlak, J. M., and Peters, D., Eds., *Proc. 4th Int. Coll. Invertebr. Pathol.,* 1986, 265.

47. **Nguyen, K. B.,** A New Nematode Parasite of Mole Crickets: Its Taxonomy, Biology and Potential for Biological Control, Ph.D. thesis, University of Florida, Gainesville, 1988.

48. **Gaugler, R.,** Ecological considerations in the biological control of soil-inhabiting insects with entomopathogenic nematodes, *Agric. Ecosys. Environ.,* 24, 351, 1988.

49. **Georgis, R., and Poinar, G. O., Jr.,** Effect of soil texture on the distribution and infectivity of *Neoaplectana carpocapsae* (Nematoda: Steinernematidae), *J. Nematol.,* 15, 308, 1983.

50. **Moyle, P. L., and Kaya, H. K.,** Dispersal and infectivity of the entomogenous nematode, *Neoaplectana carpocapsae* Weiser (Rhabditida: Steinernematidae) in sand, *J. Nematol.,* 13, 295, 1981.

51. **Georgis, R., and Poinar, G. O., Jr.,** Vertical migration of *Heterorhabditis bacteriophora* and *H. heliothidis* (Nematoda: Heterorhabditidae) in sandy loam soil, *J. Nematol.,* 15, 652, 1983.

52. **Choo, H. Y., Kaya, H. K., Burlando, T. M., and Gaugler, R.,** Entomopathogenic nematodes: host-finding ability in the presence of plant roots, *Environ. Entomol.,* 18, 1136, 1989.

53. **Kaya, H. K., Choo, H. Y., and Burlando, T. M.,** Influence of plant roots on the infectivity of *Heterorhabditis heliothidis* to insect hosts, *J. Nematol.,* 19, 537, 1987.

54. **Bird, A. F., and Bird, J.,** Observations on the use of insect parasitic nematodes as a means of biological control of root-knot nematodes, *Int. J. Parasitol.,* 16, 511, 1986.

55. **Ishibashi, N., and Kondo, E.,** Dynamics of the entomogenous nematode, *Steinernema feltiae,* applied to soil with and without nematicide treatment, *J. Nematol.,* 19, 404, 1987.

56. **Agudelo-Silva, F.,** unpublished data, 1987.

57. **Agudelo-Silva, F., Lindegren, J. E., and Valero, K. A.,** Persistence of *Neoaplectana carpocapsae* (Kapow selection) infectives in almonds under field conditions, *Fl. Entomol.,* 70, 288, 1987.

58. **Zimmerman, R. J.,** Entomogenous Nematodes for Use in Turf IPM Systems, MS thesis, Colorado State University, Fort Collins, 1988.

59. **Coppel, H. C., and Mertins, J. W.,** *Biological Insect Pest Suppression,* Springer-Verlag, New York, 1977, 7.

60. **Ishibashi, N., Choi, D. R., and Kondo, E.,** Integrated control of insects/nematodes by mixing application of steinernematid nematodes and chemicals, *J. Nematol.,* 19, 531, 1987.

61. **Quattlebaum, E. C.,** Evaluation of Fungal and Nematode Pathogens to Control the Red Imported Fire Ant, *Solenopsis invicata* Buren, Ph.D. thesis, Clemson University, Clemson, 1980.

62. **Verbruggen, D., De Grisse, A., and Heungens, A.,** Mogelijkheden voor de geintegreerde bestrijding van *Otiorhynchus sulcatus* (F.), *Meded. Fac. Landbouww. Rijksuniv. Gent,* 50, 133, 1985.

63. **Kamionek, M., Sandner, H., and Seryczynska, H.,** The combined action of *Beauveria bassiana* (Bals./Vuill) (Fungi imperfecti: Moniliales) and *Neoaplectana carpocapsae* Weiser (Nematoda: Steinernematidae), *Bull. Acad. Pol. Sci.,* 22, 253, 1974.

64. **Kamionek, M., Sandner, H., and Seryczynska, H.,** The combined action of *Paecilomyces farinosus* Dicks (Brown et Smith) (Fungi imp.: Moniliales) and *Neoaplectana carpocapsae* Weiser 1955 (Nematoda: Steinernematidae) on certain insects, *Acta Parasitol. Pol.,* 22, 357, 1974.

65. **Lam, A. B. Q., and Webster, J. M.,** Effect of the DD-136 nematode and of a β-exotoxin preparation of *Bacillus thuringiensis* var. *thuringiensis* on leather jackets, *Tipula paludosa* larvae, *J. Invertebr. Pathol.,* 20, 141, 1972.

66. **Molyneux, A. S., Bedding, R. A., and Akhurst, R. J.,** Susceptibility of larvae of the sheep blowfly *Lucilia cuprina* to various *Heterorhabditis* spp., *Neoaplectana* spp. and an undescribed steinernematid (Nematoda), *J. Invertebr. Pathol.,* 42, 1, 1983.

67. **Miller, R. W.,** Novel pathogenicity assessment technique for *Steinernema* and *Heterorhabditis* entomopathogenic nematodes, *J. Nematol.,* 21, 574, 1989.

68. **Renn, N., Barson, G., and Richardson, P. N.,** Preliminary laboratory tests with two species of entomophilic nematodes for control of *Musca domestica* in intensive animal units, *Ann. Appl. Biol.,* 106, 229, 1985.

69. **Georgis, R.,** unpublished data, 1985.

70. **Vinson, B.,** personal communication, 1986.
71. **Lindegren, J. E., Dibble, J. E., Curtis, C. E., Yamashita, T. T., and Romero, E.,** Compatibility of NOW parasite with commercial sprayers, *Calif. Agric.*, 35(3&4), 16, 1981.
72. **Giles, K.,** unpublished data, 1988.
73. **Dapsis, L.,** unpublished data, 1987.
74. **Curran, J., and Patel, V.,** Use of trickle irrigation system to distribute entomopathogenic nematodes (Nematoda: Heterorhabditidae) for the control of weevil pests (Coleoptera: Curculionidae) of strawberries, *Aust. J. Exp. Agric.*, 28, 639, 1988.
75. **Schroeder, W. J.,** unpublished data, 1988.
76. **Wright, R. J.,** unpublished data, 1989.
77. **Reed, D. K., Reed, G. L., and Creighton, C. S.,** Introduction of entomogenous nematodes into trickle irrigation systems to control striped cucumber beetle, *Acalymma vittatum* (Coleoptera: Chrysomelidae), *J. Econ. Entomol.*, 79, 1330, 1986.
78. **Meredith, J. A., Smart, G. C., and Overman, A. J.,** Evaluation of drip irrigation systems for the delivery of *Neoaplectana carpocapsae*, *IFAS, Univ. Florida Annu. Report, Mole Cricket Research*, 120, 1987-88.
79. **Bari, M., and Georgis, R.,** unpublished data, 1986.
80. **Lindegren, J. E., Agudelo-Silva, F., Valero, K. A., and Curtis, C. E.,** Comparative small-scale field application of *Steinernema feltiae* for navel orangeworm control, *J. Nematol.*, 19, 503, 1987.
81. **Bari, M.,** unpublished data, 1987.
82. **Begley, J.,** unpublished data, 1988.
83. **Roger, T.,** unpublished data, 1989.
84. **Pye, A.,** unpublished data, 1988.

Efficacy

10. Efficacy Against Soil-Inhabiting Insect Pests

Michael G. Klein

I. INTRODUCTION

Although chemical insecticides have been the primary means of controlling soil insects for many years, concerns about public safety, environmental contamination, and reduced efficacy due to possible microbial degradation or insect tolerance and resistance have created a need for alternative control strategies.[1,2] The milky disease bacterium, *Bacillus popilliae*, is the most widely available alternative against Japanese beetle larvae,[3] but greater use of *B. popilliae* and other milky disease organisms has been hampered by serious drawbacks such as a narrow host range, slow build-up in the soil, and poor growth of *in vitro* cultures. Entomopathogenic nematodes in the families Steinernematidae and Heterorhabditidae lack these limitations and possess many qualities that make them excellent biological control agents.[1] They have a broad host range, can be easily mass produced, possess the ability to seek out their host, kill their host rapidly, are environmentally safe, and have been exempted from registration by the U.S. Environmental Protection Agency (EPA). This combination of attributes has generated an intense interest in the development of these nematodes for use against soil insect pests.

The soil environment offers an excellent site for insect-nematode interactions; more than 90% of insect pests spend part of their life cycle in the soil, and soil is the natural reservoir of steinernematid and heterorhabditid nematodes.[1,4] Poinar[5,6] has presented a comprehensive review of the many entomopathogenic nematode species and has examined their relationships with insect hosts. This chapter examines recent studies on the efficacy of entomopathogenic nematodes against a number of soil-inhabiting pests.

II. WHITE GRUBS

White grubs, larvae of Scarabaeidae, are major pests of turfgrass, pastures, sugarcane, and forests throughout the world. They are the most serious pests of turf in the northeastern U.S. and are major pests throughout the Midwest.[7] Larvae of the Japanese beetle, *Popillia japonica*, alone cause over $234 million in damage annually.[8] *B. popilliae* has been the primary biological control agent for suppression of the Japanese beetle and other scarab grubs.[3] However, recent field tests with nematodes in the genera *Steinernema* (=*Neoaplectana*) and *Heterorhabditis* have shown that they can be effective biological control agents against these insects (Table 1).

TABLE 1. Field Efficacy of *Steinernema* spp. and *Heterorhabditis* spp. for Various Scarab Larvae

Insect host	Nematode species (strain)	Control	Ref.
Turf			
Popillia japonica	*S. carpocapsae*	55% 39 DATa	12
		2-19%	13
		7-50% 28 DAT	14
		84-90% 80 DAT	14
		0% 386 DAT	14
		55%	15
	S. glaseri	8-15% reduction 1st yr, significant reduction subsequent yr	9,10
		Colonization unsuccessful	11
		2-31% 14 DAT	13
	H. bacteriophora	73%	12
		64% 47 DAT	2
		35-60% 29 DAT	14
		84-96% 280 DAT	14
		93-99% 386 DAT	14
	H. bacteriophora (HP88)	100% 28 DAT and 93-97% of the next generation	14
		>70%	15
Cyclocephala borealis	*S. carpocapsae*	<40%	15
	H. bacteriophora (HP88)	>70%	15
Cyclocephala hirta	*S. feltiae*	65% 28 DAT	16
	H. bacteriophora (HP88)	47% 28 DAT	16
Potted Yews			
Popillia japonica	*S. carpocapsae*	>50%	17
	S. glaseri	>90%	17
	H. bacteriophora	>95%	17
	Heterorhabditis sp.	>95%	17
Rhizotrogus majalis	*S. carpocapsae*	<20%	17

TABLE 1 (continued). Field Efficacy of *Steinernema* spp. and *Heterorhabditis* spp. for Various Scarab Larvae

Insect host	Nematode species (strain)	Control	Ref.
Potted Yews (continued)			
	S. glaseri	<70%	17
	H. bacteriophora	<55%	17
	Heterorhabditis sp.	<65%	17
Pasture			
Popillia japonica	*S. carpocapsae* (All)	0%	18
	S. carpocapsae (Breton)	40%	18
	S. glaseri	80%	18
White grubs[b]	*Heterorhabditis* sp.	60%	19
	S. carpocapsae	43%	19
	H. bacteriophora	69%	19
Costelytra zealandica	*S. glaseri*	Infection established, but no population reduction	20
		30% (range 0-56%)	21
	H. bacteriophora	47% (range 0-92%)	21
Adoryphorus couloni	*S. glaseri*	Significant control	22
		35-52%	23
Sugarcane			
Alissonotum impressicolle	*S. glaseri*	>80% reduction	24
	Heterorhabditis sp.	<80%	24
Ligyrus subtropicus	*S. carpocapsae*	No significant control	25
	S. glaseri	No significant control	25

a   DAT = Days after treatment.
b   White grub complex — *Phyllophaga anxia, P. fusca,* and *Polyphyla comes.*

Historically, *P. japonica* has played a significant role in the development of entomopathogenic nematodes as biological insecticides.[9-11] The isolation of *Steinernema glaseri* from Japanese beetle larvae in New Jersey resulted in the first effort to use nematodes to control this pest. Initial field experiments were encouraging, and methods of mass production of the nematodes were developed. A large-scale colonization program was initiated in the 1930s in New Jersey and Maryland, but it did not enjoy the success of previous colonization efforts using the milky disease bacteria.[11] A lack of understanding of the complex biology of the nematode, particularly of its symbiotic relationship with the bacterium in the genus *Xenorhabdus*, hampered efforts to use *S. glaseri* effectively against Japanese beetle populations.[5] Although the work was conducted more than 50 years ago, optimal conditions for nematode establishment defined then — soil temperatures of >16°C, soil moisture of >20% without flooding, the presence of turf or permanent cover, and dense grub populations[11] — remain valid today.

A. Effect of Nematode Species or Strain

Considerable effort has gone into screening nematode species and strains for efficacy in killing white grubs. Table 2 shows some of the species and strains tested in the laboratory against the Japanese beetle and the northern masked chafer, *Cyclocephala borealis*. Overall, *Heterorhabditis* spp. have been more effective than *S. carpocapsae*, but often not more effective than *S. glaseri*. The superiority of the *Heterorhabditis* spp. is particularly apparent when examining this nematode's activity against *C. borealis* larvae in the laboratory. Pot tests in New York showed similar results when comparing the susceptibility of Japanese beetle and European chafer, *Rhizotrogus majalis*, larvae.[17] Field tests against scarab larvae in the U.S. have concentrated on *Heterorhabditis* species and *S. carpocapsae* because of the problems in maintaining *S. glaseri* cultures[19] and in producing sufficient material to test. Similarly, although *H. megidis* was isolated from Japanese beetle larvae[27] and was highly infective in laboratory studies (Table 2), it has not been possible to mass produce sufficient quantities for field tests.

Shetlar et al.[12] reported 73% control of Japanese beetle larvae with *H. bacteriophora* (=*heliothidis*) NC strain in Ohio turf compared with 55% for *S. carpocapsae*. *H. bacteriophora* NC and HP88 strains were also more efficacious than *S. carpocapsae*.[14] Georgis and Poinar[4] summarized the results of field tests conducted between 1984 and 1987, reporting that the HP88 strain of *H. bacteriophora* was more efficacious than the NC strain, and was equivalent to registered insecticides, for suppression of Japanese beetle and European chafer larvae. In addition, *H. bacteriophora* NC strain was superior to *S. carpocapsae*. *Heterorhabditis* species are probably superior against scarabs and other soil insects because they are active,[28] move down in the soil profile, and apparently can penetrate the soft intersegmental membrane of insects.[1,4] Since 1987, however, field tests with *H. bacteriophora* HP88 strain in the U.S. have been less successful. These negative results may reflect the influence of particularly adverse environmental factors or may be caused by a loss of virulence in the nematode-bacterial complex as a result of unfavorable mass production or storage conditions.

TABLE 2. Mortality of Japanese Beetle (JB) and Northern Masked Chafer (MC) Larvae by Various Steinernematid spp. and Heterorhabditid spp. in Laboratory Tests

Nematode species (strain)	% Mortality at indicated days after treatment						
	7a		14b	10c		10d	
	JB	MC	JB	JB	MC	JB	MC
S. carpocapsae(Agriotos)	61	33	64	-	-	-	-
S. carpocapsae(Agriotos)	61	33	64	-	-	-	-
S. carpocapsae(All)	48	50	55	80	0	20	0
S. carpocapsae(Breton)	88	50	52	-	-	-	-
S. carpocapsae(Breton x DD-136)	88	33	70	-	-	-	-
S. carpocapsae(HOP)	48	33	24	-	-	-	-
S. carpocapsae(Italian)	61	84	55	-	-	-	-
S. carpocapsae(Mexican)	75	50	24	90	0	40	0
S. carpocapsae(Pye)	-	-	24	-	-	-	-
S. carpocapsae(Ohio)	-	-	-	-	-	40	0
S. anomal	-	-	-	80	0	30	10
S. glaseri	-	-	-	90	40	70	0
H. bacteriophora(HP88)	75	67	82	-	-	-	-
H. bacteriophora(NC)	-	-	76	90	70	-	-
H. bacteriophora	-	-	55	-	-	-	-
H. megidis	-	-	-	100	70	50	0
Heterorhabditis sp.(SC)	-	-	-	0	0	10	10

a   Shetlar[13] 250 nematodes/vial.
b   Shetlar[13] 450 nematodes/vial.
c   Klein[26] 400 nematodes/vial, 25% moisture.
d   Klein[26] 400 nematodes/vial, 40% moisture.

The nematode of choice varies between countries. *S. glaseri* gave over 80% reduction of scarab larvae in sugarcane in China[24] but did not significantly reduce scarab larvae in Florida sugarcane fields.[25] *S. glaseri* has also been used for white grub control in Australian[22,23] and Azorean[18] pastures, but *H. bacteriophora* has been shown to be more effective in New Zealand.[21] These differences may be due to the availability of different strains, the environmental parameters found in each country, application techniques, susceptibility of the insect species, or the methodologies employed in rearing and storing the various nematodes.

B. Effect of Temperature and Moisture

Low soil temperatures are a limiting factor in using milky disease organisms[3] and also restrict nematode activity. Georgis and Gaugler[29] analyzed field trials with nematodes against white grubs and found that average efficacy was nearly 80% when soil temperatures were 21-30°C but was less than 40% when temperatures were 12-16°C. These results are in agreement with the 16°C threshold first noted in the early *S. glaseri* field trials. The nematodes and milky disease bacteria are most effective over the same temperature range.[11] This temperature limitation almost precludes nematode application in the northern U.S. during the spring when soil moisture is usually favorable for nematode survival, but soil temperatures are too low for infection. Thus, many applications are restricted to the fall when soil temperatures are above the threshold, but soil moisture must be supplemented by irrigation. Similar conditions exist in New Zealand where there is only a 1 to 2 month window when moisture and temperatures are adequate for application.[30] However, the recent isolation of a low temperature strain of *S. feltiae* (=*bibionis*)[30] may allow for a longer period of optimal temperature and moisture interactions for field applications. It has been used successfully against *Cyclocephala* larvae in California,[16] indicating that this strain, or other improved strains, may be useful against soil pests.

Soil moisture is often the most critical factor in the survival and movement of entomopathogenic nematodes. For example, there was ca. 50% reduction in mortality of Japanese beetle larvae in vials when the soil moisture was reduced from 40% to 25% (Table 2). Shetlar et al.[12] demonstrated that steinernematid and heterorhabditid nematodes required at least 0.64 cm of irrigation after application to turf to ensure establishment. Similarly, Jackson et al.[31] found that nematodes applied during rain or prior to irrigation were more likely to establish than those applied under dry conditions. The requirement for watering nematodes into the soil will not be new for most turf managers, because all registered turf insecticides require immediate irrigation or rain to wash them below the soil surface. Jackson et al.[31] also found that nematodes applied as a drench or with a subsurface injector provided superior control compared with those applied as a conventional spray. Because drenching requires unacceptably large quantities of water (10,000 l/ha), subsurface injection has the greater potential for insect control. More recently, Berg et al.[23] demonstrated that a commercially available pasture seeder could be modified to inject nematodes into pastures in Australia for grub control. They found that subsurface injection was as efficient as "watering-in" the nematodes while using 75% less water. Injection would solve several problems in the establishment of nematodes in turf. It would remove them from the exposed surface into the soil where they are protected from desiccation and UV inactivation. Moreover, this approach would place the nematodes beyond the barrier of thatch or organic matter and in the zone of grub activity.

C. Nematode Persistence

If nematodes are able to persist in the soil for long periods, they could be used in colonization programs similar to those for parasitoids and milky disease bacteria rather than being used for their immediate effect as a biological insecticide (e.g., *Bacillus thuringiensis*). In fact, some degree of persistence would enhance nematodes as agents for short-term control.

Fleming[11] reported that *S. glaseri* could maintain itself in the field for 14 years with a grub population of less than 54/m^2. Furthermore, it survived 24 years when the grub population was augmented periodically. Recently, Klein and Georgis[14] found that the NC and HP88 strains of *H. bacteriophora* caused significant mortality of Japanese beetle larvae in Ohio turf in the grub generations following applications (Table 3). The nematodes are probably not surviving in the free-living stage in the soil for such extended periods. Because the nematodes reproduce in larvae under field conditions (Figure 1), the nematodes are probably recycling in the white grubs or other hosts, or the cadavers are offering some degree of protection to the nematodes. Possibly the nematodes are persisting and dispersing by infection of adult beetles through the period when larvae are not available. Glaser and Farrell[10] noted that nematode-infected adults of the Japanese beetle were a factor in the persistence and natural movement of *S. glaseri*. In addition, Hatsukade[32] reported that over 90% of adult Japanese beetles became infected when held in soil infested with *S. carpocapsae*. Clearly, more information is needed about the role of adult insects in the persistence of entomopathogenic nematodes in the field.

III. WEEVILS

Many weevils (Coleoptera, Curculionidae) are pests of pastures, turf, and agricultural and horticultural crops throughout the world. They are often difficult to kill with chemical insecticides, but as a group tend to be highly susceptible to entomopathogenic nematodes (Table 4). Accordingly, they have been targeted as prime candidates for biological control by these nematodes.

A. Black Vine Weevil

Larvae of the black vine weevil, *Otiorhynchus sulcatus*, are major root pests of nursery and greenhouse plants worldwide.[36] Entomopathogenic nematodes have consistently controlled larvae in numerous tests (Table 4) and may serve as acceptable alternatives for the persistent chemical pesticides used in the past. Bedding and Miller[36] demonstrated that an isolate of *H. bacteriophora* provided up to 100% control of larvae in freshly potted yew, raspberries, and grapes, and over 87% control in cyclamens and strawberries in Australia. They noted the importance of watering the treated pots and the potential for improved activity with the use of more cold-tolerant strains of nematodes. In The Netherlands, Simons[33] also reported that a *Heterorhabditis* species provided up to 100% control in greenhouse tests. Variation in the effect against larvae

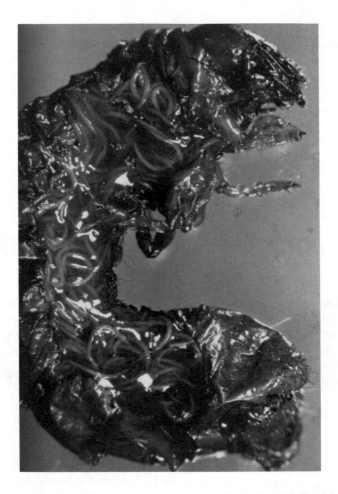

Figure 1. Reproduction of *Heterorhabditis megidis* in a field-collected infected larva of the northern masked chafer, *Cyclocephala borealis*. Host cuticle partially removed to expose nematodes.

infesting different species of plants indicated that soil moisture and structure and type of the plants had an impact on results. More recently, Simons and van der Schaaf[54] reported isolates of *Heterorhabditis* infecting black vine weevil larvae at 12°C. However, the process of infectivity and mortality of the larvae was retarded because the *Xenorhabdus* bacterium did not reproduce until temperatures reached 20°C. A cold-adapted strain of nematode-bacterial complex would be particularly useful in temperate regions where reduced soil temperatures have hindered field efficiency. Other research in The Netherlands demonstrated that *Heterorhabditis* sp. infected 70% of larvae feeding on roots of *Taxus* plants in the field.[34] This level of control may be sufficient to reduce the larval damage but is probably not satisfactory for nurseries because their plants

TABLE 3. Persistence of *Steinernema* and *Heterorhabditis* Nematodes Applied to Turf at 6.2×10^9/ha Against Japanese Beetle Larvae in Ohio

Nematode spp. (strain)	% Control of larvae			
	28 DAT[a]	Next generation[b]	28 DAT	Next generation[c]
S. carpocapsae (All)	45	0	-	-
H. bacteriophora (NC)	45	93	61	39
H. bacteriophora (HP88)	-	-	100	93

[a]   DAT = Days after treatment.
[b]   386 days after 4 September 1986 treatment.
[c]   138 days after 13 May 1987 treatment.

have a high value and a low economic threshold. Georgis and Poinar[4] summarized trials using *Steinernema* and *Heterorhabditis* species against black vine weevil larvae in potted plants in the greenhouse and outdoors and plants in field soil. This summary showed that moderate (50-80%) to high (>80%) control occurred when temperatures were above 16°C.

The black vine weevil and other root weevils are serious pests of cranberries in Massachusetts, Wisconsin, and Washington, U.S. Chemical insecticides are restricted or have not provided adequate control against the larval stage. A number of field trials demonstrated that nematodes are efficacious and may replace chemical insecticides as the preferred approach for suppression of these weevils. For example, the NC strain of *H. bacteriophora* applied to a Washington cranberry bog in April reduced larvae and pupae by >70% in the spring.[44] One year later control was still >70%. Moreover, *H. bacteriophora* HP88 strain and *S. carpocapsae* All strain reduced weevil larvae and pupae by 100% and >75%, respectively.

B. White-Fringed Beetles

White-fringed beetles, *Graphognathus* spp., are pests of field crops and pastures in South America, the southeastern U.S., Australia, and New Zealand.[47,48,55-58] In these areas entomopathogenic nematodes have been the primary biological control agents naturally associated with weevil larvae. Early work in the U.S. noted natural infections of up to 24% by species of *Steinernema*,[55] but attempts to colonize *S. glaseri* into weevil populations were not fruitful.[47] However, as with the Japanese beetle program, a lack of understanding about the nematode's biology and its symbiotic bacterial association resulted in an unsuccessful colonization effort. Young et al.[59] concluded that nematodes were

TABLE 4. Efficacy of *Steinernema* spp. and *Heterorhabditis* spp. against Various Weevil Larvae

Insect host	Nematode species	Control	Ref.
Greenhouse and Pot Tests			
Otiorhynchus sulcatus	*S. carpocapsae*	82-91%	38
	S. glaseri	82-91%	38
	H. bacteriophora	Up to 100% control in yews, raspberries, and grapes	36
		>87% in cyclamens and strawberries	36
		Better than diazinon	37
		82-91%	38
	Heterorhabditis sp.	100% in primula	33
		88-100% in strawberries	33
		50-88% in cyclamen	33
		Up to 100% in *Taxus*	34
		85-100% in azalea	35
		60-100% with other plants	35
Diaprepes abbreviatus	*S. carpocapsae*	48-92%	40
	S. glaseri	92-100% 8 DAT[a]	39
		12-34%	40
Cosmopolites sordidus	*S. carpocapsae*	88-100%	41
	S. feltiae	75-100%	41
	S. glaseri	100%	41
Field Tests			
Otiorhynchus sulcatus	*S. carpocapsae*	56%	44
	H. bacteriophora	93%	43
		56-70% in cranberries	44
	H. bacteriophora (HP88)	100%	44
	Heterorhabditis sp.	70% in *Taxus*	34
		Effective alone or with chemical insecticides	42

TABLE 4 (continued). Efficacy of *Steinernema* spp. and *Heterorhabditis* spp. against Various Weevil Larvae

Insect host	Nematode species	Control	Ref.
	Field Tests		
Otiorhynchus	*S. carpocapsae*	Not effective	37
ovatus		82-91%	38
	S. glaseri	82-91%	38
	H. bacteriophora	Not effective	37
		82-91%	38
	Heterorhabditis sp.	Not effective	37
Nemocestes	*S. carpocapsae* +	6-65%	45
incomptus	*S. glaseri*		
Diaprepes	*S. carpocapsae*	65%	40
abbreviatus		55-97%	46
	S. glaseri	35%	40
		70%	46
	H. bacteriophora	70-100%	46
Graphognathus	*S. carpocapsae*	38-50%	48
spp.			
	S. glaseri	Nematodes colonized in 5 states (U.S.)	47
Hylobius	*S. carpocapsae*	89%	49
abietis		66%	50
Hylobius	*S. carpocapsae*	Infection initiated	51
radicis			
Curculio	*S. carpocapsae*	67% 20 DAT	52
caryae			
Sphenophorus	*S. carpocapsae*	56-65% 14 DAT	53
parvulus			
	H. bacteriophora (HP88)	72-79%	53

[a]   DAT = Days after treatment.

an important factor in regulation of beetle populations only in the coastal flatlands of Mississippi where moisture, soil, and temperature conditions were particularly favorable to the nematodes. Subsequently, Harlan et al.[48] found that weevil populations could be reduced by 38-50% with application of *S. carpocapsae*, and that nematodes could be recovered 16 months after application. However, because of inconsistencies in efficacy, problems in mass-rearing nematodes, and the availability of persistent chemical insecticides, nematodes never became a tactic in white-fringed beetle control strategies.

Recent attempts to control beetle larvae in an isolated infestation in southern California by field applications of *H. bacteriophora*, *S. carpocapsae*, and *S. feltiae* (=*bibionis*) did not significantly reduce larval populations.[60] Fortunately, there was high mortality of larvae due to natural populations of *Steinernema* sp. and *Heterorhabditis* sp. A similar situation was noted in Australia. A natural level of *Heterorhabditis* and *Steinernema* infection occurred, but artificial introduction of additional nematodes did not significantly increase weevil mortality.[61] Perhaps these natural isolates from weevil populations may yet be useful in white-fringed beetle suppression programs. In spite of these results, *S. carpocapsae* and *H. bacteriophora* are being applied by the California Department of Food and Agriculture in attempts to suppress white-fringed weevil larvae. In addition, carbaryl is being used to suppress adult weevil populations. This approach is not for eradication but for containment of the weevil population within the infested area.

C. Citrus Weevils

In laboratory, greenhouse, and field tests on citrus in Florida, *S. carpocapsae*, *S. glaseri*, and *H. bacteriophora* provided significant control of a root weevil, *Diaprepes abbreviatus*.[39,40,46] Results have been encouraging with up to 100% mortality of the weevils in field plots treated with *S. carpocapsae*.[40] Nematodes can be added to irrigation systems, resulting in excellent protection of young seedling trees and reducing the emergence of *Diaprepes* adults by >90%. These findings have opened a potential major commercial market for *S. carpocapsae* in Florida and have encouraged research to increase efficacy through isolation of new nematode species or strains, genetic selection of nematodes, habitat management, and application techniques.[62]

D. Other Weevils

Entomopathogenic nematodes show potential for field control of several other weevils. Strains of *S. carpocapsae* are effective against a number of root weevils including *Hylobius*[49,50] and the pecan weevil, *Curculio caryae*.[52] However, because of a lack of production and application methods at the time of these tests, the nematodes were not adopted to control these insects.

Figueroa[41] indicated that *S. carpocapsae*, *S. glaseri*, and *S. feltiae* (=*bibionis*) provided excellent control of the banana root borer, *Cosmopolites sordidus*, in Puerto Rico. The tropical environment appears to be a logical situation for successfully using these nematodes. Thus, Jansson et al.[63] demonstrated that several entomopathogenic nematode species offer excellent potential for biological control of the sweetpotato weevil, *Cylas formicarius elegantulus*, in Florida and Puerto Rico. *S. carpocapsae* (G13) and *H. bacteriophora* (HP88) were efficacious in simulated field conditions. *H. bacteriophora* (HP88) was superior to *S. carpocapsae* (All) and chemical insecticides in protecting roots of the sweetpotato in the field.

Field tests in Ohio indicated that nematodes are equivalent to chemical insecticides for the control of the bluegrass billbug, *Sphenophorus parvulus*.[53] Georgis and Poinar[4] summarized field trials against billbugs from 1984 to 1987 and found that *H. bacteriophora* (HP88) averaged 77% control, whereas *S. carpocapsae* (All) averaged 57% and the chemical insecticides averaged 84%. These results indicated that nematodes can be considered an alternative to chemical insecticides for billbug control.

IV. TERMITES

Using entomopathogenic nematodes for control of termites in buildings has been a controversial topic. Poinar and Georgis[64] stated that pest control operators using *S. carpocapsae* against termites (*Reticulitermes* spp.) achieved a success rate of 80-87%. They emphasized the importance of high doses and adequate moisture but noted a lack of quantitative data on control that would allow an accurate evaluation of the prospects of using nematodes for property protection. Skepticism is needed because nematodes did not control termites in a simulation of soil treatments under concrete slabs or in logs.[65] Moreover, Epsky and Capinera[66] showed that large numbers of nematodes were required to kill termites ($LD_{50} = 1.5 \times 10^4$ nematodes/termite). Further, termites avoided areas with large numbers of nematodes and exploited gaps in coverage to reinfest a food source. On the other hand, success has been reported in protecting tea plants from termites in Sri Lanka.[67] Large numbers of nematodes were required, but once the termites were infected, a chain of infection led to the annihilation of a colony. However, Fujii[68] found that healthy Formosan subterranean termites (*Coptotermes formosanus*) walled off infected individuals, thus breaking any chain of infection.

Although it may be feasible to protect trees or crops from termites with nematodes, a different set of circumstances exists for protection of homes and buildings. The low economic threshold for termite damage to homes, the large colonies, the immense numbers of nematodes required to kill termites, the presence of secondary reproductives, and the behavior of healthy termites (finding gaps in nematode barriers or walling off nematode-infected individuals) preclude nematodes as a viable alternative for termite control.

V. MOLE CRICKETS

Mole crickets, *Scapteriscus* spp., are major pests of turf in the southeastern U.S. Burrowing through the top layers of soil, they uproot plants and leave ridges throughout the turf.[69] Georgis and Poinar[4] summarized field tests between 1984 and 1987 and reported that mole crickets, unlike white grubs, are more susceptible to infection by *S. carpocapsae* (average 62%) than *H. bacteriophora* (average 8%). This difference in susceptibility probably relates to the behavior of the nematodes. Infective juveniles of *Steinernema* remain closer to the soil surface where they are more likely to encounter the mole crickets.[1,4] Cobb and Georgis[69] found that two strains of *S. carpocapsae* gave 57-70% reduction of damage in the field compared with 53% for the standard chemical insecticide treatment. In addition, *S. carpocapsae* was superior to *S. feltiae* (=*bibionis*) and *H. bacteriophora*.

Nematodes are important natural biological control agents of mole crickets in South America. Fowler[70] collected 18 isolates of *S. carpocapsae* and 13 isolates of *Heterorhabditis* species which gave at least 50% mortality of mole crickets in the laboratory. Extensive investigations being conducted in Florida with *Steinernema scapterisci* from Uruguay show that 50-94% infection can be obtained under laboratory conditions. Moreover, adult mole crickets are more susceptible to infection than nymphs and serve as agents for nematode dispersal.[71] In field tests, nematodes were still recovered 36 months after release. From all appearances, nematodes are promising alternatives to standard insecticide treatments for mole cricket control.

VI. OTHER INSECTS

Entomopathogenic nematodes have shown their potential as biological control agents for many soil-inhabiting insect pests or even those that only come in contact with the soil during a brief period of their life cycle. Reviewing efforts to utilize nematodes against the red imported fire ant, Poinar and Georgis[4] reported that *S. carpocapsae* and *H. bacteriophora* infected this insect in the laboratory and field. Quattlebaum[72] obtained up to 100% mortality in the field with *S. carpocapsae*, whereas *H. bacteriophora* gave 89% mortality. If nematodes are to be used as biological control agents against fire ants, proper timing of application, the size of the ant mound, and climatic conditions are important factors to consider. Unfortunately, the low infection rate of workers and the formation of satellite mounds have discouraged their use as control agents.

Larvae of several fruit fly species are susceptible to entomopathogenic nematodes. *S. carpocapsae* caused mortality to larvae of the Mediterranean fruit fly, *Ceratitis capitata*, the melon fly, *Dacus cucurbitae*, and the oriental fruit fly, *D. dorsalis* at rates of 92, 85, and 86%, respectively.[73] Field tests showed that >85% mortality could be achieved with as few as 500 nematodes/cm^2 of soil.[74] Mass production of more efficacious strains coupled with persistence of the nematodes in the soil for 1 week or more may make nematodes viable tools for fruit fly eradication programs.[75]

S. carpocapsae, *S. glaseri*, and *H. bacteriophora* were effective in laboratory and field studies against the Colorado potato beetle, *Leptinotarsa decemlineata*.[76,77] However, these nematodes may not be used for this insect because of the recent commercial development of *Bacillus thuringiensis* var. *san diego*.

Chrysomelid larvae are among the most serious pests of corn, vegetables, and specialty crops. Rootworms (*Diabrotica* spp.) appeared to be a major potential market for entomopathogenic nematodes, but early tests (1968-1969) showed no benefit from applying *S. carpocapsae* DD-136 strain in corn field plots in Nebraska.[78] However, Poinar et al.[79] indicated that the Breton strain of *S. carpocapsae* was significantly better than an insecticide treatment for rootworm larvae in corn. Recently, in the laboratory, larvae of the western corn rootworm, *Diabrotica virgifera virgifera*, were shown to be most susceptible to the Mexican strain of *S. carpocapsae* and least susceptible to the All strain.[80] Yet *S. carpocapsae* All strain applied through center pivot irrigation in a corn field reduced rootworm larval numbers as much as the standard application of a registered chemical insecticide.[81] The reasons for the differences in response to *S. carpocapsae* All strain in the field and laboratory are unknown. Perhaps the quality of nematodes in the field tests was superior to that of the nematodes in the laboratory tests, or perhaps if the Mexican strain had been used, the efficacy would have been even better than the All strain.

Increasing difficulties in suppressing the banded cucumber beetle, *Diabrotica balteata*, has also focused attention on the possible use of nematodes. Although *S. carpocapsae* (DD-136) caused high mortality (76-94%) of this insect in the field, the nematodes did not adequately protect sweet potatoes against the total insect pest complex present in the soil.[82] Subsequent tests with *H. bacteriophora* (=*heliothidis*, SC strain) showed that this nematode was more efficacious against *D. balteata* (>95% control) and was more persistent in the soil than *S. carpocapsae*.[83] These results suggest that this nematode, in combination with insect resistant cultivars of sweet potatoes, may reduce insect damage to acceptable levels.[84]

S. carpocapsae or *H. bacteriophora* (SC) applied through a trickle irrigation system was used to control the striped cucumber beetle, a close relative of *Diabrotica*.[85] Control of this vegetable pest with nematodes may be advanced by combining the best nematodes with new and innovative application techniques.

In addition to the rootworms and their relatives, flea beetles are important chrysomelid pests of agricultural crops. Although *S. carpocapsae* (All) had no observable effect on the crucifer flea beetle, *Phyllotreta criciferae*, *H. bacteriophora*, and *S. carpocapsae* (All) gave 94% and 67% control of the mint flea beetle, *Longitarsus waterhousei*.[87] Because no chemical insecticides are registered for mint flea beetle larvae, and other members of the soil insect complex attacking mint can also be controlled with nematodes, it appears that rapid adoption of entomopathogenic nematodes will occur for this crop.

Kaya and Hara[88] examined the susceptibility of numerous lepidopterous pupae to entomogenous nematodes. Whereas almost all species were infected to some degree, those that pupated above ground were most susceptible. However, in nature, only those pupating in the litter or soil would be in a habitat suitable for nematode survival, movement, and subsequent infection. Of more practical interest, they found that prepupae of the beet armyworm, *Spodoptera exigua*, and the armyworm, *Pseudaletia unipuncta*, in the soil were highly susceptible to infection by *S. carpocapsae*.[89] Furthermore, *S. carpocapsae* could infect >90% of *S. exigua* prepupae in soil at concentrations as low as 5 nematodes/cm^2 of soil surface.[90] In field trials, *S. feltiae* (=*bibionis*), applied to soil significantly reduced larval populations of the turnip moth, *Agrotis seg-*

etum, compared with the control.[91] Some control was attained even under adverse conditions, defined as lack of rainfall, rise in air temperature, and increased solar radiation.

Adult beet armyworms were also susceptible to nematode infections during emergence from the soil.[90] More recently, Timper et al.[92] demonstrated that beet armyworm adults infected with *S. carpocapsae* dispersed up to 11 m from the site of infection, thus suggesting that infected adult insects may account, in part, for the wide distribution of entomopathogenic nematodes.

VII. CONCLUSIONS

The soil environment is the ideal location to take advantage of the interactions between entomopathogenic nematodes and some of the over 90% of insects which spend part of their life cycles in contact with the soil. Because soil is the natural reservoir for these nematodes, their potential for use as biological insecticides is great. In fact, commercial firms are selling entomopathogenic nematodes for control of black vine weevil larvae in nurseries throughout the world and in cranberry bogs in the U.S., and for citrus weevil control in Florida. Additionally, a number of turf pests are potential new markets for nematodes. In most cases, poor control by conventional chemical insecticides or the absence of available chemical controls has spurred the commercialization of nematodes. *Heterorhabditis* spp. or *S. glaseri* have often given superior efficacy when compared with *S. carpocapsae* strains for many pests, particularly the white grubs. Production and storage capabilities of the most efficacious nematode species and strains need to be developed so that consistent field results can be obtained. As new species and more motile and cold active strains are discovered through exploration and developed through genetic manipulation, they can be combined with advances in application techniques to provide reliable pest control to a broad range of consumers. As we move forward, a better understanding of the complex interactions in the soil ecosystem will enable us to maximize use of entomopathogenic nematodes for biological control.

REFERENCES

1. **Gaugler, R.,** Ecological considerations in the biological control of soil-inhabiting insects with entomopathogenic nematodes, *Agric. Ecosys. Environ.*, 24, 351, 1988.
2. **Villani, M. G., and Wright, R. J.,** Entomogenous nematodes as biological control agents of European chafer and Japanese beetle (Coleoptera: Scarabaeidae) larvae infesting turfgrass, *J. Econ. Entomol.*, 81, 487, 1988.
3. **Klein, M. G.,** Pest management of soil-inhabiting insects with microorganisms, *Agric. Ecosys. Environ.*, 24, 337, 1988.
4. **Georgis, R., and Poinar, G. O., Jr.,** Field effectiveness of entomophilic nematodes *Neoaplectana* and *Heterorhabditis*, in *Integrated Pest Management for Turfgrass and Ornamentals*, Leslie, A. R., and Metcalf, R. L., Eds., U.S. Environmental Protection Agency, Washington, D.C., 1989, 213.

5. **Poinar, G. O., Jr.**, *Nematodes for Biological Control of Insects*, CRC Press, Boca Raton, 1979.

6. **Poinar, G. O., Jr.**, Entomophagous nematodes, *in Biological Plant and Health Protection*, Franz, J. M., Ed., G. Fischer Verlag, Stuttgart, 1986, 95.

7. **Tashiro, H.**, *Turfgrass Insects of the U.S. and Canada*, Cornell University Press, Ithaca, NY, 1987.

8. **Ahmad, S., Streu, H. T., and Vasvary, L. M.**, The Japanese beetle: a major pest of turfgrass, *Am. Lawn Appl.*, 4(2), 2, 1983.

9. **Glaser, R. W.**, Studies on *Neoaplectana glaseri*, a nematode parasite of the Japanese beetle (*Popillia japonica*), *N. J. Dept. Agric. Cir.*, 211, 1932.

10. **Glaser, R. W., and Farrell, C. C.**, Field experiments with the Japanese beetle and its nematode parasite, *J. N. Y. Entomol. Soc.*, 43, 345, 1935.

11. **Fleming, W. E.**, Biological control of the Japanese beetle, *USDA Tech. Bull. No.* 1383, 1968.

12. **Shetlar, D. J., Suleman, P. E., and Georgis, R.**, Irrigation and use of entomogenous nematodes *Neoaplectana* spp. and *Heterorhabditis heliothidis* (Rhabditida: Steinernematidae and Heterorhabditidae) for control of Japanese beetle (Coleoptera: Scarabaeidae) grubs in turfgrass, *J. Econ. Entomol.*, 81, 1318, 1988.

13. **Shetlar, D. J.**, Entomogenous nematodes for control of turfgrass insects with notes on other biological control agents, in *Integrated Pest Management for Turfgrass and Ornamentals*, Leslie, A. R., and Metcalf, R. L., Eds., U. S. Environmental Protection Agency, Washington, D.C., 1989, 225.

14. **Klein, M. G., and Georgis, R.**, unpublished data, 1988.

15. **Klein, M. G., Moyseenko, J. J., and Georgis, R.**, unpublished data, 1987.

16. **Kaya, H. K., and Klein, M. G.**, unpublished data, 1989.

17. **Wright, R. J., Villani, M. G., and Agudelo-Silva, F.**, Steinernematid and heterorhabditid nematodes for control of larval European chafers and Japanese beetles (Coleoptera: Scarabaeidae) in potted yew, *J. Econ. Entomol.*, 81, 152, 1988.

18. **Simoes, N., and Laumond, C.**, unpublished data, 1989.

19. **Kard, B. M. R., Hain, F., and Brooks, W. M.**, Field suppression of three white grub species (Coleoptera: Scarabaeidae) by the entomogenous nematodes *Steinernema feltiae* and *Heterorhabditis heliothidis*, *J. Econ. Entomol.*, 81, 1033, 1988.

20. **Jackson, T. A., and Trought, T. E.**, Progress with the use of nematodes and bacteria for the control of grass grub, in *Proc. 35th N.Z. Weed Pest Control Conf.*, 103, 1982.

21. **Hoy, J. M.**, The use of bacteria and nematodes to control insects, *N.Z. Sci. Rev.*, 13, 56, 1955.

22. **Berg, G. N., Bedding, R. A., Williams, P., and Akhurst, R. J.**, Developments in the application of nematodes for the control of subterranean pasture pests, *Proc. 4th Aust. Appl. Entomol. Res. Conf.*, Adelaide, Bailey, P., and Swincer, D., Eds., 1984, 352.

23. **Berg, G. N., Williams, P., Bedding, R. A., and Akhurst, R. J.**, A commercial method of application of entomopathogenic nematodes to pasture for controlling subterranean insect pests, *Plant Prot. Quart.*, 2, 174, 1987.

24. **Wang, J. X., and Li, L. Y.**, Entomogenous nematode research in China, *Rev. Nématol.*, 10, 483, 1987.

25. **Sosa, O., Jr., and Hall, D. G.**, Mortality of *Ligyrus subtropicus* (Coleoptera: Scarabaeidae) by entomogenous nematodes in field and laboratory trials, *J. Econ. Entomol.* 82, 740, 1989.

26. **Klein, M. G.**, unpublished data, 1985.

27. **Poinar, G., Jr., Jackson, T., and Klein, M.**, *Heterorhabditis megidis* sp. n. (Heterorhabditidae: Rhabditida) parasitic in the Japanese beetle, *Popillia japonica* (Scarabaeidae: Coleoptera), in Ohio, *Proc. Helminthol. Soc. Wash.*, 54, 53, 1987.

28. **Choo, H. Y., Kaya, H. K., Burlando, T. M., and Gaugler, R.**, Entomopathogenic nematodes: host-finding ability in the presence of plant roots, *Environ. Entomol.*, 18, 1136, 1989.

29. **Georgis, R., and Gaugler, R.**, unpublished data, 1989.
30. **Wright, P. J., and Jackson, T. A.**, Low temperature activity and infectivity of a parasitic nematode against porina and grass grub larvae, *Proc. 41st N.Z. Weed Pest Control Conf.*, 138, 1988.
31. **Jackson, T. A., Todd, B. W., and Wouts, W. M.**, The effect of moisture and method of application on the establishment of the entomophagous nematode *Heterorhabditis bacteriophora* in pasture, *Proc. 36th N.Z. Weed Pest Control Conf.*, 195, 1983.
32. **Hatsukade, M.**, Application of entomogenous nematodes for control of some turfgrass insects, in *Recent Advances in Biological Control of Insect Pests by Entomogenous Nematodes in Japan*, Ishibashi, N., Ed., Ministry of Education, Japan, Grant No. 59860005, 1987, 81.
33. **Simons, W. R.**, Biological control of *Otiorhynchus sulcatus* with heterorhabditid nematodes in the glasshouse, *Neth. J. Plant Pathol.*, 87, 149, 1981.
34. **Dolmans, N. G. M.**, Biological control of the black vine weevil (*Otiorhynchus sulcatus*) with a nematode (*Heterorhabditis* sp.), *Meded. Fac. Landbouww. Rijksuniv. Gent*, 48, 417, 1983.
35. **Klingler, J.**, Einsatz und Wirksamkeit insektenparasitischer Nematoden gegen den Gerfurchten Dickmaulrussler, *Gartnermeister*, 89, 277, 1986.
36. **Bedding, R. A., and Miller, L. A.**, Use of a nematode, *Heterorhabditis heliothidis*, to control black vine weevil, *Otiorhynchus sulcatus*, in potted plants, *Ann. Appl. Biol.*, 99, 211, 1981.
37. **Rutherford, T. A., Trotter, D., and Webster, J. M.**, The potential of heterorhabditid nematodes as control agents of root weevils, *Can. Entomol.*, 119, 67, 1987.
38. **Georgis, R., Poinar, G. O., Jr., and Wilson, A. P.**, Susceptibility of strawberry root weevil *Otiorhynchus sulcatus* to neoaplectanid and heterorhabditid nematodes, *IRCS Med. Sci.*, 10, 42, 1982.
39. **Beavers, J. B.**, Susceptibility of *Diaprepes abbreviatus* to the parasitic nematode (*Steinernema glaseri*), *IRCS Med. Sci.*, 12, 480, 1984.
40. **Schroeder, W. J.**, Laboratory bioassays and field trials of entomogenous nematodes for control of *Diaprepes abbreviatus* (Coleoptera: Curculionidae) in citrus, *Environ. Entomol.*, 16, 987, 1989.
41. **Figueroa, W.**, unpublished data, 1989.
42. **Verbruggen, D., De Grisse, A., and Heungens, A.**, Mogelijkheden voor de geintegreerde bestrijding van *Otiorhynchus sulcatus* (F.), *Meded. Fac. Landbouww. Rijksuniv. Gent*, 50, 133, 1985.
43. **Barratt, B. I. P., Ferguson, C. M., and Jackson, T. A.**, Control of black vine weevil (*Otiorhynchus sulcatus* [F.]) larvae with parasitic nematodes and fungal pathogens, *Proc. 42nd N.Z. Weed Pest Control Conf.*, 1989, in press.
44. **Shanks, C. H., Jr., and Agudelo-Silva, F.**, Field pathogenicity and persistence of heterorhabditid and steinernematid nematodes (Nematoda) infecting black vine weevil larvae (Coleoptera: Curculionidae) in cranberry bogs, *J. Econ. Entomol.*, 83, 107, 1990.
45. **Georgis, R., and Poinar, G. O., Jr.**, Field control of the strawberry root weevil, *Nemocestes incomptus* by neoaplectanid nematodes (Steinernematidae: Nematoda), *J. Invertebr. Pathol.* 43, 130, 1984.
46. **Schroeder, W. J.**, Control of *Diaprepes abbreviatus* (Coleoptera: Curculionidae) in citrus with entomogenous nematodes, *Environ. Entomol.*, 1990, in press.
47. **Swain, R. B., Littig, K. S., Gordon, M. F., and Saul, I. A.**, Studies of the nematodes and bacterial diseases found associated with the white fringed beetle, *White-fringed Beetle Investigations*, 2nd Quart. Rep., Gulfport, 1943.
48. **Harlan, D. P., Dutky, S. R., Padgett, G. R., Mitchell, J. A., Shaw, Z. A., and Bartlett, F. J.**, Parasitism of *Neoaplectana dutkyi* in white-fringed beetle larvae, *J. Nematol.*, 3, 280, 1971.

49. **Pye, A., and Pye, N.**, Different applications of the insect parasitic nematode *Neoaplectana carpocapsae* to control the large pine weevil, *Hylobius abietis*, *Nematologica*, 31, 109, 1985.

50. **Burman, M., Pye, A. E., and Nojd, N. O.**, Preliminary field trials of the nematode *Neoaplectana carpocapsae* against larvae of the large pine weevil, *Hylobius abietis* (Coleoptera, Curculionidae), *Ann. Entomol. Fennici*, 45, 88, 1979.

51. **Schmiege, D. C.**, The biology and host-parasite relationship of a neoaplectanid nematode parasitic on some forest insect pests, Ph.D. thesis, University of Minnesota, St. Paul, 1962.

52. **Tedders, W. L., Weaver, D. J., and Wehunt, E. J.**, Pecan weevil: suppression of larvae with the fungi *Metarhizium anisopliae* and *Beauveria bassiana* and the nematode *Neoaplectana dutkyi*, *J. Econ. Entomol.*, 66, 723, 1973.

53. **Niemczyk, H. D., Klein, M. G., and Power, K. T.**, unpublished data, 1988.

54. **Simons, W. R., and van der Schaaf, D. A.**, Infectivity of three *Heterorhabditis* isolates for *Otiorhynchus sulcatus* at different temperatures, in *Fundamental and Applied Aspects of Invertebrate Pathology*, Samson, R. A., Vlak, J. V., and Peters, D., Eds., Proc. 4th Int. Coll. Invertebr. Pathol., Wageningen, 1986, 285.

55. **Swain, R. B.**, Nematode parasites of the white-fringed beetles, *J. Econ. Entomol.*, 36, 671, 1943.

56. **Ahmad, R.**, Studies on *Graphognathus leucoloma* (Boh.) Col: Curculionidae and its natural enemies in the central provinces of Argentina, *Tech. Bull. Commonwealth Inst. Biol. Control*, 17, 19, 1974.

57. **Sexton, S. B., and Williams, P.**, A natural occurrence of parasitism of *Graphognathus leucoloma* (Bohheman) by the nematode *Heterorhabditis* sp., *J. Aust. Entomol. Soc.*, 20, 253, 1981.

58. **Wouts, W. M.**, Biology, life cycle and redescription of *Neoaplectana bibionis* Bovien, 1937 (Nematoda: Steinernematidae), *J. Nematol.*, 12, 62, 1980.

59. **Young, H. C., App, B. A., Gill, J. B., and Hollingsworth, H. S.**, White-fringed beetles and how to combat them, *USDA Circ. 850*, 1950.

60. **Klein, M. G., Kaya, H. K., and Paine, T.**, unpublished data, 1989.

61. **Akhurst, R. J.**, personal communication, 1989.

62. **Schroeder, W. J.**, personal communication, 1989.

63. **Jansson, R. K., Lecrone, S. H., Gaugler, R., and Smart, G. C.**, Potential of entomopathogenic nematodes as biological control agents of sweetpotato weevil (Coleoptera: Curculionidae), *J. Econ. Entomol.*, 1990, in press.

64. **Poinar, G. O., Jr., and Georgis, R.**, Biological control of social insects with nematodes, in *Integrated Pest Management for Turfgrass and Ornamentals*, Leslie, A. R., and Metcalf, R. L., Eds., U. S. Environmental Protection Agency, Washington, D.C., 1989, 255.

65. **Mauldin, J. K., and Beal, R. H.**, Entomogenous nematodes for control of subterranean termites, *Reticulitermes* spp. (Isoptera: Rhinotermitidae), *J. Econ. Entomol.*, 82, 1638, 1989.

66. **Epsky, N., and Capinera, J. L.**, Efficacy of the entomogenous nematode *Steinernema feltiae* against a subterranean termite, *Reticulitermes tibialis* (Isoptera: Rhinotermitidae), *J. Econ. Entomol.*, 81, 1313, 1988.

67. **Danthanarayana, W., and Vitarana, S.**, Control of the live-wood tea termite *Glyptotermes dilatatus* using *Heterorhabditis* sp. (Nemat.), *Agric. Ecosys. Environ.*, 19, 333, 1987.

68. **Fujii, J. K.**, Effects of an entomogenous nematode *Neoaplectana carpocapsae* Weiser, on the Formosan subterranean termite, *Coptotermes formosanus*, Ph.D. thesis, University of Hawaii, Honolulu, 1975.

69. **Cobb, P., and Georgis, R.**, unpublished data, 1989.

70. **Fowler, H. G.**, Occurrence and infectivity of entomogenous nematodes in mole crickets in Brazil, *Int. Rice Res. Newsl.*, 13, 34, 1988.

71. **Frank, J. H., Castner, J. L., and Hudson, W. G.,** Update on biological control of *Scapteriscus* mole crickets, *Florida Turf Digest*, 5, 1988.
72. **Quattlebaum, E. C.,** Evaluation of fungal and nematode pathogens to control the red imported fire ant, *Solenopsis invicta* Buren, Ph.D. thesis, Clemson University, Clemson, SC, 1980.
73. **Lindegren, J. E., and Vail, P. V.,** Susceptibility of Mediterranean fruit fly, melon fly, and oriental fruit fly (Diptera: Tephritidae) to the entomogenous nematode *Steinernema feltiae* in laboratory tests, *Environ. Entomol.*, 465, 1986.
74. **Lindegren, J. E., Wong, T. T., and McInnis, D. O.,** Response of Mediterranean fruit fly (Diptera: Tephritidae) to the entomogenous nematode *Steinernema feltiae* in field tests (Hawaii, 1983 and 1984), *Environ. Entomol.*, 1990, in press.
75. **Lindegren, J. E.,** unpublished data, 1988.
76. **Toba, H. H., Lindegren, J. E., Turner, J. E., and Vail, P. V.,** Susceptibility of the Colorado potato beetle and the sugarbeet wireworm to *Steinernema feltiae* and *S. glaseri*, *J. Nematol.*, 15, 597, 1983.
77. **Wright, R. J., Agudelo-Silva, F., and Georgis, R.,** Soil application of steinernematid and heterorhabditid nematodes for control of Colorado potato beetles, *Leptinotarsa decemlineata* (Say), *J. Nematol.*, 19, 201, 1987.
78. **Munson, J. D., and Helms, T. J.,** Field evaluation of a nematode (DD-136) for control of corn rootworm larvae, *Proc. N. Cent. Br. Entomol. Soc. Am.*, 25, 97, 1970.
79. **Poinar, G. O., Jr., Evans, J. S., and Schuster, E.,** Field tests of the entomogenous nematode, *Neoaplectana carpocapsae*, for control of corn rootworm larvae (*Diabrotica* sp., Coleoptera), *Prot. Ecol.* 5, 337, 1983.
80. **Jackson, J. J., and Brooks, M. A.,** Susceptibility of immune response to western corn rootworm larvae (Coleoptera: Chrysomelidae) to the entomogenous nematode, *Steinernema feltiae* (Rhabditida: Steinernematidae), *J. Econ. Entomol.*, 82, 1073, 1989.
81. **Georgis, R.,** personal communication, 1990.
82. **Creighton, C. S., Cuthbert, F. P., Jr., and Reid, W. J., Jr.,** Susceptibility of certain coleopterous larvae to the DD-136 nematode, *J. Invertebr. Pathol.*, 10, 368, 1968.
83. **Creighton, G. S., and Fassuliotis, G.,** *Heterorhabditis* sp. (Nematoda: Heterorhabditidae): a nematode parasite isolated from the banded cucumber beetle *Diabrotica balteata*, *J. Nematol.*, 17, 150, 1985.
84. **Schalk, J. M., and Creighton, C. S.,** Influence of sweet potato cultivars in combination with a biological control agent (Nematoda: *Heterorhabditis heliothidis*) on larval development of the banded cucumber beetle (Coleoptera: Chrysomelidae), *Environ. Entomol.*, 18, 897, 1989.
85. **Reed, D. K., Reed, G. L., and Creighton, C. S.,** Introduction of entomogenous nematodes into trickle irrigation systems to control striped cucumber beetle (Coleoptera: Chrysomelidae), *J. Econ. Entomol.*, 79, 1130, 1989.
86. **Morris, O. M.,** Evaluation of the nematode, *Steinernema feltiae* Filipjev, for the control of the crucifer flea beetle, *Phyllotreta cruciferae* (Goeze) (Coleoptera: Chrysmelidae), *Can. Entomol.*, 119, 95, 1987.
87. **Morris, M., Agudelo-Silva, F., and Berry, R.,** unpublished data, 1989.
88. **Kaya, H. K., and Hara, A. H.,** Susceptibility of various species of lepidopterous pupae to the entomogenous nematode *Neoaplectana carpocapsae*, *J. Nematol.*, 13, 291, 1981.
89. **Kaya, H. K., and Hara, A. H.,** Differential susceptibility of lepidopterous pupae to infection by the nematode *Neoaplectana carpocapsae*, *J. Invertebr. Pathol.*, 36, 389, 1980.
90. **Kaya, H. K., and Grieve, B. J.,** The nematode *Neoaplectana carpocapsae* and the beet armyworm *Spodoptera exigua*: infectivity of prepupae and pupae in soil and of adults during emergence from soil, *J. Invertebr. Pathol.*, 39, 192, 1982.
91. **Theunissen, J., and Fransen, J. J.,** Biological control of cutworms in lettuce by *Neoaplectana bibionis*, *Meded. Fac. Landbouww. Rijksuniv. Gent*, 49, 771, 1984.
92. **Timper, P., Kaya, H. K., and Gaugler, R.,** Dispersal of the entomogenous nematode *Steinernema feltiae* (Rhabditida: Steinernematidae) by infected adult insects, *Environ. Entomol.*, 17, 546, 1988.

11.   Efficacy Against Insects in Habitats Other than Soil

Joe W. Begley

I. INTRODUCTION

Entomopathogenic nematodes in the families Steinernematidae and Heterorhabditidae offer an alternative to chemical insecticides for a number of insect pests. They have been used against diverse pests, including those found in the soil, in cryptic habitats, on foliage, in manure, and in aquatic habitats. Greater successes have been recorded against insects in the first two habitats than the latter three. For most insects, initial susceptibility tests have been conducted in petri dishes where high levels of mortality are typically recorded. This high mortality occurs because conditions are favorable for nematode survival and infectivity, and ecological barriers to nematode infection are absent. When taken to the field, however, nematode applications have often yielded inconsistent results. Although several reasons have been cited to explain these inconsistencies, placement of nematodes in a foreign environment, soil being their natural habitat, has frequently had a deleterious effect on their ability to produce adequate control.

Because a number of reviews have been published in recent years,[1-7] especially against soil insects,[1-4] I will examine the recent efficacy data, supplemented with some of the older literature, on the use of entomopathogenic nematodes against insects occurring on or in foliar, cryptic, manure, and aquatic habitats.

II. FOLIAR HABITAT

Field trials using nematodes as biological insecticides against foliage-feeding insects, including the nonfeeding prepupal and pupal stages, have primarily involved *Steinernema carpocapsae* and, to a limited extent, *S. feltiae* (=*S. bibionis*) and *Heterorhabditis bacteriophora* (=*H. heliothidis*). With a few notable exceptions, many problems have been encountered in using nematodes on foliage. Nematodes require adequate moisture for infectivity so that high ambient humidity (>90%) and free water on the leaves are important prerequisites for infection; however, the foliar environment often exposes the nematodes to unfavorable moisture conditions that result in their rapid desiccation and death.[8-10] Moreover, high temperatures[10] and sunlight[11] are fatal to nematodes exposed on the leaf surface. Accordingly, nematodes applied to foliage must be protected from these detrimental environmental effects. The tropics, regions with monsoon seasons, and glasshouses (i.e., areas or situations with high humidities and moderate temperatures) are preferred environments for foliar application of nematodes. In addition, evening applications, the incorporation of antidesiccants into aqueous nematode suspensions, oil formulations, or anhydrobiotic nematodes, offer possibilities for increasing efficacy against foliar insects.

A. Lepidoptera

In a few cases, entomopathogenic nematodes have successfully reduced foliar insect populations to an acceptable level. For example, application of *S. carpocapsae* (Mexican strain) on cherry trees effectively controlled third-instar fall webworm, *Hyphantria cunea*.[12] In three field trials, nematodes applied at 3000 nematodes/milliliter under conditions of high humidity and moderate or low temperatures persisted on the foliage between 11 and 40 hr and caused 43 to 100% mortality. Because fall webworm larvae are gregarious and form silken webs at the feeding site, the webs may provide some protection to the nematodes by slowing the rate of nematode desiccation and reducing exposure to solar radiation. In other examples with *S. carpocapsae*, wet foliage for 24 hr in the field was a key to infecting >80% of the armyworm, *Mythimna separata*, and high humidity for the 15 hr following evening applications and the addition of a "viscous material" enhanced mortality of the imported cabbageworm, *Pieris rapae*.[13]

Successful application of *S. carpocapsae* against the beet armyworm, *Spodoptera exigua,* has been attained in a commercial nursery producing chrysanthemum stock in Florida. The susceptibility of the beet armyworm to entomopathogenic nematodes has been well documented.[14-17] Although targeting the prepupal or pupal stage in the soil eliminates nematode desiccation problems and increases success of infection, this tactic does not prevent larval feeding damage to the leaves. Thus, an approach designed to control the larval stage on foliage was initiated. Under optimal laboratory conditions, armyworm larvae were exposed to *S. carpocapsae* for different lengths of time to determine the interval required to obtain high armyworm mortality. These exposures, conducted in petri dishes with high humidity and free water, showed that 1 hr resulted in 52% mortality at 1 day and 80% mortality 4 days after treatment.[17] Four hours exposure resulted in 97% mortality after 4 days.

Grown under shade-cloth covered fields, chrysanthemum stock plants at a Florida nursery are often wet each morning from dewfall. Along with the dense canopy, this moist situation creates favorable conditions for nematode survival and infectivity. Thus, application at dawn resulted in 52-84% survival of the infective juveniles after 3 hr (Table 1).[17] More significantly, field trials showed that a tolerable rate of damaged chrysanthemum cuttings could be obtained with a single application of nematodes applied as a foliar spray. Nematode-killed beet armyworm larvae were recovered from foliage in large numbers 72 hr posttreatment. Although the nematode-treated plots had 8% damage to the cuttings compared with 30% in the control plots at 10 days posttreatment, no live larvae were found in the treated plots.

These efficacious results, which were equal to chemical insecticides,[18] led to the development of a beet armyworm control program with *S. carpocapsae* and eliminated the use of three conventional chemical insecticides. The critical points in the program were as follows. (1) Applications were timed to ensure a minimum of 4 hr contact between insect and nematode. Both the time of day and wind had to be considered because of their affect on humidity and thus nematode survival. (2) The volume of water in applications was adjusted to provide good coverage and avoid significant runoff loss of nematodes. (3) The nematode application rate was calibrated to maintain an effective dose after the expected 50% nematode mortality during a 3 hr drying time.[19] The incorporation of an evaporetardant that enhances nematode survival and efficacy is an economic consideration that may allow use of a lower nematode concentration.

Field efficacy of nematodes against the beet armyworm can be affected by certain insecticides used on chrysanthemums for other pests. Therefore, the impact on *S. carpocapsae* by chemical insecticide residues on foliage was assessed (Table 2).[17] Abamectin residues one day posttreatment adversely affected nematode survival, whereas, endosulfan and chlorpyrifos appeared to have minimal effect, especially during the first two days posttreatment. Clearly, nematodes can be used in conjunction with a conventional insecticide program, but information about the impact of insecticide residues on the plants will be needed for timing their application. For the interim, nematodes can probably be best integrated into a control program with systemic and short residual insecticides.

TABLE 1. Survival of Infective Juvenile *Steinernema carpocapsae* Applied to Foliage at Three Dosages[a,b]

Time on foliage (hr)	Percent nematode survival at		
	5.0×10^9/ha	2.5×10^9/ha	1.25×10^9/ha
1	86.1a	66.5a	74.2a
2	62.0a	53.4a	57.2a
3	83.8a	48.5ab	52.0a
4	40.4b	37.0b	30.3a
5	16.2c	10.3c	11.9b

[a] Application time was 0600 hours (EST).
[b] Based on average recovery of live nematodes per chryanthemum cutting at 0.5 hr after application. Data transformed to arcsine before analysis. Means in a column followed by the same letter are not significantly different (P=0.05) according to Newman-Keuls multiple range test.

TABLE 2. Effect of Foliar Residues of Pesticides on Survival of Infective Juvenile *Steinernema carpocapsae*[a,b]

| | Percent nematode survival | | |
| | Days between insecticide and nematode applications | | |
Pesticide treatment	1	2	3
Chlorpyrifos	17.2b	22.2b	14.1b
Endosulfan	23.8a	18.7b	6.3c
Abamectin	4.3c	11.3b	14.7b
Control	16.1b	17.7b	5.8a

[a]   Means in a column followed by the same letter are not significantly different (P = 0.05) according to Duncan's new multiple range test.

[b]   Application rate of 2500 million nematodes/hectare.

In most cases, foliar applications of nematodes have not reduced insect populations to an economically acceptable level. A few examples are documented here to illustrate reasons for the failures. When treated with *S. carpocapsae,* corn plants infested with the European corn borer, *Ostrinia nubilalis,* had a lower damage index compared with untreated plots.[20] However, in another study, damage to the whorl of the corn plant was not decreased by the nematode treatment.[21] The nematodes settled in water accumulated at the bottom of the whorl, which essentially removed them from the larval feeding area. In this case, the presence of free water was detrimental in obtaining efficacious results.

The DD-136 strain of *S. carpocapsae* was tested against five different foliage-feeding pests of apple in the laboratory and field.[22] This nematode caused high mortality in the laboratory, but field application of the nematode against a dense population of the winter moth, *Operophtera brumata*, did not result in larval suppression. The discrepancy between laboratory and field performance was attributed to rapid nematode desiccation and to the application method. A contributing factor may have been the low nematode concentration. Because 5×10^3 infective juveniles/milliliter provided 95% larval mortality in the laboratory, nematodes were applied at the rate of 2.5×10^3 infective juveniles/milliliter in the field, a concentration that was probably too low for effective control.

Laboratory studies indicated that *S. carpocapsae* (All)[23] and *Heterorhabditis bacteriophora* (NC)[24] caused high mortality against the western spruce budworm, *Choristoneura occidentalis*, and the spruce budworm, *C.*

fumiferana, respectively. With western spruce budworm field tests, despite rates of up to 8000 infective juveniles/ml and the addition of 2% Volck oil to the aqueous nematode suspension as an antidesiccant, nematodes did not reduce larval populations.[23] The addition of oils and antidesiccants did increase survival of the nematodes on pine foliage.[25] The reasons for the failure of *S. carpocapsae* were not clear, although the low temperature (15°C) the night after application, the poor nematode persistence (16 hr or less), and rain washing nematodes from the foliage after the application probably played significant roles.[23]

S. carpocapsae and *S. feltiae* (=*bibionis*) caused >90% mortality of the gypsy moth, *Lymantria dispar*, in the laboratory.[26] To take advantage of the migratory behavior of the older instars of this insect, which seek resting places under bark flaps and crevices during the day, nylon pack cloth bands lined with Pellon fleece or terrycloth were placed on tree boles and treated with nematodes. Larvae migrated under the bands and were infected. Especially high mortality was observed for those larvae that moved under the bands shortly after treatment. Increasing nematode survival within the bands should enhance infection among the gypsy moth larvae. Because a small proportion of the larval population occur under the bands at any given time, this approach may be feasible only in conjunction with other control tactics.

B. Hymenoptera

Many sawfly species are serious forest defoliators and are susceptible to nematode infection under laboratory conditions.[27,28] In field tests, the larch sawfly, *Pristiphora erichsonii,* was not suppressed by applications of *S. carpocapsae* (DD-136) to young larvae on foliage or fifth instar larvae entering sphagnum on the ground to form cocoons.[27] *S. carpocapsae* (UK) applied to control the web-spinning larch sawfly, *Cephalcia lariciphila,* caused ca. 29% infection of larvae on the trees and 61% infection of prepupae entering the soil.[29] The nematodes persisted in the soil and were recovered from prepupae 1 year later. In Czechoslovakia, *S. kraussei* is an important natural mortality agent of *Cephalcia* sawfly prepupae and pupae in soil and may be useful for biological control of other sawfly species.[30]

C. Coleoptera

Entomopathogenic nematodes do not appear to be feasible biological control agents for coleopterous foliage-feeding insects such as the elm leaf beetle, *Pyrrhalta luteola,*[23] and Colorado potato beetle, *Leptinotarsa decemlineata.*[31,32] Although the addition of antidesiccants to the nematode suspension increased mortality of Colorado potato beetle larvae feeding on foliage to 30-60% as compared with 10% without antidesiccants, the increased mortality was not sufficient to provide economic control.[31] *S. carpocapsae* applied to the soil to suppress the fourth-stage larvae produced 59-71% mortality.[33] These data suggest that an integrated approach employing a chemical or biological insecticide (i.e., *Bacillus thuringiensis* var. *sandiego*) for the foliar-feeding stages and nematodes for the soil-inhabiting stages (i.e., fourth-stage larva and pupa) may offer successful population reduction.

D. Other Insects

Controlling grasshoppers with *S. carpocapsae* has been attempted in the field. Deployment of alginate or alfalfa-wheat pellets containing nematodes resulted in 41 and 33% mortality, respectively, 3 days posttreatment.[34] This level of suppression is unsatisfactory in most situations.

The squash bug, *Anasa tristis,* is primarily diurnal in habit, hiding around the base of squash plants or the upper leaves or vines. Field application of *S. carpocapsae* against this insect caused 24-71% infection.[35] Although the results appeared encouraging, Wu[35] concluded that the low infection rate precludes the use of this nematode in Fresno, California, an area characterized by hot, dry summers. Generally, the same limitations which inhibit the nematodes against other foliage-feeding insects prevail. Interestingly, this is one of the few reports of nematodes infecting insects with sucking mouthparts.

III. CRYPTIC HABITATS

Although sensitivity to low moisture, high temperatures, and ultraviolet radiation has limited nematode use against foliage-feeding insects, cryptic habitats are generally characterized by conditions more favorable for nematode survival and infectivity. In fact, the most consistent, efficacious results with nematodes have been obtained in cryptic habitats, especially against insects that bore into plants.

A. Lepidoptera

A number of insect pests spend a portion of their life cycle within a host plant, in particular lepidopterous insects in the families Cossidae (carpenterworms) and Sesiidae (clearwing moths). In addition to being in a favorable habitat, most species have large gallery openings which are useful sites for nematode entry. These large gallery openings occur because the borers frequently clear their galleries of frass. A decrease in frass activity can actually be used as an index of successful control. Extensive studies with cossids and sesiids have shown that they can be effectively controlled by *S. carpocapsae* or *S. feltiae (=bibionis).*[36-46] In China, *S. carpocapsae* has been effective in controlling the borer, *Holcocercus insularis*, which has interconnecting galleries containing several hundred individuals.[45] When infected, many borers leave the galleries and die outside the tree. Those remaining in the galleries die and produce infective juveniles which can infect the borers that escaped earlier infection.

A major problem in controlling borers has been finding an economically feasible method of nematode application. Locating the gallery openings of these insects in trees can be time consuming and expensive. Often, the galleries are scattered throughout the tree, and for large trees, special equipment may be needed to treat the upper bole and branches. Other factors affecting efficacy include the small gallery openings of the early instars,[38] the dry galleries of some species that only bore into the bark,[44] and gums and resins which cover and restrict the size of the gallery opening. In addition, treatment is often initiated only on the larger insects, after the damage has been done, because the young instars go undetected.

Besides cossids and sesiids, entomopathogenic nematodes have been effective in controlling other lepidopterous species. The larvae of the navel orangeworm, *Amyelois transitella*, a serious pest on almonds, enter through splits in the husk and destroy the nut meat. The protected environment within the almond nut interior is ideal for both the development of the navel orangeworm larvae and the survival of entomopathogenic nematodes.[47] Field trials with *S. carpocapsae* have demonstrated that the level of suppression is rate dependent.[48] At levels of ca. 500 infective juveniles per nut, the mortality rate was 100%. However, depending on almond variety, hull split occurs over a period of a few days to several weeks, thus necessitating frequent nematode applications to attain effective control. Presently, the high nematode rates needed to obtain economic control limit their use as an acceptable control agent for this insect.

Codling moth, *Cydia pomonella*, is one of the most important pests of apples. They pupate and overwinter as prepupae under the bark of their host trees. When corrugated cardboard bands were placed around the trunks of apple trees as an artificial bark substrate, prepupae moved into them.[49] Subsequent applications of *S. carpocapsae* (All) resulted in 23-73% infection, with the best mortality occurring in the winter. Although producing insufficient control on their own, nematodes could be used as one element of an integrated management program for codling moth.

Field tests with the artichoke plume moth, *Platyptilia carduidactyla*, have shown effective control of third and fourth instars 15 days after treatment with *S. carpocapsae*.[50] The residual effect of nematode treatment was greater than that of the standard chemical pesticide. This insect bores into the leaf stalk and flower bud of artichokes, which occur in a cool, foggy climate of coastal California. Both the cryptic habitat and the humid climate make plume moth a prime candidate for control with entomopathogenic nematodes. The nematodes are used against the plume moth larvae when plants are young because relatively few nematodes are needed to maintain the insect population below the economic threshold of 5% infestation. On large, older plants, the numbers of nematodes needed for economic control is not feasible.

B. Coleoptera

Many pestiferous beetles, especially scolytids and cerambycids, occur in galleries in trunks or branches of trees. It is not surprising, therefore, that attempts have been made to control these insects with entomopathogenic nematodes. Bark beetles are susceptible to steinernematids[51-55] and heterorhabditids.[55] Applications of *S. carpocapsae* to logs infested with bark beetles demonstrated that the nematodes can enter the galleries and infect beetle larvae and adults.[51,53,54] However, entomopathogenic nematodes do not appear promising as biological insecticides for bark beetles. These insects usually infest dead or dying forest trees which have low or no economic value. When high value trees are attacked, the trees are probably already under stress, and mass attack by bark beetles requires a control agent that can act quickly to ward off tree death. Even with fast and effective beetle control, the trees may not be saved.

A research area which may prove fruitful is the initiation of epizootics in trees infested with bark beetles. Finney and Walker[53] noted that control logs which were not sprayed with nematodes contained beetles infected with *S. carpocapsae*. This observation implied that adult beetles served as carriers from treated to untreated logs. Epizootics may reduce beetle populations below outbreak levels. One caveat is that bark beetle galleries contain nematophagous mites,[56] and probably other organisms, that may shorten or prevent an epizootic of nematode disease.

S. carpocapsae is a promising control agent for the cerambycid, *Monochamus alternatus*, a vector of the pine wilt nematode in Japan.[57] Applications made by spraying the pine logs with nematodes or injecting them into galleries under field conditions provided 50-80% mortality of larvae and adults. In comparison, *S. glaseri, S. feltiae* (=*bibionis*), and *H. bacteriophora* (=*heliothidis*) resulted only in 46, 46, and 7% mortality, respectively.[57] *S. kushidai* was ineffective against this insect.[58]

C. Other Insects

The leafminers, *Liriomyza* spp., are major pests of ornamentals and vegetable crops in the tropics, subtropics, and greenhouses. The larval stages of these insects actively feed within the leaf tissue (i.e., mines) where they are protected from most insecticides. Laboratory[59] and greenhouse studies[60] with *S. carpocapsae* against *L. trifolii* suggest that this nematode may be an effective control agent. In the greenhouse, infective juveniles applied in an aqueous suspension located and infected the maggots within the mines. Apparently, the nematodes enter the mines through small tears on the epidermal leaf surface or ovipositional punctures made by the adult leafminer. The nematode treatment was as effective as abamectin, the only efficacious insecticide registered for use on ornamentals against leafminers.

Biological control of cockroaches is appealing because these nocturnal insects are major pests of human habitations where use of chemical insecticides is restricted. Cockroaches are susceptible to entomopathogenic nematodes. For example, the German cockroach, *Blatella germanica*, is susceptible to *S. carpocapsae,* and traps containing an attractant and nematodes are being tested as a means of control.[61] This approach appears feasible, although much more information is needed on nematode persistence and cockroach behavior in the presence of the nematodes. In another study, *Heterorhabditis zealandica* (T327) was more effective in killing the first three instars (>44% mortality) of the American cockroach, *Periplaneta americana*, than the fourth instar or adults (<15% mortality).[62] Triatomid bugs have also been shown to be susceptible to entomopathogenic nematodes.[63]

IV. MANURE HABITAT

Filth-breeding insects would appear to be good candidates for control with entomopathogenic nematodes. Many abiotic factors such as moisture, temperature (where insect activity occurs), and low ultraviolet radiation would seem to favor nematode survival. Moreover, many of these insects have become resistant to chemical insecticides and, as very few alternatives are available, the nematodes are considered potential control agents.

A. Diptera

In general, laboratory petri dish tests indicate that filth-breeding fly larvae are susceptible to steinernematids and heterorhabditids.[64-67] However, the addition of manure or fly medium into the tests reduces the efficacy of the nematodes.[64-67]

Field tests of the ability of *S. carpocapsae* and *H. bacteriophora* to control maggots in poultry houses have produced mixed results. Mullens et al.[68] demonstrated no larval or adult reduction in four fly species when *S. carpocapsae* and *H. bacteriophora* were applied. In contrast, Belton et al.[69] showed a significant reduction of house fly (*Musca domestica*) adults in a poultry house treated with *H. bacteriophora* as compared with the control. In the former study, manure moisture was considered favorable for fly development and, presumably, favorable for nematode survival. Failure was attributed to the poor survival of the nematodes in manure, probably because of the ammonia and salts which were toxic to the nematodes[68] or to predatory mites.[70] Moreover, the upper range of ambient temperatures (38°C) of the manure was within the nematodes' thermal inactivation zone. In the latter study, success was attributed to favorable differences in the manure environment, especially temperature and moisture. The manure temperature varied from 11 to 19°C and the manure was moist. Moisture was not quantified, but laboratory observations showed that the nematodes survived best in moist, not wet, manure. Although both groups of investigators used *H. bacteriophora* (NC), the poultry houses in the Mullens et al.[68] tests were open, whereas those in the Belton et al.[69] tests were closed. The results obtained by Belton et al.[69] are encouraging, but nematodes seem unlikely candidates for controlling filth fly larvae in poultry houses because of the hostile manure habitat.

Control of adult flies in nematode-baited traps shows more promise. Offering *S. carpocapsae* on cotton balls with 5% sucrose to house flies resulted in 63-67% fly mortality.[66] Another baiting system for the house fly consisted of a sex pheromone, fly medium, and nematodes.[65] Using this bait, *S. carpocapsae* killed 100% and *H. bacteriophora* killed 94% of the flies. A trap containing a bait that prevents nematode desiccation for a long period of time (1-3 months) would seem to be essential for making this system commercially viable.

Grown on pasteurized compost in cool, moist, dark houses, mushrooms are produced under conditions favorable for nematode survival and development. Major pests in mushroom cultivation include larvae of phorid, sciarid, and cecidomyiid flies, all of which can be infected by steinernematids and heterorhabditids.[71,72] In field trials with the sciarid *Lycoriella auripila*, *S. feltiae* (=*bibionis*) was the most effective and gave larval reduction comparable to the standard chemical, diflubenzuron;[72] *S. carpocapsae* and *H. bacteriophora* were less effective. An attractive feature of the nematodes is their ability to recycle. Thus, *S. feltiae* (=*bibionis*) and *H. bacteriophora* persisted in the compost and showed a marked population increase at 3 to 4 weeks posttreatment when a new generation of nematodes was released from the sciarid maggots.

B. Coleoptera

The lesser mealworm, *Alphitobius diaperinus*, lives in the litter of poultry houses, where the beetles breed in spilled feed and manure. Laboratory trials indicate that *S. carpocapsae* is superior to *H. bacteriophora* and *S. glaseri* for control of late instar lesser mealworm.[73] Field trials with *S. carpocapsae* showed that beetle populations initially increased more slowly in treated than untreated houses.[74] By the end of 10-13 weeks, however, the adult populations were about equal in treated and untreated houses. Although *S. carpocapsae* persisted in the soil (which contained manure and feed) for up to 15 weeks, bioassay of soil showed a marked decline 7 weeks after treatment. Geden et al.[74] speculated that the reason for the temporary control was the relatively short period when high populations of insects coincided with high nematode populations. They also noted that the time interval of greatest nematode loss (5 to 9 weeks posttreatment) occurred when outdoor temperatures exceeded 36°C for 1 week.

V. AQUATIC HABITAT

The aquatic habitat offers an excellent environment for nematode survival. A seemingly logical extension would be to use these nematodes for control of pestiferous insects in this habitat. However, steinernematids and heterorhabditids are soil organisms and are not adapted for directed motility in the aquatic environment.

A. Diptera

Mosquitoes and black flies would appear to be prime candidates for control with nematodes because they readily ingest nematodes. However, a number of factors reduce efficacy, including damage to the nematode during ingestion,[75,76] host immune response,[77-79] and spatial separation of host and nematode.[77,80] In the case of black flies, Gaugler and Molloy[76] demonstrated that the nematodes were physically excluded during feeding of the first through third instars, rendering the host resistant to infection. Older instars (fourth through seventh) were susceptible to infection, with the oldest instar being the most susceptible. In the fourth through sixth instars, the principal factor regulating susceptibility was nematode injury caused by the larval mouthparts during ingestion. In mosquitoes, Dadd[75] also observed that larval size excluded nematode ingestion by early instars and that some nematode injury occurred during ingestion. In some cases, nematodes that were ingested appeared viable but were rapidly degraded in the gut.

Nematodes that do reach the hemocoel of mosquito larvae can be encapsulated.[77-79] If one or only a few infective juveniles enter the hemocoel, the mosquito may survive by virtue of its immune response. However, such larvae may take longer to develop and often die at pupation or adult emergence.[77]

Mosquito feeding behavior and spatial separation of the nematodes also affect efficacy.[77,80] The substrate type influences the uptake of nematodes by mosquito larvae. Nematodes settle quickly to the bottom. Mosquitoes will easily remove them from a smooth surface, but when debris is added, the nematodes are less available to the mosquitoes.

Using *S. carpocapsae* and *H. bacteriophora*, Molta and Hominick[81] found that larval mortality of *Aedes aegypti* was a positive linear function of nematode dosage and exposure time. Inundation of the system with nematodes was necessary to promote larval mortality. They speculated that the nematodes might increase the fitness of a mosquito population because the larvae readily consumed the nematodes, thereby gaining nutrition, whereas only a small proportion of the larvae were killed.

In field trials, 50% mortality of late instar black fly larvae, *Simulium* spp., was observed in a New York stream treated with *S. carpocapsae*.[82] Further trials with the same nematode species against *Simulium* vectors of onchoceriasis in Mexico produced no significant mortality.[83] Gaugler and Molloy[76] had demonstrated that a postgenal length of >481 μm is needed for ingestion of *S. carpocapsae* infective juveniles without consistently injuring the nematodes with the larval mouthparts. The postgenal length of the oldest instar of the Mexican black flies was <481 μm. Probably, the nematodes were not ingested or were damaged by the larval mouthparts.

B. Coleoptera

Larvae of the rice water weevil, *Lissorhoptrus oryzophilus*, feed on the roots of flooded rice during the critical early vegetative period. In laboratory tests, 100% of the adult beetles confined in a tube with moist filter paper were killed by *S. carpocapsae*, but <6% of the adults in an open environment were infected when sprayed with nematodes.[84] Although flooded soil would seem adverse to nematode movement and host-finding, preliminary trials using *Steinernema* spp. to control rice water weevil larvae appear encouraging.[85]

VI. CONCLUSIONS

Entomopathogenic nematodes have been highly successful against a number of insect species in cryptic habitats, particularly plant borers. Less encouraging results have been obtained for control efforts on pests in foliar, aquatic, and manure habitats. The success with plant borers is attributed to favorable conditions in the galleries, the nematodes' ability to seek a host, and the high susceptibility of the target pests. In the foliar habitat, many species are also highly susceptible to nematode infection, but the narrow window of opportunity for treatment often restricts application to the evening or to times when adequate moisture is available. Attempts to extend nematode survival with the addition of evaporetardants to aqueous suspensions of nematodes have been partially successful, and the quest for the ideal evaporetardant is an ongoing endeavor.

The manure habitat appears to be too hostile for nematode survival. Moreover, the insects in this habitat are generally less susceptible to the nematode, especially in the presence of manure. As an exception, adult house flies may be controllable with nematode baits, but there is a need to close the gap between expected and observed mortality.

The aquatic habitat is poorly suited to nematode-based management strategies. The rice water weevil, which occurs in flooded soil, might be susceptible to nematode treatment, although much more data are needed to ascertain the nematodes' biological control potential for this insect.

Innovative approaches are needed to provide the impetus to use nematodes in various habitats. Some, such as baits or attractants for grasshoppers, house fly adults, and cockroaches, have been tried with encouraging results. Fine tuning of these approaches may provide a breakthrough in insect control. Adaptation of new methodologies and technologies from other fields to research on entomopathogenic nematodes is needed. We also must develop innovative approaches which will be useful for nematodes and other biological control agents.

Perhaps too much emphasis is being placed here on using nematodes as the sole agent for insect control. In fact, nematodes may be only one component of a pest management strategy. The life cycle of the target pest must be examined, its most vulnerable stage determined in relation to its environment and to nematode infection, the most virulent and persistent nematode species or strains identified, the economics of nematode usage ascertained, and the management strategy defined.

REFERENCES

1. **Klein, M. G.,** Efficacy against soil-inhabiting insect pests, in *Entomopathogenic Nematodes in Biological Control*, Gaugler, R., and Kaya, H. K., Eds., CRC Press, Boca Raton, FL, 1990, chap. 10.
2. **Gaugler, R.,** Ecological considerations in the biological control of soil-inhabiting insects with entomopathogenic nematodes, *Agric. Ecosys. Environ.*, 24, 351, 1988.
3. **Shetlar, D. J.,** Entomogenous nematodes for control of turfgrass insects with notes on other biological control agents, in *Integrated Pest Management for Turfgrass and Ornamentals*, Leslie, A. R., and Metcalf, R. L., Eds., U. S. Environmental Protection Agency, Washington, D.C., 1989, 225.
4. **Kaya, H. K.,** Entomogenous nematodes for insect control in IPM systems, in *Biological Control in Agricultural IPM Systems*, Hoy, M. A., and Herzog, D. C., Eds., Academic Press, New York, 1985, 283.
5. **Kaya, H. K.,** Entomopathogenic nematodes in biological control of insects, in *New Directions in Biological Control: Alternatives for Suppressing Agricultural Pests and Diseases*, Baker, R. R., and Dunn, P. E., Eds., Alan R. Liss, New York, 1990, 189.
6. **Georgis, R., and Poinar, G. O., Jr.,** Field effectiveness of entomophilic nematodes *Neoaplectana* and *Heterorhabditis*, in *Integrated Pest Management for Turfgrass and Ornamentals*, Leslie, A. R., and Metcalf, R. L., Eds., U. S. Environmental Protection Agency, Washington, D.C., 1989, 213.
7. **Poinar, G. O., Jr., and Georgis, R.,** Biological control of social insects with nematodes, in *Integrated Pest Management for Turfgrass and Ornamentals*, Leslie, A. R., and Metcalf, R. L., Eds., U. S. Environmental Protection Agency, Washington, D.C., 1989, 225.
8. **Kaya, H. K.,** *Steinernema feltiae*: use against foliage feeding insects and effects on nontarget insects, in *Fundamental and Applied Aspects of Invertebrate Pathology*, Samson, R. A., Vlak, J. M., and Peters, D., Eds., Proc. 4th Int. Coll. Invertebr. Pathol., Wageningen, 1986, 268.
9. **Moore, G. E.,** The bionomics of an insect-parasitic nematode, *J. Kansas Entomol. Soc.*, 38, 101, 1965.
10. **Kamionek, M., Maslana, I., and Sandner, H.,** The survival of invasive larvae of *Neoaplectana carpocapsae* weiser in a waterless environment under various conditions of temperature and humidity, *Zesz. Probl. Postepow Nauk Roln.*, 154, 409, 1974.
11. **Gaugler, R., and Boush, G. M.,** Effects of ultraviolet radiation and sunlight on the entomogenous nematode, *Neoaplectana carpocapsae*, *J. Invertebr. Pathol.*, 32, 291, 1978.
12. **Yamanaka, K., Seta, K., and Yasuda, M.,** Evaluation of the use of entomogenous nematode, *Steinernema feltiae* (Str. Mexican) for the biological control of the fall webworm, *Hyphantria cunea*, (Lepidoptera: Arctiidae), *Jpn. J. Nematol.*, 16, 26, 1986.
13. **Wang, J. X., and Li, Y. L.,** Entomogenous nematode research in China, *Rev. Nématol.*, 10, 403, 1987.
14. **Kaya, H. K., and Hara, A. H.,** Differential susceptibility of lepidopterous pupae to infection by the nematode *Neoaplectana carpocapsae*, *J. Invertebr. Pathol.*, 36, 389, 1980.
15. **Kaya, H. K., and Hara, A. H.,** Susceptibility of various species of lepidopterous pupae to the entomogenous nematode *Neoaplectana carpocapsae*, *J. Nematol.*, 13, 291, 1981.
16. **Kaya, H. K., and Grieve, B. J.,** The nematode *Neoaplectana carpocapsae* and the beet armyworm *Spodoptera exigua*: infectivity of prepupae and pupae in soil and of adults during emergence from soil, *J. Invertebr. Pathol.*, 39, 192, 1982.
17. **Begley, J. W.,** Use of the Nematode *Steinernema feltiae* to Control the Beet Armyworm, Ph.D. thesis, West Virginia University, Morgantown, 1986.
18. **Morris, J.,** personal communication, 1987.
19. **Begley, J. W.,** unpublished data, 1987.
20. **Welch, H. E., and Brian, L. J.,** Field experiment on the use of a nematode for the control of vegetable crop insects, *Proc. Entomol. Soc. Ontario*, 91, 197, 1961.

21. **York, G. T.**, European corn borer research, *Station Annual Report, USDA, ARS, Ankeny, Iowa*, 121, 1957.

22. **Jaques, R. P.**, Mortality of five apple insects induced by the nematode DD-136, *J. Econ. Entomol.*, 60, 741, 1967.

23. **Kaya, H. K., Hara, A. H., and Reardon, R. C.**, Laboratory and field evaluations of *Neoaplectana carpocapsae* (Rhabditida: Steinernematidae) against the elm leaf beetle (Coleoptera: Chrysomelidae) and the western spruce budworm (Lepidoptera: Tortricidae), *Can. Entomol.*, 113, 787, 1981.

24. **Finney, J. R., Lim, K. P., and Bennett, F.**, The susceptibility of the spruce budworm, *Choristoneura fumiferana* (Lepidoptera: Tortricidae), to *Heterorhabditis heliothidis* (Nematoda: Heterorhabditidae) in the laboratory, *Can. J. Zool.*, 60, 958, 1982.

25. **Kaya, H. K., and Reardon, R. C.**, Evaluation of *Neoaplectana carpocapsae* for biological control of the western spruce budworm, *Choristoneura occidentalis*: ineffectiveness and persistence of tank mixes, *J. Nematol.*, 14, 595, 1982.

26. **Reardon, R. C., Kaya, H. K., Fusco, R. A., and Lewis, F. B.**, Evaluation of *Steinernema feltiae* and *S. bibionis* (Rhabditida: Steinernematidae) for suppression of *Lymantria dispar* (Lepidoptera: Lymantriidae) in Pennsylvania, U.S.A., *Agric. Ecosys. Environ.*, 15, 1, 1986.

27. **Drooz, A. T.**, The larch sawfly, its biology and control, *USDA Tech. Bull. No.* 1212, 1960.

28. **Finney, J. R., and Bennett, G. F.**, The susceptibility of some sawflies (Hymenoptera: Tenthredinidae) to *Heterorhabditis heliothidis* (Nematoda: Heterorhabditidae) under laboratory conditions, *Can. J. Zool.*, 61, 1177, 1983.

29. **Georgis, R., and Hague, N. G. M.**, Field evaluation of *Steinernema feltiae* against the web-spinning larch sawfly *Cephalcia lariciphila, J. Nematol.*, 20, 317, 1988.

30. **Mrácek, Z.**, Hortizonal distribution in soil, and seasonal dynamics of the nematode, *Steinernema kraussei*, a parasite of *Cephalcia abietis, Z. Ang. Entomol.*, 94, 110, 1982.

31. **MacVean, C. M., Brewer, J. W., and Capinera, J. L.**, Field test of antidesiccants to extend the infection period of the entomogenous nematode *Neoaplectana carpocapsae* against the Colorado potato beetle, *J. Econ. Entomol.*, 75, 97, 1982.

32. **Welch, H. E., and Brian, L. J.**, Tests of the nematode DD-136 and an associated bacterium for control of the Colorado potato beetle, *Leptinotarsa decemlineata* (Say), *Can. Entomol.*, 43, 759, 1961.

33. **Toba, H. H., Lindegren, J. E., Turner, J. E., and Vail, P. V.**, Susceptibility of the Colorado potato beetle and the sugarbeet wireworm to *Steinernema feltiae* and *S. glaseri, J. Nematol.*, 15, 597, 1983.

34. **Capinera, J. L., and Hibbard, B. E.**, Bait formulations of chemical and microbial insecticides for suppression of crop-feeding grasshoppers, *J. Agric. Entomol.*, 4, 337, 1987.

35. **Wu, H-J.**, Biocontrol of squash bug with *Neoaplectana carpocapsae* (Weiser), *Bull. Inst. Zool. Acad. Sin.*, 27, 195, 1988.

36. **Lindegren, J. E., Yamashita, T. T., and Barnett, W. W.**, Parasitic nematode may control carpenterworm in fig trees, *Calif. Agric.*, 35(1&2), 25, 1981.

37. **Bedding, R. A., and Miller, L. A.**, Disinfesting blackcurrant cuttings of *Synanthedon tipuliformis*, using the insect parasitic nematode, *Neoaplectana bibionis, Environ. Entomol.*, 10, 449, 1981.

38. **Lindegren, J. E., and Barnett, W. W.**, Applying parasitic nematodes to control carpenterworms in fig orchards, *Calif. Agric.*, 36(11&12), 7, 1982.

39. **Miller, L. A., and Bedding, R. A.**, Field testing of the insect parasitic nematode, *Neoaplectana bibionis* (Nematoda: Steinernematidae) against currant borer moth, *Synanthedon tipuliformis* (Lep.: Sesiidae) in black currants, *Entomophaga*, 27, 109, 1982.

40. **Foschi, S., and Deseö, K. V.**, Risultati di lotta con nematodi entomopatogenic su *Zeuzera pyrina* L. (Lepidopt.; Cossidae) Nel 1982, *La Difesa delle Plante*, 3-4, 153, 1983.

41. **Kaya, H. K., and Lindegren, J. E.**, Parasitic nematode controls western poplar clearwing moth, *Calif. Agric.*, 37(3&4), 31, 1983.

42. **Deseö, K. V., and Miller, R.**, Efficacy of entomogenous nematodes, *Steinernema* spp., against clearwing moths, *Synanthedon* spp., in north Italian apple orchards, *Nematologica*, 31, 100, 1985.

43. **Capinera, J. L., Cranshaw, W. S., and Hughes, H. G.,** Suppression of raspberry crown borer, *Pennisetia marginata* (Harris) (Lepidoptera: Sesiidae) with soil applications of *Steinernema feltiae* (Rhabditida: Steinernematidae), *J. Invertebr. Pathol.*, 48, 257, 1986.

44. **Kaya, H. K., and Brown, L. R.,** Field application of entomogenous nematodes for biological control of clear-wing moth borers in alder and sycamore trees, *J. Arboricult.*, 12, 150, 1986.

45. **Qin, X., Kao, R., Yang, H., and Zhang, G.,** Study on application of entomopathogenic nematodes of *Steinernema bibionis* and *S. feltiae* to control *Anoplophora glabripennis* and *Holcocerus insularis, Forest Res.*, 1, 179, 1988.

46. **Forschler, B. T., and Nordin, G. L.,** Comparative pathogenicity of selected entomogenous nematodes to the hardwood borers, *Prionoxystus robiniae* (Lepidoptera: Cossidae) and *Megacyllene robiniae* (Coleoptera: Cerambycidae), *J. Invertebr. Pathol.*, 52, 343, 1988.

47. **Agudelo-Silva, F., Lindegren, J. E., and Valero, K. A.,** Persistence of *Neoplectana carpocapsae* (kapow selection) infectives in almonds under field conditions, *Florida Entomol.*, 70, 288, 1987.

48. **Lindegren, J. E., Curtis, C. E., and Poinar, G. O., Jr.,** Parasitic nematode seeks out navel orangeworm in almond orchards, *Calif. Agric.* 32(6), 10, 1978.

49. **Kaya, H. K., Joos, J. L., Falcon, L. A., and Berlowitz, A.,** Suppression of the codling moth (Lepidoptera: Olethreutidae) with the entomogenous nematode, *Steinernema feltiae* (Rhabditida: Steinernematidae), *J. Econ. Entomol.*, 77, 1240, 1984.

50. **Bari, M. A., and Kaya, H. K.,** Evaluation of the entomogenous nematode *Neoplectana carpocapsae* (=*Steinernema feltiae*) Weiser (Rhabditida: Steinernematidae) and the bacterium *Bacillus thuringiensis* Berliner var. *kurstaki* for suppression of the artichoke plume moth (Lepidoptera: Pterophoridae), *J. Econ. Entomol.*, 77, 225, 1984.

51. **Moore, G. E.,** *Dendroctonus frontalis* infection by the DD-136 strain of *Neoplectana carpocapsae* and its bacterium complex, *J. Nematol.*, 2, 341, 1970.

52. **Finney, J. R., and Mordue, W.,** The susceptibility of the elm bark beetle *Scolytus scolytus* to the DD-136 strain of *Neoplectana* sp., *Ann. Appl. Biol.*, 83, 311, 1976.

53. **Finney, J. R., and Walker, C.,** The DD-136 strain of *Neoplectana* sp. as a potential biological control agent for the European elm bark beetle, *Scolytus scolytus, J. Invertebr. Pathol.*, 29, 7, 1977.

54. **Finney, J. R., and Walker, C.,** Assessment of a field trial using the DD-136 strain of *Neoplectana* sp. for the control of *Scolytus scolytus, J. Invertebr. Pathol.*, 33, 239, 1979.

55. **Poinar, G. O., Jr., and Deschamps, N.,** Susceptibility of *Scolytus multistriatus* to neoaplectanid and heterorhabditid nematodes, *Environ. Entomol.*, 10, 85, 1981.

56. **Kinn, D. N.,** Mutualism between *Dendrolaelaps neodisetus* and *Dendroctonus frontalis*, *Environ. Entomol.*, 9, 756, 1980.

57. **Mamiya, Y.,** Application of entomogenous nematode on pine logs infested with pine sawyer, *Monochamus altenatus*, in *Recent Advances in Biological Control of Insect Pests by Entomogenous Nematodes in Japan*, Ishibashi, N., Ed., Ministry of Education, Japan, Grant No. 59860005, 1987, 31.

58. **Mamiya, Y.,** Comparison of the infectivity of *Steinernema kushidai* (Nematode: Steinernematidae) and other steinernematid and heterorhabditid nematodes for three different insects, *Appl. Entomol. Zool.*, 24, 302, 1989.

59. **Harris, M., and Warkentin, D.,** Leafminers? Send in the nematodes, *Grower Talks*, 52(1), 74, 1988.

60. **Harris, M. A., Begley, J. W., and Warkentin, D. L.,** Control of *Liriomyza trifolii* (Burgess) (Diptera: Agromyzidae) suppression with foliar applications of *Steinernema carpocapsae* (Weiser) (Rhabditida: Steinernematidae) and abamectin, *J. Econ. Entomol.*, 1990, in press.

61. **Georgis, R.,** Formulation and application technology, in *Entomopathogenic Nematodes in Biological Control*, Gaugler, R., and Kaya, H. K., Eds., CRC Press, Boca Raton, FL, 1990, chap. 9.

62. **Zervos, S., and Webster, J. M.,** Susceptibility of the cockroach *Periplaneta americana* to *Heterorhabditis heliothidis* (Nematoda: Rhabditoidea) in the laboratory, *Can. J. Zool.*, 67, 1609, 1989.

63. **Minter, D. M., and Oswald, W. J. C.,** The nematode *Neoaplectana* in the biological control of some trypanosomatid vectors, *Trans. R. Soc. Trop. Med. Hyg.*, 74, 679, 1980.

64. **Renn, N., Barson, G., and Richardson, P. N.,** Preliminary laboratory tests with two species of entomophilic nematodes for control of *Musca domestica* in intensive animal units, *Ann. Appl. Biol.*, 106, 229, 1985.

65. **Geden, C. J., Axtell, R. C., and Brooks, W. M.,** Susceptibility of the house fly, *Musca domestica* (Diptera: Muscidae), to the entomogenous nematodes *Steinernema feltiae, S. glaseri* (Steinernematidae), and *Heterorhabditis heliothidis* (Heterorhabditidae), *J. Med. Entomol.*, 23, 326, 1986.

66. **Mullens, B. A., Meyer, J. A., and Cyr, T. L.,** Infectivity of insect-parasitic nematodes (Rhabditida: Steinernematidae, Heterorhabditidae) for larvae of some manure-breeding flies (Diptera: Muscidae), *Environ. Entomol.*, 16, 769, 1987.

67. **Georgis, R., Mullens, B. A., and Meyers, J. A.,** Survival and movement of insect parasitic nematodes in poultry manure and their infectivity against *Musca domestica*, *J. Nematol.*, 19, 292, 1987.

68. **Mullens, B. A., Meyer, J. A., and Georgis, R.,** Field tests of insect-parasitic nematodes (Rhabditida: Steinernematidae, Heterorhabditidae) against larvae of manure-breeding flies (Diptera: Muscidae) on caged-layer poultry facilities, *J. Econ. Entomol.*, 80, 438, 1987.

69. **Belton, P., Rutherford, T. A., Trotter, D. B., and Webster, J. M.,** *Heterorhabditis heliothidis*: a potential biological control agent of house flies in caged-layer poultry barns, *J. Nematol.*, 19, 263, 1987.

70. **Wicht, M. C., Jr., and Rodriguez, J. S.,** Integrated control of muscid flies in poultry houses using predator mites, selected pesticides and microbial agents, *J. Med. Entomol.*, 7, 687, 1970.

71. **Richardson, P. N.,** Susceptibility of mushroom pests to the insect-parasitic nematodes *Steinernema feltiae* and *Heterorhabditis heliothidis*, *Ann. Appl. Biol.*, 111, 433, 1987.

72. **Richardson, P. N.,** Nematode parasites of mushroom flies: their use as biological control agents, in *Cultivating Edible Fungi*, Wuest, P. J., Royse, D. I., and Beelman, R. B., Eds., Elsevier, Amsterdam, 1986, 385.

73. **Geden, C. J., Axtell, R. C., and Brooks, W. M.,** Susceptibility of the lesser mealworm, *Alphitobius diaperinus* (Coleoptera: Tenebrionidae) to the entomogenous nematodes *Steinernema feltiae, S. glaseri* (Steinernematidae) and *Heterorhabditis heliothidis* (Heterorhabditidae), *J. Entomol. Sci.*, 20, 331, 1985.

74. **Geden, C. J., Arends, J. J., and Axtell, R. C.,** Field trials of *Steinernema feltiae* (Nematoda: Steinernematidae) for control of *Alphitobius diaperinus* (Coleoptera: Tenebrionidae) in commercial broiler and turkey houses, *J. Econ. Entomol.*, 80, 136, 1987.

75. **Dadd, R. H.,** Size limitations on the infectibility of mosquito larvae by nematodes during filter-feeding, *J. Invertebr. Pathol.*, 18, 246, 1971.

76. **Gaugler, R., and Molloy, D.,** Instar susceptibility of *Simulium vittatum* (Diptera: Simuliidae) to the entomogenous nematode *Neoaplectana carpocapsae*, *J. Nematol.*, 13, 1, 1981.

77. **Welch, H. E., and Bronskill, J. F.,** Parasitism of mosquito larvae by the nematode DD-136 (Nematoda: Neoaplectanidae), *Can. J. Zool.*, 40, 1263, 1962.

78. **Poinar, G. O., Jr., and Leutenegger, R.,** Ultrastructural investigation of the melanization process in *Culex pipiens* (Culicidae) in response to a nematode, *J. Ultrastruct. Res.*, 30, 149, 1971.

79. **Poinar, G. O., Jr., and Kaul, H. N.,** Parasitism of the mosquito *Culex pipiens* by the nematode *Heterorhabditis bacteriophora*, *J. Invertebr. Pathol.*, 39, 382, 1982.

80. **Finney, J. R., and Harding, J. B.,** Some factors affecting the use of *Neoaplectana* sp. for mosquito controls, *Mosquito News*, 41, 798, 1981.

81. **Molta, N. B., and Hominick, W. M.,** Dose- and time-response assessments of *Heterorhabditis heliothidis* and *Steinernema feltiae* (Nematoda: Rhabditida) against *Aedes aegypti* larvae, *Entomophaga*, 34, 485, 1989.

82. **Gaugler, R., and Molloy, D.,** Field evaluation of the entomogenous nematode, *Neoaplectana carpocapsae*, as a biological control agent of black flies (Diptera: Simuliidae), *Mosquito News*, 41, 459, 1981.

83. **Gaugler, R., Kaplan, B., Alvarado, C., Montoya, J., and Ortega, M.,** Assessment of *Bacillus thuringiensis* serotype 14 and *Steinernema feltiae* (Nematoda: Steinernematidae) for control of the *Simulium* vectors of onchocerciasis in Mexico, *Entomophaga*, 28, 309, 1983.

84. **Kishimoto, R.,** The entomogenous nematode *Steinernema feltiae* as a biological control agent of the rice water weevil, the brown rice planthopper, and the diamondback moth, in *Recent Advances in Biological Control of Insect Pests by Entomogenous Nematodes in Japan*, Ishibashi, N., Ed., Ministry of Education, Japan, Grant No. 59860005, 1987, 111.

85. **Smith, K., and Georgis, R.,** unpublished data, 1989.

12. Logistics and Strategies for Introducing Entomopathogenic Nematode Technology Into Developing Countries

Robin Bedding

I. INTRODUCTION

Entomopathogenic nematodes have great potential for the biological control of many important insect pests. They are already being used for the control of the black vine weevil, *Otiorhynchus sulcatus*, in Australia and Europe,[1] currant borer moth, *Synanthedon tipuliformis*, in Australia,[1] and a tree-boring cossid, *Holcocercus insularis*, and the peach borer moth, *Carposina nipponensis*, in China.[2] Environmentally safe and likely to induce minimal problems of insect resistance, these nematodes are sometimes more effective than insecticides. They are also ideally suited economically to developing countries. Whereas the cost of imported chemical insecticides is high, nematode-based control is relatively inexpensive and production can be adapted as a local cottage industry. Furthermore, developing countries usually have the tropical, humid conditions conducive to control using entomopathogenic nematodes, and their farmers often have a close relationship with their carefully and intensively tended land.

This chapter is based largely on experience gained from a comprehensive collaborative program between CSIRO, Australia, and two Chinese institutes, with some insights derived from several small programs with various other countries. Many of the logistics and strategies suggested have been developed and used in the above programs; others are being introduced with the benefit of hindsight. Some of the strategies involved are also applicable to the transfer of entomopathogenic nematode technology within industrialized countries.

II. INITIATION

Obviously for any comprehensive program, the industrialized country's organization must itself have a substantial commitment to entomopathogenic nematode research and a broad background in nematode technology. Full cooperation between the developing and industrialized countries' organizations will be best achieved if both clearly realize the potential mutual advantages. The developing country may benefit from: (1) a source of diverse nematode isolates, (2) access to sufficient nematodes for large field trials at an early stage, (3) financial and material assistance from the industrialized country, (4) training and research opportunities for their personnel in the industrialized country, and (5) research collaboration with an industrialized country.

The advantages to the organization of the industrialized country may include: (1) grants for mutually beneficial research within the industrialized country, (2) expanded research opportunities, such as extensive laboratory and field trials against a variety of insects made possible by a large inexpensive labor force, (3) research collaboration with scientists from the developing country, and (4) access to new isolates of entomopathogenic nematodes from the developing country.

III. PRELIMINARY ASSISTANCE

Before substantial technological assistance can be offered to groups in a developing country, a source of outside funding is likely to be required. However, early assistance should include helping to determine the appropriateness of using entomopathogenic nematodes in the developing country and providing the background necessary to obtain outside funding for more comprehensive programs. This stage, likely to require 2 to 3 years, should not be overly expensive for either the industrialized or developing country's organization. The following steps are suggested:

1. Identify pest problems likely to be amenable to nematode-based control.
2. Visit to developing country by an expert to further assess prospects for success.
3. Prepare a circular containing a general summary and an overview of entomopathogenic nematodes and technology.
4. Provide relevant literature. This packet should contain 10 to 20 key publications on entomopathogenic nematodes including reviews, important technique and strategy papers, and articles relevant to controlling insect pests of specific importance to the developing country.
5. Establish the credentials of the industrialized country's organization by sending relevant nematode publications written by personnel.
6. Supply a full publication list (about 750 references) which encompasses the scope of work in the discipline.
7. Assist in providing information to the developing country's quarantine authorities to facilitate import permits.
8. Supply full instructions on extracting nematodes from storage medium, maintaining them in suspension, counting them, and performing infectivity testing.
9. Make available at least eight species of entomopathogenic nematodes for importation.

IV. SUBSTANTIAL ASSISTANCE

Having established a sound basis from which to proceed, both the industrialized and developing country's organizations will now need to procure substantial funding either from within the developing country or from appropriate organizations within the industrialized country. This funding is likely to be forthcoming only if a strong probability of success can be demonstrated. Therefore, at the end of the preliminary phase, a detailed report should be prepared concerning the overall progress made, the feasibility of controlling various insect pests important to the developing country, and the financial and environmental advantages of using entomopathogenic nematodes instead of insecticides. A detailed budget should be included. Funding success will be enhanced if the project has direct relevance to both countries and if each will be committing significant resources. In the case of the developing country, the number of personnel devoted to the project will be of particular relevance.

Entomopathogenic nematology is becoming increasingly commercialized and therefore competitive (e.g., in the U.S., U.K., Europe, and Australia). Where the industrialized country's organization is involved or likely to be involved with a commercial company, there may be concern over the confidentiality of proprietary interests and the possibility of competition from the developing country in international markets (a problem generally recognized in the transfer of biotechnology[3]). This may result in a reluctance or even in legal barriers to transferring up to date technology which has often been developed at considerable expense. These issues can be addressed by drafting and signing the appropriate documents.

Where it appears worthwhile to enter into a major project, funding over a 3 year period will be required for the following:

1. Short-term visits (about 2 weeks) by personnel from the developing country to the industrialized country.
2. At least one person from the developing country to come to the industrialized country's entomopathogenic nematode laboratory for 6 to 12 months to learn general techniques and work on one or more collaborative research projects. Other visiting personnel could specialize in bacteriology, mass rearing, ecology, application methods, or taxonomy for approximately 3 month periods.
3. Development of mutually relevant projects in the industrialized country's organization. These projects may involve the appointment of research scientists and technical assistants (one of the major benefits to the industrialized country's organization).
4. Annual or biannual visits of 1 to 2 weeks to the developing country by one or more scientists from the industrialized country. The scientist(s) will give seminars and workshops on entomopathogenic nematode technology to local scientists, end users, and government officials, and provide advice on the establishment of laboratory facilities.
5. Purchase of equipment and supplies unavailable in the developing country.
6. Provision of sufficient quantities of appropriate nematode species for field trials while local production technology is being developed.
7. Provision of monoxenic cultures of appropriate nematodes and healthy phase one cultures of their symbiotic bacteria.

8.   Provision of a full library of reprints and books on entomopathogenic nematodes and a library computer search system with comprehensive key wording.
9.   Supply of slides and videos of important processes.
10. Compilation and publication of a comprehensive manual which describes all techniques in considerable detail.

V. THE TECHNOLOGY

The technology associated with entomopathogenic nematodes ranges from simple to advanced. In general the simple technology should be introduced during the phase of preliminary assistance, whereas introduction of the more complex technology should coincide with the substantial assistance phase.

A. Simple Technology

The results of preliminary laboratory and field testing against insect pests must provide reliable evidence for either proceeding with or terminating the project. Thus, the industrialized country's organization must ensure that healthy infective juveniles were applied at recommended dosages against the target pests and that the results can be compared with similar trials in the industrialized country (i.e., they have been obtained using the same methodology). Consequently, instructions must be very explicit and more detailed than those included in a scientific publication. Specifically, these instructions should include information on the following activities:

1.   Extracting nematodes from consignments and storing them in water.
2.   Counting nematodes. Principles and procedures for nematode quantification are important because it is involved in most aspects of entomopathogenic nematode research from mass rearing, storage, infectivity testing, and ecology to large-scale field testing.
3.   Standardizing infectivity tests or quality control. Methodologies employed must be similar in both countries so that comparisons can be made and personnel can readily estimate the feasibility of controlling for particular insects by particular nematodes.
4.   Screening of insect pests in laboratory tests.
5.   Rearing of the greater wax moth, *Galleria mellonella*, where this insect is permitted by quarantine authorities.
6.   Culturing nematodes in *Galleria* larvae, or if this is not permitted by quarantine regulations, in another insect.
7.   Using *Galleria* larvae or other suitable insect larvae as bait for indigenous nematodes.
8.   Conducting small-scale field trials.

B. Advanced Technology

Once the simple technology has been introduced, the more advanced technology training can commence. A major constraint is adequate instruction in aseptic techniques. Training in advanced technology should include several topics:

1. General sterile techniques with special attention to the handling of large cultures.
2. Culture of symbiotic bacteria. In particular, the maintenance of phase one cultures and use of freeze-dried symbiotic bacteria should be covered.
3. Subculture of monoxenic entomopathogenic nematodes on polyether-polyurethane foam in tubes and flasks. Some of the necessary steps are isolation of bacterial symbionts, surface sterilization of infective juveniles, and establishment of monoxenic cultures.
4. Large-scale production methods (once flask culture has been thoroughly mastered). These methods comprise large-scale culture and harvesting and processing techniques.
5. Storage techniques, including large-scale nematode storage protocols, and liquid nitrogen storage of nematode and bacterial isolates.
6. Entomopathogenic nematode taxonomy. Teach morphology and cross-breeding for steinernematids, and morphology and electrophoretic techniques for heterorhabditids.
7. Assessment of bacterial presence in infective juveniles.
8. Basic bacterial taxonomy.
9. Experimental design. This training is best implemented during long-term visits of the developing country's personnel to the industrialized country's organization. This training should be supplemented by detailed discussions of the research plans during visits to the developing country, by an exchange of written research plans, and by frequent telephone contacts.
10. Modification of methodology to suit local conditions. Collaborative research may be necessary to develop appropriate improvements.

VI. LARGE SCALE DEVELOPMENT

When field tests and grower trials establish that one or more important insect pests can be economically controlled with nematodes, provision must be made for introducing this means of control into normal agricultural practices. Importing nematodes on a large scale from industrialized country's commercial concerns will be far too expensive. Thus, nematodes must be produced in the developing country. Establishment of facilities can be undertaken by government or local commercial concerns, but may best be achieved in joint venture with an established commercial entomopathogenic nematode company from the industrialized country.

VII. PROVISION OF ENTOMOPATHOGENIC NEMATODES

No one species of entomopathogenic nematode is the best control agent for all or even most insect species.[4] For example, there can be a several hundred-fold difference in LD_{50}s between two species. *Steinernema carpocapsae*, the most commonly available species, is not usually the best for any particular insect. Consequently, unless interest is in only one insect pest already known to be susceptible to a given nematode species, a minimum of eight nematode species must be sent initially. It is preferable to examine different species rather than various strains of one or more species because species usually exhibit greater differences in infectivity.[1]

Unfortunately, predicting the most pathogenic nematode species for an untested insect is difficult. Known efficacy against related pests may be a loose guide to potentially effective species. Most entomopathogenic nematodes have a low LD_{50} for Lepidoptera; however, *Steinernema glaseri*, *Steinernema* (513 strain), *S. anomali*, *S. kushidai*, and various *Heterorhabditis* species are more pathogenic against scarabs.[1] Some *Steinernema* species (e.g., *S. carpocapsae* and *S. feltiae* [=*bibionis*]) may never reach the LD_{50} for scarabs even when dosages exceed 100,000 per insect.[1,4] Even for closely related insects, relative efficacy can vary. Whereas *Heterorhabditis zealandica* (T310, T327) and *H. bacteriophora* (C1) are excellent for controlling larvae of the weevil, *Otiorhynchus sulcatus*, in potted plants,[1,5] they are poor for controlling another weevil, *Phlyctinus callosus*, in the same environment. The reverse is true for *S. feltiae* (=*bibionis*).[1] *H. bacteriophora* and *S. glaseri* are best, although not good, against larvae of another weevil *Grapthognathus leucoloma*.[1] Lastly, *S. carpocapsae* is best against larvae[6] but not the adults[7] of the weevil, *Cosmopolites sordidus*, in banana corms.

The environmental conditions under which control will be attempted should also be considered when determining the best entomopathogenic nematodes to send. For example, to test *S. kraussei*, which is active at 2°C, but cannot function at temperatures above 26°C[8] in the tropics, or *Heterorhabditis* sp. (Q380), which has a minimum temperature of activity at 16°C[8] in cool climates, is obviously inappropriate. Where application against foliage-feeding insects will be attempted, *S. carpocapsae* would be the main species sent because of its superior capability to withstand a certain degree of desiccation.[1]

The availability of various species of entomopathogenic nematodes will obviously be a factor in determining which one to supply to a developing country. However, attention is drawn to the database developed by Dr. R. J. Akhurst (CSIRO, Canberra, Australia) which contains information on the availability and location of some 250 isolates. These isolates represent 30 species of entomopathogenic nematodes held by 20 laboratories.

VIII. COLLECTION OF INDIGENOUS ISOLATES

Collection of indigenous nematodes has several merits. A relatively simple procedure, it may provide isolates more suitable for inundative release against local pest insects because of adaptation to local climate and population regulators. Collection provides information on indigenous fauna prior to possible introductions of exotic species. Lastly, new isolates are potentially beneficial to collaborators from the industrialized country.

The *Galleria* trap method[9] is recommended for general collection. If isolates are required for a particular insect, this host can be used to bait soil samples. All isolates should be stored in liquid nitrogen[10] to ensure preservation and conserve genetic diversity. Details of all isolates should be sent to the international entomopathogenic nematode database compiled by Dr. R. J. Akhurst.

IX. SCREENING OF INSECT PESTS

It may be tempting for the developing country's organization to plunge directly into field tests. However, an efficient screening procedure should eliminate many unsuitable nematode species. Moreover, many insects which cannot feasibly be controlled with nematodes because of natural resistance or environmental constraints will be eliminated. This procedure will save time while minimizing the possibility of erroneous rejection.

Standardized techniques are used to determine the relative infectivities of different species and batches of nematodes. Petri dish/filter paper assays are not recommended because this approach is far removed from the natural situation. Such assays are likely to favor those nematodes more able to nictate,[1] and there are likely to be problems with establishment of an adequate attraction gradient.[4] Using more than one insect per container is misleading because of individual variation in attractiveness and susceptibility between individual insects of the same species and stage.[1] For many soil-inhabiting insects, exposing individual insects in a standard sand type and moisture content to five logarithmic dosages (e.g., 2^1 to 2^5 or $10^1, 10^{1.5}, 10^2, 10^{2.5}, 10^3$) of nematodes[4] is perhaps the most satisfactory means of obtaining consistent LD_{50}s.

With insects that bore into plant tissues (often among the most promising for control with entomopathogenic nematodes), use of a sand assay is not so appropriate as sawdust or fine vermiculite. Whereas a high LD_{50} should preclude consideration of a particular nematode species against a given borer, a low LD_{50} may be relatively meaningless. Other factors may affect the access of infective juveniles to the target pest. For example, with many cerambycid beetles, the frass is so tightly packed in the galleries that infective juveniles cannot penetrate it.[1] With other borers, nematode entry may be blocked by callus tissue, and unless it is feasible to artificially insert nematodes into the plant interior, prospects for control may be poor. For these reasons, those borer-nematode combinations that show a low LD_{50} in laboratory assays should be further tested *in situ* on excised portions of the plant maintained at 100% humidity. The nematodes should be denied access to the cut surfaces, and, if this is not possible, small-scale tests should be conducted in the field.

Recommended procedure for soil-inhabiting insects is to test one very high dosage (i.e., 1000 infective juveniles per insect) against each of 20 individual insects. If less than 10 of these are killed, reject this nematode species for further testing. However, if at least 10 insects are killed, then the LD_{50} should be determined. Each stage of the insect that is likely to be present in the soil for a significant time should be tested. The two or three nematode species with low LD_{50}s should be selected for pot trials using soil types and host plants similar to those found in the field. If one or more species are better than others, test various strains of the best species.

Small-scale field trials can be conducted if pot trials have produced satisfactory results. Where natural insect populations are sparse, insects from other areas can be supplemented, provided they are undamaged and can be introduced without unduly disturbing the plots. However, effective nematode dosage appears to be proportional to host density,[11] so this aspect should be considered during evaluation. When small scale-field trials prove encouraging, plan and conduct larger trials after consultation with a statistician.

After a range of successful field tests in diverse situations, five to ten commercial growers in different localities should be encouraged to adopt the nematode as a control measure on part of their crop. This stage should only be attempted when all concerned are confident that nematodes are an effective means of control. The commercial growers must be well-trained in nematode handling and application.

Evaluation of stem-boring insects may be attempted in the same general manner as described for soil-inhabiting insects. However, in many cases, experiments in the field should begin at an earlier stage. Field trials avoid the difficulty of extracting insects from the plant tissue and take advantage of the ease with which many borers can be evaluated by the presence or absence of fresh frass outside the plant.

X. MASS REARING AND PROCESSING

For rearing small numbers (a few million) of nematodes, use of *Galleria* larvae and White traps[12] is adequate. When the stage requiring substantial assistance is reached, the developing country's organization should begin the adoption of methods for large-scale rearing. This rearing can be achieved aseptically by liquid or solid culture. Solid culture, which has been used successfully to produce hundreds of billions of infective juveniles of a variety of nematode species,[1] may be the most cost effective method for developing countries.

The use of polyether-polyurethane foam coated with homogenated solid medium in 500 ml flasks, followed in later stages in larger containers, forms a proven process of mass rearing for all species of entomopathogenic nematodes.[13-15] The developing country's workers must be well trained in this process because maintaining monoxenicity during subculture requires experience. Ensuring the quality of the bacterial symbiont and nematode inoculum is also of great importance. Flask culture should readily provide sufficient nematodes for small-scale field trials. For larger scale trials and for commercial development, a system using self-aerating trays has been developed.[15] Such a system allows for almost unlimited nematode production in solid culture. Finally, for long-term storage of large quantities of nematodes, milled, calcined attapulgite clay combined with a nematode cream[16] has proven effective. Storing anhydrobiotic nematodes in superabsorbent gels[16] also appears to be promising.

XI. A CASE STUDY — CHINA

A comprehensive collaborative program between the Division of Entomology, CSIRO, Australia, and the Guangdong Entomological Institute (GEI) in Guangzhou and the Biological Control Laboratory (BCL), Chinese Academy of Sciences in Beijing was supported from 1985 until the present by the Australian Centre for International Agricultural Research (ACIAR).

The collaboration commenced in 1979 with visits to the CSIRO entomopathogenic nematode laboratory in Hobart, Tasmania, by scientists from GEI. Over the next few years, literature, advice, and nematode and bacterial symbiont cultures were sent from CSIRO to GEI. Scientists at GEI and then at BCL commenced evaluations against important insect pests. In collaboration with the Pomology Institute, Zhengzhou, field evaluations against *Carposina nipponensis* (Lepidoptera, Carposinidae), the most serious pest of apples, were initiated.

At the start of the funded phase of the project, a rigorous training program for the collaborating Chinese scientists was implemented. There can be no doubt that funding of short- and long-term visits by Chinese scientists to Australia was of considerable benefit. Each of the long-term (1 year) visitors is now a scientific leader of his/her group, and each is doing and directing excellent research in several fields of nematode research, as well as helping to organize the work of other institutes.

Dr. R. J. Akhurst, Dr. J. Curran, and I visited laboratories and field trial areas and gave seminars and laboratory workshops at various stages of the project. Moreover, extensive written and telephone communication occurred throughout the project.

Both GEI and BCL have established satisfactory small-scale mass rearing facilities. Full use is now being made of self-aerating trays which, unlike previous techniques, make possible the rearing of large quantities of nematodes. Mass rearing media comprising only dry ingredients readily and cheaply obtainable in China and Australia have been developed for both *Steinernema* and *Heterorhabditis* species. Procedures for extracting, cleaning, mass storage, and transport have been implemented. Although these procedures need to be improved further, they are adequate at this stage for the production and processing of hundreds of kilograms of nematodes per year.

Substantial research effort has been made during the project to identify and test those economically important pest insects which are amenable to control by nematodes. Progress has been made on many fronts, and preliminary results have led to the targeting of a number of pest species for special attention. The research has been conducted in three phases: (1) laboratory trials on susceptibility of the target insect to a range of nematode species, (2) laboratory experiments and small-scale trials to determine efficacy under field conditions, and (3) extensive field trials and evaluations leading to utilization of nematodes as biological control agents.

The entomopathogenic nematode surveys made by both BCL and GEI scientists have yielded many unique isolates and greatly increased the armory of potentially useful agents. New and interesting bacterial symbionts have been found associated with the nematode isolates obtained from China. Important advances in liquid nitrogen storage[10] have helped to conserve this genetic diversity by enabling indefinite, low maintenance storage of hundreds of isolates.

A noteworthy development has been the degree of collaboration between BCL and GEI; both groups are continually in touch with each other, hold joint meetings, and exchange information and cultures. This cooperation extends to many other groups working in China, and ranges from supplying others with BCL and GEI nematode cultures to conducting a major training seminar in 1988. The seminar was attended by 34 scientists from 12 provinces. This cooperation has established the administrative and technical infrastructure needed within China to support the extensive research projects currently underway.

The most important result of this collaboration has been the world's first wide-scale use of nematodes for pest control becoming a reality in China, which is an effective method for the control of the fruit borer, *C. nipponensis*. This insect is the major pest threatening 70% of the large and rapidly expanding Chinese apple industry (1 million hectares producing 4 million metric tons of apples per year).[17,18] With further collaboration, it is hoped that inundative applications of entomopathogenic nematodes will soon become the common agricultural practice for controlling *C. nipponensis*. The cost is expected to be comparable or lower than that of chemical insecticides. To put the scale of potential nematode use into perspective, some 360 metric tons of infective juveniles will be needed each year to control *Carposina* alone. Unfortunately, the situation is complicated because all of these nematodes will be required during one month of the year.

Having a wealth of expertise on *C. nipponensis* has been a great advantage. Chinese scientists had earlier determined the precise biology of the insect.

They were using pheromone trapping and damage indices to assess population levels, and economic injury and threshold levels were established. Development of an entomopathogenic nematode control system for general use requires such extensive information about the pest insect.

Research has identified the appropriate application time (monitored with soil temperature and moisture data), determined dosage rates (1.2-$2.4 \times 10^9/$ ha), and compared different application methods and spray equipment.[17,18] Field trials in the autumn achieved >90% larval mortality and <3% fruit damage (considered an acceptable level of control) over a 3-year period. However, treatments are now applied in the late spring because peak emergence of the larvae from overwintering hibernacula occurs at a temperature of 19°C and a soil moisture content of 10%, conditions found to be ideally suited to the nematodes. Furthermore, the orchard soils are weed free and well watered, and farm labor is not involved with other operations at this time of year. Nematodes are applied directly to the soil surface, which is protected by the tree canopy, and are concentrated around the base of the tree. In large orchards nematodes are applied with a spray nozzle fitted to a bamboo pole; in small orchards they are applied by skillfully casting a nematode suspension from a seed sowing bowl. Large-scale (3.3-26.6 ha) areas have been treated with nematodes. Field trials have been completed by each of seven institutes over periods of several years. All these trials have given excellent results, achieving consistently better control of fruit damage by *Carposina* than the chemical insecticide treatments (Table 1).

TABLE 1. Example of Nematode vs. Chemical Control of *Carposina nipponenesis* in Apple Orchards[a]

Year	Treatment[b]	% Larval mortality	% Fruit damaged
1983	Nematode	94	1.9
	Phoxim	77	2.7
1984	Nematode	100	2.3
	Phoxim	89	2.4
1985	Nematode	99	0.1
	Phoxim	96	0.2
1986	Nematode	92	2.8
	Parathion	86	2.6
	Diazinon	78	3.5

[a]   Data from field trials undertaken by Dr. Lee Yan, Zhengzhou Pomology Institute.

[b]   Nematodes = 1-2 billion/ha; Insecticides (phoxim, parathion, and diazinon) = 7.5 kg/ha.

In China, *S. carpocapsae* is presently being used commercially for the control of the tree-boring cossid moth, *Holcocercus insularis*. Larvae of this moth are responsible for killing tens of thousands of shade trees in China's northern cities every year and constitute a serious pest in gingko, maple, and flowering crab apple plantations. To indicate the scale of this problem, Beijing has 14 million and Tianjin 2 million shade trees with up to 5% of the trees infested with larvae of this cossid moth. In Tianjin, authorities have been battling unsuccessfully against this insect for 15 years, losing 10,000 mature trees annually. Previous control has relied on chemical pesticides within the city (30 metric tons in 1986). The use of chemical fumigants for shade trees in the streets has posed health hazards to citizens. During 1988, 14,000 shade trees were treated with *S. carpocapsae* in Tianjin and Shi Jiazhang, and it is expected that another 100,000 trees will have been treated in five cities in northern China by the end of 1989.[17]

The adult cossid moth lays its eggs on the tree bark, the eggs hatch, and the larvae rapidly crawl under the bark and bore into the wood of the tree. They form extensive, interconnecting galleries containing 50-1000 larvae. The galleries seriously weaken the tree and eventually lead to its death. Because larvae are only exposed on the tree surface for a brief period, surface spraying of chemical insecticides or the use of contact biological control agents such as the insect pathogenic fungus *Beauveria* are ineffective. Larvae deep in the wood have been killed with a chemical fumigant (i.e., DDVP; O,O-dimethyl-O-2,2-dichlorovinylphosphate); however, to be effective the numerous entry and exit holes of the larvae have to be blocked, an extremely time consuming and expensive procedure. There is a high probability that missed holes will allow escape of the noxious fumigant into the city streets. The ability of entomopathogenic nematodes to seek their target insect by dispersing through the galleries[19,20] means that injecting nematodes into the top-most gallery opening on the tree will be effective and plugging of other openings is unnecessary. Indeed, injection of *S. carpocapsae* Agriotos strain into the infested tree trunk is followed within days by a mass exodus of hundreds of dying larvae, their behavior modified by parasitization with the nematode.[17] These insects fall to the ground and ring the tree with infested cadavers. Moreover, larvae dying within the tree are a source of hundreds of thousands of new infective nematodes which can attack uninfected insects. Thus, nearly 100% of the insects are eventually killed.

Field control of the lychee bark borer, *Inoarbela dea*, using *S. carpocapsae* recently commenced, with 2300 trees from three orchards being treated, and pest mortality ranging from 91 to 98%.[18] Investigations into the possibilty of controlling several other insect pests in China are at various stages of development.

To fully exploit the potential for nematode control of insects in China, certain areas require further development and consolidation. These include bringing nematode-based control of *Carposina* and *Holcocercus* into standard practice, continuing research on other pests, improving in culture and quality control procedures, and establishing and economically evaluating pilot production plants. Because of the highly seasonal demand for nematodes for *Carposina* control, perhaps the most important research area is long-term mass storage.

XII. CONCLUSIONS

Several critical considerations for potential collaboration have been identified. These include the selection of competent and enthusiastic partners, clear and frequent communication, reciprocal visits to maximize the mutual benefit of the cooperation, and the evaluation of a variety of entomopathogenic nematode species against target pests. The development of programs of collaboration on entomopathogenic nematodes between developing and industrialized countries will do much to advance the discipline and hopefully help to foster good relations between the countries concerned. It benefits both countries by providing additional funds and stimulus for the development of an important alternative to chemical control measures.

ACKNOWLEDGMENTS

Drs. Ray Akhurst, John Curran, Li Liying, Li Pingxu, Mr. Wang Jin Xian, and Mrs. Yang Huaiwen have played a considerable part in developing the strategy for the ACIAR funded collaborative project on entomopathogenic nematodes between Australian and Chinese institutions. I thank Drs. Akhurst and Curran for their suggestions and comments on this paper.

REFERENCES

1. **Bedding, R. A.**, unpublished data, 1983-87.
2. **Bedding, R. A., and Curran, J.**, Utilization of entomopathogenic nematodes to control insect pests, ACIAR Project No. 8451, Report 1985-1988, 1988.
3. **Persley, G.**, Agricultural biotechnology opportunities for international development, World Bank-ISNAR-AIDAB-*Heterorhabditis* spp., *Neoaplectana* spp. and *Steinernema* ACIAR, Synthesis Report, 1989.
4. **Bedding, R. A., Molyneux, A. S., and Akhurst, R. J.**, *Heterorhabditis* spp., *Neoaplectana* spp. and *Steinernema kraussei*: interspecific and intraspecific differences in infectivity for insects, *Exp. Parasitol.*, 55, 249, 1983.
5. **Bedding, R. A., and Miller, L. A.**, Use of a nematode, *Heterorhabditis heliothidis*, to control black vine weevil, *Otiorhynchus sulcatus*, in potted plants, *Ann. Appl. Biol.*, 99, 211, 1981.
6. **Bedding, R. A., and Pinese, B.**, unpublished data, 1982.
7. **Treverrow, N., Parniski, P., and Bedding, R. A.**, unpublished data, 1989.
8. **Molyneux, A. S.**, *Heterorhabditis* spp. and *Steinernema* (=*Neoaplectana*): temperature, and aspects of behaviour and infectivity, *Exp. Parasitol.*, 62, 169, 1986.
9. **Bedding, R. A., and Akhurst, R. J.**, A simple technique for the detection of insect parasitic rhabditid nematodes in soil, *Nematologica*, 21, 109, 1975.

10. **Popiel, I., Holtenmann, K. D., Glazer, I., and Womersley, C.,** Commercial storage and shipment of entomogenous nematodes, Int. Patent WO 88/01134, 1988.
11. **Bedding, R. A., Akhurst, R. J., and Molyneux, A. S.,** unpublished data, 1983.
12. **Dutky, S. R., Thompson, J. V., and Cantwell, G. E.,** A technique for the mass propagation of the DD-136 nematode, *J. Insect Pathol.*, 6, 417, 1964.
13. **Bedding, R. A.,** Low cost in vitro mass production of *Neoaplectana* and *Heterorhabditis* species (Nematoda) for field control of insect pests, *Nematologica*, 27, 109, 1981.
14. **Bedding, R. A.,** Large scale production, storage and transport of the insect parasitic nematodes, *Neoaplectana* spp. and *Heterorhabditis* spp., *Ann. Appl. Biol.*, 104, 117, l984.
15. **Bedding, R. A.,** Apparatus and method for rearing and harvesting nematodes, Aust. Patent Appl. PJ 0630/88, 1988.
16. **Bedding, R. A.,** Storage of entomopathogenic nematodes, Int. Patent Appl. PCT/AU88/00127, 1987.
17. **Yang, H.,** unpublished data, 1989.
18. **Wang, J. X.,** unpublished data, 1989.
19. **Lindegren, J. E., Yamashita, T. T., and Barnett, W. W.,** Parasitic nematode may control carpenterworm in fig trees, *Calif. Agric.*, 35(1), 25, 1981.
20. **Deseö, K. V., Grassi, S., Foschi, F., and Rovesti, L.,** Un sistema di lotta biologica contro il rodilegno giallo (*Zeuzera pyrina* L.; Lepidoptera, Cossidae), *Atti Giornate Fitopathol.*, 2, 403, 1984.

Biotechnology and Genetics

13. *Caenorhabditis elegans* as a Model for the Study of Entomopathogenic Nematodes

András Fodor, Gabriella Vecseri, and Tibor Farkas

I. INTRODUCTION

Nematodes have been model organisms for developmental biologists for a century. The nematode *Caenorhabditis elegans* (Rhabditidae) occupies a unique place in developmental genetics because it is the only animal for which the complete cell lineage is known, from the single cell of the zygote to the thousand or so differentiated cells of the adult. Largely as a result of the work of Sydney Brenner[1] and his followers,[2,3] *C. elegans* has become the most completely understood metazoan in terms of molecular and classical genetics, development, behavior, and anatomy.[1,3] Moreover, *C. elegans* is one of the most well-known model organisms for molecular, developmental, and neuro-genetic studies on eukaryotic animals.[3] The methods developed for and the information obtained from this nematode should be widely applicable to heterorhabditid and steinernematid nematodes. The families Rhabditidae, Heterorhabditidae, and Steinernematidae are included in the Superfamily Rhabditoidea; the species within these families are closely related.[3-5]

Brenner chose *C. elegans* as his experimental animal with the objective of understanding the genetic program of development and the organization of a complex organism. *C. elegans* possesses a limited number of cells, a good reproductive capacity, rapid development, and suitability for genetic analysis and gene manipulations.[1,3,6] The first articles by Brenner[1,6] and Sulston and Brenner[2] pioneered this program, and about 60 laboratories throughout the world are currently conducting research on *C. elegans*.

The unique feature of Brenner's concepts in *C. elegans* research is that almost all complex phenomena (e.g., timing of cell lineages, dauer formation and recovery, aging, behavior, and drug resistance) have been described in genetic and molecular terms. More than 75% of the nematode genome has been physically mapped and is kept in DNA libraries; the whole cell lineage chart has been reconstructed and described at phenological and genetic levels.[3]

We do not intend to review recent *C. elegans* research. Our aim is rather to discuss this research from the aspect of its potential for studying entomopathogenic nematodes. We have focused on some selected features of *C. elegans* genetics and developmental biology (with special emphasis on dauer formation and recovery), and on a few aspects of the molecular biology and gene manipulations which, we believe, might be adopted for use in research on entomopathogenic nematodes. When comparing the genetics of *C. elegans*,

Steinernema spp., and *Heterorhabditis* spp., we must realize how much is known about *C. elegans* and how little about entomopathogenic nematodes.

II. COMPARISON OF THE DEVELOPMENTAL BIOLOGY OF FREE-LIVING AND ENTOMOPATHOGENIC RHABDITIDA

A. Alternative Life Cycles

1. Embryonic and Postembryonic Development

The life cycles of the Rhabditida (including *Caenorhabditis, Heterorhabditis*, and *Steinernema* genera) are similar. Embryonic development lasts about 1 hr and can be separated into a cell proliferation and a morphogenetic (cell migrating) stage. Hatching is followed by various juvenile stages (J_1-J_4) preceding the adult stage. Each developmental stage is followed by a short quiescent state (called lethargus) and molt. The lengths of the different developmental stages are temperature dependent. In *C. elegans*, at 25°C, the average lengths of the different developmental stages are as follows: J_1, 11.5 hr; J_2, 7 hr; J_3, 7.5 hr; and J_4, 11.5 hr. At 25°C, a hermaphrodite produces about 320 self-fertilized eggs over 4 days. However, if permanently mated by males, the hermaphrodite may produce over a thousand cross-fertilized eggs during 5-6 days.[7] The life span extends up to 3 weeks. *C. elegans* growth and reproduction occurs between 15 and 25°C, whereas *C. briggsae* and *C. remanei* grow and reproduce between 15 and 27.5°C. The optimal temperature for entomopathogenic nematodes, just like their average life span at different temperatures, is not definitely known.

The duration of each juvenile stage at different temperatures can easily be determined for entomopathogenic species by applying the "pumping curve" method of Swanson and Riddle.[8] A well-synchronized population is essential, therefore eggs should be obtained from gravid females or hermaphrodite nematodes by using hypochlorite.[9] Eggs surviving the treatment are incubated in sterile M9 buffer[1,3] for 24 hr, during which time the J_1 hatch, but do not grow and develop until receiving food. Both BIOSYS, Inc. (Palo Alto, California) and we have used this technique for entomopathogenic nematodes. A J_1 population obtained from hypochlorite-treated eggs of *S. carpocapsae* grows highly synchronously either in an artificial liquid medium or in hemolymph.[10]

Caenorhabditid development can be monitored by scoring the rate of pharyngeal pumping of individual animals on an agar (NGM)[1,3] plate using a stereomicroscope. Pumping continues until the premolt lethargus starts and resumes after molting. Fifty to 100 nematodes should be scored at each time point, and the cultures held in an incubator controlled at ±0.5°C. The observation period should not exceed 5 min at ambient temperature.

A typical pharyngeal pumping curve of *C. briggsae* (G16 strain) is shown in Figure 1.[11] Development was initiated by placing the synchronized J_1 on

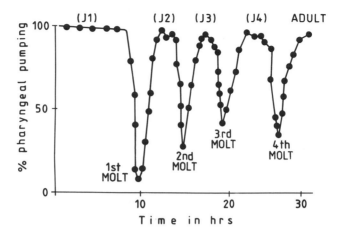

Figure 1. A typical pumping curve of a growing population of *C. briggsae* (G16) synchronized at hatch.[14] This curve is based upon the scores of 87 individual *C. briggsae* juveniles.

food. The abscissas of the lowest points of the curve denote the average time for the respective molt.

2. Alternative Developmental Pathway: Dauer Formation

a. Physiological Aspects of the Dauer State

Under certain conditions, the third developmental stage of some species of Rhabditoidea, including all known *Caenorhabditis*, *Steinernema*, and *Heterorhabditis* spp., is facultatively replaced by a unique, morphologically distinct, nonaging, nonfeeding, alternative developmental variant called a dauer[1,12,14,15] juvenile.

The dauer juvenile of entomopathogenic nematodes is capable of tolerating stresses fatal to other developmental forms, and of infecting new hosts, and is therefore also called an infective juvenile or infective third-stage juvenile. Embryogenesis, stages J_1-J_4 of postembryonic development, and the adult stage of entomopathogenic nematodes occur within the insect host. The infective juvenile is free-living, leaving the host to infect a new host.[5]

The dauer juveniles of caenorhabditids and the infective juveniles of steinernematids and heterorhabditids are capable of active movement, display an altered energy metabolism, and are arrested in development and aging.[12,13] The *C. elegans* dauer can survive four to eight times longer than the 3 week life span of the nematode that has by-passed the dauer stage.[14,15] *Caenorhabditis* dauers cannot survive long periods at 5°C, whereas *Steinernema* infective juveniles can survive for months or years at 5°C. The consumption of stored energy may be a major factor limiting life expectancy.

Environmental factors influencing the development of the caenorhabditid dauers include a pheromone, food supply, and temperature.[11,15,16,19] The dauer-inducing pheromone results in an altered developmental path at the second molt. The proportion of a population that can be induced to form dauers in response to the pheromone depends upon the presence of food.[15] Temperature-shift and pheromone-shift experiments on synchronous cultures of *C. elegans* proved that discrimination between dauer formation and continued growth begins no later than the first molt.[17] *C. elegans* acquires the full complement of morphological, physiological, and behavioral properties characteristic of dauers at the end of the second postembryonic stage, a morphogenetic process that takes about 12 hr at 25°C.[18] Dauer formation can be reversed by changing one or more of the three environmental parameters prior to the second molt.[19]

Some arrested developmental forms of numerous plant and animal parasitic nematodes exist as obligatory stages in their life cycle (Figures 2B, C).[15,20] In contrast, the dauer juvenile form of many Rhabditoidea is an example of a facultative stage (Figure 2A).[15]

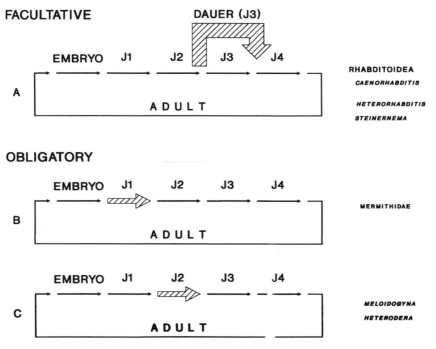

Figure 2. Dauer formation. The dauer stage ( ) of many Rhabditoidea occurs facultatively after the second molt instead of the normal third developmental stage, in response to unfavorable environmental conditions (i.e., crowding, absence of food), mediated by a genus-specific pheromone.

The infective juvenile of entomopathogenic nematodes and the dauer juvenile of caenorhabditids are comparable in many respects. Both exhibit behavior not observed in other developmental stages. Both dauer and infective juveniles tend to crawl up objects that project from the substrate, stand on their tails, and wave their heads back and forth (nictation).[21] In the soil environment, this behavior by steinernematids may permit attachment to passing insects.

Morphologically, both the caenorhabditid and entomopathogenic dauers are thin and dense as a consequence of radial shrinkage of the body at the dauer-specific molt.[12] The body wall cuticle is thick and, when viewed in transverse section, contains a radially striated inner layer not found in other stages.[22] The *C. elegans* dauer cuticle also contains a dauer-specific collagen.[23,24]

After 1 hr of radial shrinkage, the caenorhabditid dauer acquires resistance to nonionic detergent treatment,[12] presumably as a result of cuticle modification and sealing of the buccal cavity by a cuticular block.[15,23,25] Both caenorhabditid and entomopathogenic dauers are resistant to 1% sodium dodecyl sulfate (SDS).[12] Unlike entomopathogenic rhabditoids, *Caenorhabditis* dauers are extremely sensitive to 0.5% of Hyamine 2389 (SERVA).[26] Electron microscopy reveals that the intestinal lumen of the nonfeeding dauer is shrunken, and the microvilli are small and indistinct.[25,27] The ventricular portion of the intestine of the steinernematid infective juvenile is specifically modified for storage of the symbiotic bacteria.[28,29] Symbiotic bacteria are located in the isthmus of the pharynx and in the ventricular portion of the intestine in infective stage *H. bacteriophora*.[30]

Unlike other developmental forms,[31] *C. elegans* dauers do not respond promptly to chemical stimulus[27] apart from cyclic AMP.[32] No systematic chemotaxis studies have been made on the different developmental stages of entomopathogenic nematodes. Orientation[33] and host-finding behavior[34] of *S. carpocapsae* infective juveniles have been demonstrated. Detectable genetic variability among steinernematid strains[34] may make it possible to study the behavior of entomopathogenic infective stages by genetic means. Enhanced host-finding, for instance, proved a selectable trait for *S. carpocapsae*.[35]

Sunlight and ultraviolet irradiation adversely affect entomopathogenic nematodes.[36] *C. elegans* dauers are less sensitive to ultraviolet than other developmental forms, but a relatively small dose of radiation, which is not fatal for other postembryonic developmental forms, is sufficient to inhibit their recovery (i.e., molting to the fourth-stage juvenile). Irradiated dauers can survive for many days without recovery in the presence of food.[37] X-ray irradiation doses used in *C. elegans* research (up to 7500 R)[41-43] do not significantly influence the recovery of *S. carpocapsae* infective stages.[41]

The developmental pathway leading to the dauer stage in *C. elegans* is induced by the Dauer Recovery Inhibiting Factor (DRIF) pheromone discovered by Golden and Riddle.[17,19] DRIF is secreted by the nematodes throughout their life, including the dauer stage,[15] even though the dauer excretory gland is inactive.[42] The pheromone is thought to mediate unfavorable external conditions (e.g., starvation and crowding) to a receptive developmental variant (late first postembryonic stage). Some alterations in the second postembryonic developmental stage result in predauer[15] and dauer formation. We presume that the mechanism of infective juvenile formation of entomopathogenic nematodes follows a comparable pathway, using a pheromone chemically different from that of caenorhabditids. Maintaining single caenorhabditid or steinernematid J_1s in complete or partial starvation does not result in dauer induction.[43] Dauers always occur in crowded conditions, sometimes in the presence of the food,[11] indicating that starvation itself does not induce dauer formation, but crowding is an essential environmental precondition.

As for the specificity of DRIF, Fodor et al.[11] extracted DRIF from both *C. elegans* and two strains of *C. briggsae* and found cross-reactivity, indicating that DRIF was genus-specific. Extracts from other free-living rhabditids were inactive in caenorhabdids.[11] Ohba and Ishibashi[16] showed that DRIF isolated from *C. elegans* was inactive in *S. carpocapsae*. Subsequently, Fodor et al.[43] extracted a chemical fraction from *in vivo* and *in vitro* cultures of *S. carpocapsae* which induced infective stage formation. This extract proved genus- and species-specific.[41] The chemical nature and the purification conditions of the *Caenorhabditis* and *Steinernema* pheromones are obviously different.[43] The most reproducible bioassay of each pheromone is based upon its capability to induce dauer formation in the presence of a limited amount of competitive "food signal". The "food signal" for caenorhabditids is an unknown component of the yeast extract.[17-19] We suppose that insect hemolymph contains a component acting as a "food signal" for steinernematids and heterorhabditids.

The dauer stage may be terminated in response to conditions that are more favorable for growth and reproduction. In caenorhabditids, recovery from the dauer stage occurs within 1 hr after the animal is placed in a fresh environment[15] at an extremely low level of the pheromone, and in the presence of food or "food signal".[11] After 2-3 hr the dauer juvenile begins to feed, resumes development irreversibly, loses its resistance to nonionic detergents, and after an additional 8-10 hr molts to the fourth stage. There is a slight difference between *Caenorhabditis* and entomopathogenic rhabditidid species in this respect. The *C. elegans* dauer and to some extent the heterorhabditid infective juvenile, recover in an empty pheromone-free agar plate, even in the absence of food. *S. carpocapsae* infective juveniles, however, never recover in an empty plate.[43]

b. Developmental Genetics of Dauer Formation and Recovery

The interactions of epistatic genes involved in the developmental pathway of *C. elegans* dauers have been discovered by Riddle and co-workers.[15,18,44] There are wild type genes for determining function in dauer formation and recovery. Mutations of these genes result in disturbances of dauer formation and recovery.

One class of recessive mutants shows a dauer-defective (*daf-def*) phenotype. This means that animals homozygous for a *daf-def* mutation are unable to form dauers. There is another class of recessive, thermosensitive mutations of dauer-constitutive (*daf-const*) phenotypes. This means that nematodes homozygous for a *daf-const* mutant develop to the dauer at a nonpermissive temperature (above 24°C) either in the presence or absence of the pheromone and/or the food. All developmental aspects of these mutants have been described in detail.[14-18,44] While constructing double *def/def*, *const/const* mutant strains, Riddle and co-workers[14,15,19,44] found several epistatic relations between them, on the basis of which they reconstructed the genetically determined developmental pathway (Figure 3). Mutant phenotypes suggest that the pathway corresponds to neural processing of environmental stimuli.[27] If the constitutive mutants were simply blocked in entry to the third postembryonic developmental stage, while the defective ones were blocked in a separate sequence of the dauer, the constitutive-defective double mutants could not continue further development. Instead, the data show that the defective mutants were blocked in a pathway to the dauer stage, while the dauer-constitutives generated a false signal, causing the mutants to form dauers even in the absence of the environmental cues. If the pathway is blocked after the false signal, a double mutant will be dauer-defective. If the false signal is generated after the block, a double mutant will be dauer-constitutive.[15]

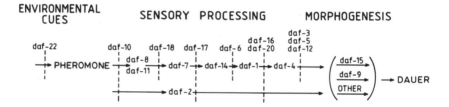

Figure 3. Riddle's scheme on the genetically determined developmental pathway of dauer formation and recovery of *C. elegans*; based upon the analysis of different dauer-defective/dauer-constitutive double mutants. *daf*=gene playing a role in dauer development. *daf-2*, *daf-8*, *daf-11*, *daf-14*, and *daf-4* are "dauer constitutive" (*daf-const*) genes: if any of them do not function, the nematodes grow to dauers spontaneously even in the absence of the DRIF pheromone at high temperature (25°C). *daf-22*, *daf-10*, *daf-18*, *daf-17*, *daf-6*, *daf-16*, *daf-20*, *daf-3*, *daf-5*, and *daf-12* are dauer-defective (*daf-def*) genes. Their normal function is essential for dauer formation. Other *daf* genes play a role in the morphogenesis of dauer formation (courtesy of D. L. Riddle).

The number of entomopathogenic infective juveniles formed in nature is dependent on the number of nematodes present in the insect cadaver (i.e., on the pheromone concentration present).[43] In the first generation, the progeny of those initiating the host infection are relatively few in number; consequently, there is a low concentration of the pheromone and few infective juveniles are produced. In the second and particularly in the third generation, the majority of the progeny develop to infective juveniles. By using a different size of inoculum (5-500), we can produce an increasing number of infective juveniles in the second or even the first generation,[45] suggesting that a pheromone must play a determining role in infective juvenile formation.

A maternal-effect dauer constitutive mutation (such as several alleles of the *daf-1* gene in *C. elegans*)[12,14,15,44] in an entomopathogenic nematode species might permit the synchronous production of infective stages on a large scale.

B. Genetics

1. Nematode Genetics in General

Conventional and molecular genetics have great potential for improving entomopathogenic nematode/bacterial symbiont complexes to yield powerful biological pest control agents.

a. Genetic Systems and Sex Determination

The genetic consequences of the various sex determination mechanisms differ, and the strategy for any kind of genetic analysis depends upon the mode of reproduction of the organism. The success of *C. elegans* research is partly due to the concepts and strategies of analysis of single gene mutants, and to the sex determination of *C. elegans*. In the latter case, use was made of the automictic (self-fertilizing) reproduction of the hermaphrodite and amphimictic crosses between males and hermaphrodites.

The formal or classical genetics of *C. elegans* is based upon the mutagenesis of a self-fertilizing hermaphrodite of XX+5AA karyotype. The mutagenized nematode produces progeny by selfing. A certain fraction of these F_1 progeny are heterozygous for some mutant allele for which a Mendelian fraction of their F_2 self-progeny become homozygous. Through a series of selfing, the "noises" of the genetic background can be eliminated and an isogenized pure Mendelian line carrying only a single mutation in homozygous form is obtained. For further genetic studies (e.g., transferring genes, studying gene and/or allelic interactions, mapping, constructing multiple mutant strains, testing complementation, etc.), the homozygote mutant hermaphrodite is usually crossed out with a wild type (or some heterozygous) X0 male.

There is great variation in the reproductive modes shown in the Phylum Nematoda.[46] Current studies based upon genetic analysis of interacting hypomorph (loss-of-function) and neomorph (gain-of-new function) mutant alleles of genes playing key roles in sex determination and/or dosage compensation in *C. elegans*[47] could promote an understanding of this reproductive variation.

b. Difficulties in Genetics of Entomopathogenic Nematodes

Steinernematids and heterorhabditids have different and more complex types of sex determination than *C. elegans*. These differences limit the options for genetic study in these nematodes.

Most nematodes, including steinernematids, are gonochoristic,[47] meaning that they have morphologically different male and female sexes, and reproduction occurs exclusively by conventional sexual (amphimictic) cross-fertilization. In genetic studies on a *Steinernema* species, we may benefit from the information and experience which have accumulated during the 70-year history of *Drosophila* genetic investigations. Moreover, *C. remanei* is a nematode species with a male/female sex system and may be a model organism for the study of steinernematids.

Autotokous (uniparental) reproduction has arisen independently in different species throughout the Nematoda. Autotoky may take the form of self-fertilization of a hermaphrodite or parthenogenesis.[46] The two mechanisms are completely different from the aspect of genetics. Classical genetic studies on an exclusively parthenogenic (or pseudogamic) organism are impossible, whereas the self-fertilization system, especially if combined with occasional male/hermaphrodite crosses, is a paradise for a geneticist.

If the genetic determination of *Heterorhabditis* spp. were the same as that of *C. elegans*, *Heterorhabditis* genetic research would be easy. As it is, there are problems. Heterorhabditid nematodes exhibit a generation dimorphism: infective juveniles infecting the hosts give rise to large automictic hermaphrodites, where as their progeny consist of smaller males and females.[46] In addition, *Heterorhabditis* species are heterogamic (i.e., alteration of amphimictic and autotokous generations). The hermaphrodites are automictic and also deuterotokic (producing male and female progeny), producing sperm in the ovotestis. *Heterorhabditis bacteriophora* shows protandrous development, with accumulated sperm in the gonoduct of the hermaphrodite.[48] The female vulva serves mainly for insemination and, rarely, for oviposition.[48]

There are several *Heterorhabditis* strains of outstanding practical importance (e.g., HP88, HL81, HW79) which have not been identified taxonomically or cytogenetically. A systematic analysis including both cytology, cross-fertility, and sex determination of these isolates is an essential precondition for any serious genetic program.

The frequency and role of heterorhabditid males in the second and third generations are unclear. It remains to be clarified whether a real amphimictic cross or merely pseudogamic mating is going on in these generations. One feature of this problem is whether the large hermaphrodites from the first generation are different from the second and third generation females in biological terms (i.e., whether the second generation females produce sperm). A detailed comparative anatomy of the dimorphic females and a cell lineage study of vulval development, and of Z and P cell lineages[49,50] in general, would probably resolve this question. Moreover, the exact frequency of the males in the dimorphic generation must be clarified, and their karyotype should be determined. If they turn out to be real X0 males, the mechanism of their origin needs to be determined.

2. How to Benefit from the Results of C. elegans Genetics

a. Genetic Analysis in a Male/Hermaphrodite Sex-Determination System

The free-living *C. elegans* and *C. briggsae* have hermaphrodite and male sexes, but no wild type female sex. Males are rarely found in natural or laboratory populations. For example, in our laboratory stock of *C. briggsae* (G16), the first male was found more than 2 years after the strain was isolated from soil. The spontaneous males are invariable X0 and must arise during meiosis.[51] Males are capable of cross-fertilizing hermaphrodites to yield equal numbers of male and hermaphrodite cross-progeny. The higher growth rate of the hermaphrodite population results in the frequency of males in most populations remaining close to the rate of X-chromosome loss (less than 0.5%).[47] Whether the situation is comparable in *Heterorhabditis* grown under *in vitro* conditions remains to be determined.

The chromosomes of *C. elegans* can be seen best by using fluorescent microscopy. Fluorescent dyes such as Hoechst 3358 have been used for viewing both meiotic[38] and mitotic[52,53] chromosomes. Albertson[54,55] has developed methods for localizing genes in the chromosomes by *in situ* hybridization of cloned probes to mitotic chromosomes.

The chromosomes of oocytes at diakinesis are highly condensed, and the six bivalents present in wild type diploid hermaphrodites are generally indistinguishable cytologically. Mitotic chromosomes are best seen in early (less than 50 cells) embryos,[52] when they appear as stiff rods 1-2 μm in length.

Following publication of Nigon's data,[56,57] Madl and Herman[58] used heat shock to establish tetraploid stocks of *C. elegans* (Bristol). Tetraploids might be useful tools in *Heterorhabditis* genetics in order to (1) overcome fertility problems between strains, (2) construct stable and productive tetraploid strains, and (3) determine the ploidy levels of existing isolates. Trisomics[57,58] with a higher resistance to stress might also be isolated.

The question of whether *Heterorhabditis* spp. are tetraploid is still open. If they are, a strategy for the genetic analysis of tetraploid nematodes as elaborated by Herman[58,59] on *C. elegans* should be adopted. With regard to "ploidity-level polymorphism" of some nematode genera, such as *Meloidogyne*,[60] the ploidy levels of the different *Heterorhabditis* species and strains should be determined as well.

C. elegans chromosomes, and probably those of entomopathogenic nematodes, have diffuse kinetochores. The metaphase chromosomes lack any visible constriction, which commonly marks the position of the centromere or kinetochore of a monocentric chromosome. Albertson and Thomson[52] have made reconstructions of *C. elegans* kinetochores using electron micrographs of dividing nuclei in serially sectioned embryos. The classically localized centromere serves two meiotic functions: (1) it provides sites for the attachment of spindle fibers, and (2) it plays a role in the orderly dysjunction of the meiotic chromatids of bivalents by keeping sister chromatids joined during meiosis I and splitting in meiosis II. Although the spindle fibers can attach throughout the length of a diffuse kinetochore, and crossovers can also occur at variable positions along the chromosome's length, it is extremely difficult to observe how the sister chromatids remain attached throughout their lengths during meiosis I. As the chromosomes of *C. elegans* become very small and condensed during late diakinesis, their bivalent structure cannot easily be interpreted. Whether the first meiotic division of the nematode is equational or reductional is unknown. A diffuse kinetochore has been reported for another nematode species,[61] but nothing has been published about heterorhabditids or steinernematids.

Brenner[1] first induced *C. elegans* mutants with ethyl methanesulfonate (EMS) and used different mutant hunt protocols to search for nonessential genes. He applied a series of recessive EMS-induced mutants of visible phenotypes as markers to map chromosomes. His total estimate, based upon EMS-induced lethals on the X chromosome, which is regarded as $1/6$ of the genome, was 1800 essential genes per genome. When considering steriles, 3000 genes with indispensable functions were estimated.[58] There is no reason to suppose a significantly higher number of genes in the genome of *Heterorhabditis* and *Steinernema* species, but the question remains of how to perform genetic studies on these organisms. A detailed genetic (linkage) map analogous to that of *C. elegans*[1,62] will be the first important step in the genetics of entomopathogenic nematodes. Isolation of entomopathogenic nematodes homozygous for a morphological mutation expressed in the infective juvenile state is not easy. There are few visible mutants of *C. elegans* which can be recognized in the dauer state. Drug-resistant[63,64] and behavioral[31,65-68] mutants, however, would be extremely useful tools for elucidating the genetics of entomopathogenic nematode species. Another useful mutant would be a maternal-effect dauer-constitutive.

b. Molecular Approaches to Some Problems in Nematode Genetics

It may be speculated that some conserved genes of *C. elegans* must be very homologous to those in entomopathogenic nematodes, allowing progress in the molecular genetics of entomopathogenic nematodes before their formal genetics has been elaborated. Many of these genes have been identified by direct molecular analysis rather than mutation. There are recessive actin mutants of wild type phenotype[70] and dominant ones of uncoordinated phenotype.[71]

In 235 kb cloned DNA surrounding the five X-linked vitellogenin genes, Heine and Blumenthal[69] identified 12 transcriptionally active genes. If this spacing of 20 kb per gene were typical of the entire genome, *C. elegans* would have a total of about 4000 genes (essential and nonessential).[59] Many genes of indispensable function are present in multiple copies in the genome. Further, several indispensable functions are coded by more than one gene. Actin, major sperm protein, acetylcholinesterase, vitellogenins, collagen, and tRNA are the best examples.

The Bristol and Bergerac strains of *C. elegans* exhibit numerous restriction fragment length differences.[9] Such DNA polymorphisms, which may be found either in *Heterorhabditis* or *Steinernema* strains, have been treated as standard phenotypic markers in two- and three-factor crosses[72-74] to localize cloned genes on the genetic map.

C. elegans is the only eukaryotic organism in which suppressor mutant genes are known. The existence of the *sup-7* amber suppressor[75] and the possibility of its integrated transformation[76] in *C. elegans*, allow the isolation of amber (UAG) mutant alleles of different genes. Whether *sup–7* can be expressed in entomopathogenic nematodes is unknown, but if it can, it would be an excellent molecular genetic tool.

The Carnegie group of *C. elegans* researchers in cooperation with the Cambridge and St. Louis groups have developed an excellent transformation system for *sup-7* and for several other important genes.[76] They have made both the cloned genes and the detailed protocol available.

The recently discovered allele-specific *smg* class of suppressors[77] seem to affect mRNA processes and might be an excellent tool for isolating mutants defective in the noncoding terminator sequences of important genes.

c. Chromosome Mechanics as a Tool of Nematode Genetics

We have recently started X-ray mutagenesis in entomopathogenic nematodes to help elaborate their formal genetics. In *C. elegans*, sets of overlapping deficiencies (i.e., deletions) at different breakpoints can be isolated and used to localize subsegments at a given region of the linkage map.[1,62] If a recessive homozygote lethal allele fails to complement at least two closely linked genes, it is assumed to be a deletion.

The most efficient method of selecting deficiencies makes use of dominant or semidominant mutants with a visible phenotype. Inactivation of such a mutant by deficiency formation can lead to an altered phenotype.[71] Other means of obtaining deletions in different segments of the genome have recently been reviewed by Herman.[59]

The balancer chromosomes are the most important tools in the genetics of essential genes whose mutations cannot be kept in homozygous strains. The balancers suppress recombination in a given segment of the genome. Translocations in the heterozygous state are routinely used as balancers in *Drosophila* genetics. In *C. elegans* genetics, Herman et al.[38] isolated an (X;V) translocation and used it to keep X-linked lethals in heterozygote strains.[59] The same translocation was used to balance lethals on linkage groups (LGV).[76]

The general scheme for isolating a balancer chromosome is to look for the absence of recombinant progeny of an ab+/++c heterozygote hermaphrodite first generation progeny of an irradiated parent(s). In order to isolate an X-linked balancer we irradiated *unc-8, +, unc3/unc-8, + unc-3* hermaphrodites, crossed them out with +, *lon-, +/0* males, transferred single *dpy-8, +, unc-3/ +, lon-2, +* hermaphrodite progeny and looked for the absence of *dpy* and *unc* recombinants of their self-progeny, but without success. However, when *dpy-8, +, unc-3/+, lon-2* hermaphrodites were irradiated, we isolated an excellent balancer szT1.[78] The most frequently used balancers are mnT1(II;X),[39] eT1,[79,80] szT1,[78] sT1 (III;X), and sT2 (IV;V).[59,81] For LGII, mnC1, an inversion balancer[39,40] is the most useful.

Translocations have been used to mark particular linkage groups cytologically.[54] The isolation and analysis of translocations during meiosis may help to elucidate meiotic chromosome pairing and segregation in nematodes such as *Heterorhabditis*. Free duplications permit the construction of segmental triploids for the study of gene dose effects and the generation of mosaics in nematodes.[38]

d. Genetic Analysis in the C. elegans System

The best examples of genetic analysis in the *C. elegans* system relate to sex determination and dose compensation,[47] muscle organization,[82] the dauer pathway,[15] and the embryonic and postembryonic cell lineage.[49,50]

The E stem cell (Figure 4) lineage of steinernematid nematodes may be of interest, for the bursa intestinalis, where the symbionts are located in the infective juvenile, is probably comprised of E cell progeny. The differences between the E cell lineages of different entomopathogenic nematodes might shed light on the genetic and cellular basis of the species-specific symbiosis between nematodes and *Xenorhabdus* spp. bacteria. Although the genetics of entomopathogenic nematodes is far from that level, the most unique features of the methods of genetic analysis in *C. elegans* are worth emphasizing. The first step is to identify those genes which are involved and occupy a key role in the genetic pathway. The hypomorphic and O-alleles (loss-of-function alleles), either themselves, or in heterozygotic combination with a deletion, or with another allele of the same gene, give information on the consequences of the partial or total loss of the gene concerned. The overproducing or neomorph (gain-of-new-function) alleles, either alone or in combination with another allele or deletion, provide information about the consequences of the loss of regulation of the gene in question.

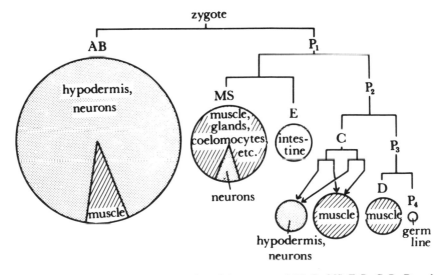

Figure 4. Sulston's scheme showing the fate of the progeny of AB, P_1, MS, E, P_2, C, P_3, D, and P_4 embryonic stem cells of *C. elegans* (courtesy of J. F. Sulston).

Epistatic relations between different alleles of different genes participating in a physiological or developmental function (as *daf* genes do)[15] form a cascade of genes that yield information about the pathway concerned.

Partial triploids can be produced by using free duplications.[39,59] The free duplications are inclining to be lost during mitosis, producing triploid/diploid genetic mosaics for the chromosome region covered by the duplication. In order to determine the function of a gene in some cells or in a special tissue, partial triploids are produced by crossing a hermaphodite homozygous (-/-) for a hypomorph recessive (-) allele of the given gene to a male carrying the wild type allele (+) of the same gene on a free duplication. The progeny will be of Dp(+)/-/- genotype. Those cells and their mitotic progeny from which Dp(+) were lost, will also lack the gene function. This so-called mosaic analysis is used to determine the cells and developmental stages within which a special gene is active. With the help of the transposon-tagging method,[82-84] the gene can readily be located, cloned, and analyzed in molecular terms.

There is hardly any other genetic system in which such a repertoire of sophisticated classical and molecular genetic and cellular biological methods can be used in combination to solve complex biological problems *in situ* at the cellular level.

e. Transposons in Nematode Genetics

The transposable elements found in the *C. elegans* genome or in other nematode species have recently been discussed in detail.[84]

In hybrids of two strains of *C. briggsae*, we found an elevated rate of spontaneous mutations, which might be caused by a new transposable element.[85] Tc1 has not been found in this species,[86] and we are searching for the presumed *C. briggsae* transposon.

It would be useful to cross different strains of entomopathogenic nematodes reciprocally, looking for hybrid dysgenesis (i.e., an elevated rate of mutations or sterility in one of the reciprocal crosses). Hybrid dysgenesis is usually caused by mobile genetic elements (transposons). Transposons might be extremely powerful tools in the genetics of entomopathogenic nematodes.

The most straightforward way to isolate a gene of interest is to take advantage of a transposon-induced allele. In *C. elegans* genetics, genes can be tagged by transposons in a mutator strain (containing a transposon of one strain and cytoplasm from the other). Most spontaneous mutants of these strains are due to a mobile genetic element in the gene concerned. A suitable selection or screen is required to isolate a transposon-induced mutant allele of the gene of interest. The newly inserted element, and some sequences of the DNA segment linked to it, can be visualized by Southern blot hybridization.

III. FIRST APPROACHES TO THE GENETICS OF ENTOMOPATHOGENIC NEMATODES

Perhaps the most exciting problem in entomopathogenic nematology which can be approached by genetic methods is the specificity of the symbiotic relation between the bacterial symbiont and the nematode. A genetic analysis of the symbiotic relations between Leguminosae and nitrogen-fixing bacteria produced spectacular results of both scientific and commercial value.

In entomopathogenic nematodes, host-finding behavior,[35] thermoadaptation (and its membrane-biochemical conditions),[87] resistance to partial desiccation, specialized host-preference behavior, and osmotic avoidance behavior are traits which might be approached by single-step mutations with a realistic hope of success. The first convincing results on entomopathogenic nematode genetics will probably relate to the establishment of mutant strains with monogenically determined resistance to different chemical pesticides.

Genetic transformation techniques might be adopted for entomopathogenic nematodes. Transferring and expressing useful mutant genes such as some *daf-1*,[15] *sup-7*,[75,76] or *smg*[77] alleles from *C. elegans* to entomopathogenous nematodes is challenging. Genetic transformation, however, involves more than microinjecting DNA into the nematodes. A systematic analysis of the genome of entomopathogenic nematodes (like that carried out by Sulston and Brenner[2] and Coulson and Sulston)[88] must come first.

A. EMS and X-Ray Mutagenesis

In EMS mutagenesis, *C. elegans* hermaphrodites are washed from the agar (NGM) plates where they had been growing and are mutagenized in a small volume (4 ml) of buffer (M9) containing the chemical mutagen (usually 20 µl EMS). The F_2 generation can be scored for mutants within a week among the adults. The nematodes are kept in transparent agar plates and can be examined easily with a stereomicroscope.

EMS also works well with entomopathogenic nematodes when the young synchronous adult population in a host larva is mutagenized.[89] The infected insect should be placed onto a damp filter paper and transferred to a desiccator containing air saturated with the vapor of 5 µl EMS/ml M9 buffer solution for 8 hr. Both in *C. elegans* and entomopathogenic nematodes the effectiveness of the mutagenesis can be tested by transferring the F_1 generation into a 1% solution of nicotine or 5×10^{-4} M levamizole. Under such conditions, nematodes heterozygous for mutations of some genes exhibit a "twitcher" phenotype, whereas the others soon become paralyzed. Irradiation with X-ray of dose 7500 R proved at least as effective as EMS. Young J_4 juveniles or adults of entomopathogenic nematodes can be irradiated either within an insect or in an agar plate.[89] UV-mutagenesis also has potential for isolating mutant nematodes.[90]

Steinernema spp. are males and females, and so lack the genetic advantages of working with a hermaphrodite/male system. Steinernematid embryos produced by fertilization of oocytes before mutagenesis are laid by the females. Twelve to 24 hr after mutagenesis, gravid females are collected and embryos are obtained by Emmons's et al.[9] technique. A significant proportion of these embryos putatively carry mutations. Eggs are hatched in M9 buffer and juveniles are grown *in vitro*. Young males should be crossed individually in microtiter vials (containing a liquid culture of artificial medium or hemolymph) with one to three young untreated virgin females obtained as late J_4s from insect larvae infected 24 hr earlier. Progeny should be scored for mutant phenotypes (see below).

B. Mutant Phenotypes

Mutants resistant to different antihelminthics can be isolated by "brute force" selection. Mutagenized nematodes are reproduced either in an *in vitro* culture or *in vivo* for three generations. When a significant proportion of the population become infective juveniles, they are washed in 0.5% Hyamine and placed on one edge of an agar plate containing the antihelminthic in selective concentration (e.g., 5×10^{-4} *M* levamizole, 1 µg/ml avermectin). Sensitive nematodes become paralyzed or die (depending upon the drug and its concentration) and cannot move far from the starting point. Resistant juveniles move away. After a few hours the agar plates are cut in half, and those nematodes from the half furthest from the placement spot of the drug are collected.

If the mutagenesis has been successful, one can always find several resistant nematodes. They are probably of different genotypes, carrying mutant alleles of different genes. If the researcher's aim is to have a resistant population, resistant nematodes are transferred into an *in vitro* medium containing the drug for mass rearing. If the aim is to identify gene loci which have been mutagenized for resistance, then single pair crosses[91] are needed and not a homozygous stock (any sensitive offspring should be segregated by individual selection from outcross progeny). Briefly, one resistant infective-stage juvenile together with one from a control population are injected into an insect larva. The infective juveniles emerging from the third generation of the pair crosses should be tested and progeny-tested again in drug-containing agar plates. This test means a number of pair crosses between the most motile nematodes. Crosses producing 100% uniformly resistant infective juveniles are considered to be founders of a homozygous resistant stock. Since the infective juveniles do not show sexual dimorphism, well below 50% of the pair crosses are successful. The crosses can also be performed in liquid culture in microtiter vials. An advantage of using *in vitro* cultures is that the F_2 generations can be scored directly, but it is more difficult to get an appropriate number of infective juveniles from *in vitro* drop cultures. Using the above methods, we have so far isolated and outcrossed seven levamizole-resistant and one avermectin-resistant mutant of *S. carpocapsae* (Mexican). Complementation tests will clarify whether they carry mutations, the same gene, or different loci. We will need morphological mutants for genetic mapping of the chromosomes.[1]

IV. CONCLUSIONS

A precondition for further biological improvement of entomopathogenic nematodes as effective biological pest control agents is elaboration of their genetics. The spectacular results achieved in *C. elegans* genetics over the past 20 years suggest that enormous benefit can be attained from this scientific information. We list below a few guidelines which may be followed:

1. It is essential to elaborate the classical genetics of steinernematids and heterorhabditids through the strategy used by Brenner.[1] Certain types of mutants (e.g., dauer-constitutive, drug-resistant, chemotactic, thermotactic, and osmotactic) comparable to those of *C. elegans* might also be isolated in entomopathogenic nematode species and provide a basis for further genetic studies. Furthermore, understanding of the genetics of the nematodes and their symbionts would make it possible to study the genetics of the symbiosis.
2. A systematic investigation of the cytogenetics and taxonomy of *Heterorhabditis* species and strains should be performed in order to learn what genetic studies are possible. The methods of Herman,[59] Albertson,[54,55] and Goldstein[61] promise to be readily adoptable. Overcoming some of the cross-fertility problems between strains would allow us to benefit from the advantages of different strains through classical breeding, by heterosis, or the obtaining of new genetic variants by hybrid dysgenesis.
3. Systematic work on the DNA of the most frequently used strains would assist their gene manipulation. This might be started by isolating conserved genes (known in *C. elegans*) by molecular biological tools.
4. A detailed cell lineage analysis of vulval, gonadal, germ line, and E-line development would lead to a better understanding of the reproduction of heterorhabditids and the symbiotic relations of bacteria and nematodes.

ACKNOWLEDGMENTS

We thank D. Durham for correcting the English text; G. O. Poinar, Jr., for encouraging us to switch to this field; A. Morris, V. Sohoni, R. Georgis, and M. Friedman of BIOSYS for nematodes and other assistance; Gy. Saringer for his cooperation in applied projects; R. Gaugler and H. Kaya for helpful consultations; and M. Kopp and G. Kalicka for technical assistance. Our work has been supported by the Alkaloida Chemical Co., Tiszavasvari, and by the OMFB grant committee of Hungary.

REFERENCES

1. **Brenner, S.,** The genetics of *Caenorhabditis elegans, Genetics,* 77, 71, 1974.
2. **Sulston, J. E., and Brenner, S.,** The DNA of *Caenorhabditis elegans, Genetics,* 77, 95, 1974.
3. **Wood, W. B., Ed.,** *The Nematode Caenorhabditis elegans,* Monograph 17, Cold Spring Harbor Laboratory, Cold Spring Harbor, 1989.

4. **Andrassy, I.,** *Evolution as a Basis for the Systematization of Nematodes*, Akademiai Kiado, Budapest, 1976.

5. **Poinar, G. O., Jr.,** Nematodes, in *Laboratory Guide to Insect Pathogens and Parasites*, 2nd ed., Poinar, G. O., Jr., and Thomas, G. M., Eds., Plenum Press, New York, 1984, 235.

6. **Brenner, S.,** The genetics of behavior, *Br. Med. Bull.*, 29, 269, 1973.

7. **Wood, W. B.,** Introduction to *Caenorhabditis elegans*, in *The Nematode Caenorhabditis elegans*, Wood, W. B., Ed., Monograph 17, Cold Spring Harbor Laboratory, Cold Spring Harbor, 1989, chap. 1.

8. **Swanson, M. M., and Riddle, D. L.,** Critical periods in the development of the *Caenorhabditis elegans* dauer larva, *Dev. Biol.*, 84, 27, 1981.

9. **Emmons, S. W., Klass, M. R., and Hirsh, D.,** Analysis of the constancy of DNA sequences during development and evolution of the nematode *Caenorhabditis elegans*, *Proc. Natl. Acad. Sci.*, 76, 1333, 1979.

10. **Fodor, A.,** unpublished data, 1987.

11. **Fodor, A., Riddle, D. L., Nelson, F. K., and Golden, J. W.,** Comparison of a new wild type *Caenorhabditis briggsae* with laboratory strains of *Caenorhabditis briggsae* and *Caenorhabditis elegans*, *Nematologica*, 29, 203, 1983.

12. **Cassada, R. C., and Russell, R. L.,** The dauer larva, a post-embryonic developmental variant of the nematode *Caenorhabditis elegans*, *Dev. Biol.*, 46, 326, 1975.

13. **Klass, M., and Hirsh, D.,** Non-aging developmental variant of *Caenorhabditis elegans*, *Nature*, 260, 523, 1976.

14. **Riddle, D. L.,** A genetic pathway for dauer larva formation in *Caenorhabditis elegans*, *Stadler Genet. Symp.*, 9, 101, 1977.

15. **Riddle, D. L.,** The dauer larva, in *The Nematode Caenorhabditis elegans*, Monograph 17, Wood, W. B., Ed., Cold Spring Harbor Laboratory, Cold Spring Harbor, NY, 1989, chap. 12.

16. **Ohba, K., and Ishibashi, N.,** A factor inducing dauer juvenile formation in *Caenorhabditis elegans*, *Nematologica*, 28, 318, 1982.

17. **Golden, J. W., and Riddle, D. L.,** A pheromone influences larval development in the nematode *Caenorhabditis elegans*, *Science*, 218, 578, 1984.

18. **Golden, J. W., and Riddle, D. L.,** The *Caenorhabditis elegans* dauer larva: developmental effects of pheromone, food and temperature, *Dev. Biol.*, 102, 368, 1984.

19. **Golden, J. W., and Riddle, D. L.,** A pheromone induced developmental switch in *Caenorhabditis elegans*: temperature sensitive mutants reveal a wild type temperature-dependent process, *Proc. Natl. Acad. Sci.*, 81, 819, 1984.

20. **Riddle, D. L., and Bird, E. F.,** Responses of *Anguina agrostis* to detergent and anesthetic treatment, *J. Nematol.*, 17, 165, 1985.

21. **Croll, N. A.,** *Behaviour of Nematodes*, Edward Arnold, London, 1977, chap. 4.

22. **Singh, R. N., and Sulston, J. E.,** Some observations on molting in *Caenorhabditis elegans*, *Nematologica*, 24, 63, 1978.

23. **Cox, G. N., Straprans, S., and Edgar, R. S.,** The cuticle of *Caenorhabditis elegans*. II. Stage-specific changes in ultrastructure and protein post embryonic development, *Dev. Biol.*, 86, 456, 1981.

24. **Cox, G. N., Carr, S., Kramer, J. R., and Hirsh, D.,** Genetic mapping of *Caenorhabditis elegans* collagen genes using DNA polymorphisms as phenotypic markers, *Genetics*, 109, 513, 1985.

25. **Popham, J. D., and Webster, J. M.,** Aspects of the fine structure of the dauer larva of the nematode *Caenorhabditis elegans*, *Can. J. Zool.*, 57, 794, 1979.

26. **Fodor, A.,** unpublished data, 1988

27. **Albert, P. S., and Riddle, D. L.,** Developmental alterations in sensory neuroanatomy of the *Caenorhabditis elegans* dauer larva, *J. Comp. Neurol.*, 219, 461, 1983.

28. **Poinar, G. O., Jr., and Leutenegger, R.,** Anatomy of the infective and normal third stage juveniles of *Steinernema carpocapsae* Weiser (Steinernematidae, Nematoda), *J. Parasitol.*, 54, 340, 1968.

29. **Akhurst, R.,** *Xenorhabdus poinarii*: its interaction with insect pathogenic nematodes, *Sys. Appl. Microbiol.*, 8, 142, 1986.
30. **Poinar, G. O., Jr., Thomas, G. R., and Hess, R.,** Characteristics of the specific bacterium associated with *Heterorhabditis bacteriophora*, *Nematologica*, 23, 97, 1977.
31. **Dusenbery, D. B., Sheridan, R. E., and Russell, R. L.,** Chemotaxis-defective mutants of the nematode *Caenorhabditis elegans*, *Genetics*, 80, 297, 1975.
32. **Ambros, V.,** personal communication, 1989.
33. **Gaugler, R., LeBeck, L., Nakagaki, B., and Boush, G. M.,** Orientation of the entomogenous nematode *Neoaplectana carpocapsae* to carbon dioxide, *Environ. Entomol.*, 9, 649, 1980.
34. **Gaugler, R., McGuire, T. R., and Campbell, J. F.,** Genetic variability among strains of the entomopathogenic nematode *Steinernema feltiae*, *J. Nematol.*, 21, 247, 1989.
35. **Gaugler, R., Campbell, J. F., and McGuire, T. R.,** Selection for host-finding in *Steinernema feltiae*, *J. Invertebr. Pathol.*, 54, 363, 1989.
36. **Gaugler, R., and Boush, G. M.,** Effects of ultraviolet radiation and sunlight on the entomopathogenic nematode, *Neoaplectana carpocapsae*, *J. Invertebr. Pathol.*, 32, 291, 1978.
37. **Fodor, A., and Riddle, D. L.,** unpublished data, 1984.
38. **Herman, R. K., Albertson, D. G., and Brenner, S.,** Chromosome rearrangements in *Caenorhabditis elegans*, *Genetics*, 83, 91, 1976.
39. **Herman, R. K.,** Crossover suppressors and balanced recessive lethals in *Caenorhabditis elegans*, *Genetics*, 88, 49, 1978.
40. **Sigurdson, D. C., Spanier, G. J., and Herman, R. K.,** *Caenorhabditis elegans* deficiency mapping, *Genetics*, 108, 331, 1984.
41. **Fodor, A.,** unpublished data, 1989.
42. **Nelson, F. K., and Riddle, D. L.,** Functional study of the *Caenorhabditis elegans* secretory-excretory system using laser microsurgery, *J. Exp. Zool.* 231, 45, 1984.
43. **Fodor, A., Farkas, T., Reape, T., Timar, T., and Galamb, V.,** unpublished data, 1989.
44. **Riddle, D. L., Swanson, M. M., and Albert, P. S.,** Interacting genes in nematode dauer larva formation, *Nature*, 290, 668, 1981.
45. **Fodor, A., and Virag, E.,** unpublished data, 1983.
46. **Poinar, G. O., Jr., and Hansen, E.,** Sex and reproductive modifications in nematodes, *Helminthol. Abstr. Ser. B.*, 5, 145, 1983.
47. **Hodgkin, J.,** Sexual dimorphism and sex determination, in *The Nematode Caenorhabditis elegans*, Monograph 17, Wood, W. B., Cold Spring Harbor, Cold Spring Harbor, NY, 1989, chap. 9.
48. **Poinar, G. O., Jr.,** Description and biology of a new insect parasitic rhabditoid, *Heterorhabditis bacteriophora*, *Nematologica*, 21, 463, 1975.
49. **Sulston, J. E.,** Cell lineage, in *The Nematode Caenorhabditis elegans*, Monograph 17, Wood, W. B., Ed., Cold Spring Harbor Laboratory, Cold Spring Harbor, NY, 1989, chap. 5.
50. **Horvitz, H. R.,** Genetics of cell lineage, in *The Nematode Caenorhabditis elegans*, Monograph 17, Wood, W. B., Ed., Cold Spring Harbor Laboratory, Cold Spring Harbor, NY, 1989, chap. 6.
51. **Hodgkin, J.,** Males, hermaphrodites and females: sex determination in *Caenorhabditis elegans*, *Trends Genet.*, 1, 85, 1985.
52. **Albertson, D. G., and Thomson, N. J.,** The kinetochores of *Caenorhabditis elegans*, *Chromosoma*, 86, 409, 1982.
53. **Ellis, H. M., and Horvitz, H. R.,** Genetic control of programmed cell death in the nematode *Caenorhabditis elegans*, *Cell*, 44, 87, 1986.
54. **Albertson, D. G.,** Localization of the ribosomal genes in *Caenorhabditis elegans* chromosomes by *in situ* hybridization using biotin-labelled probes, *EMBO J.*, 3, 1227, 1984.
55. **Albertson, D. G.,** Mapping muscle protein genes by *in situ* hybridization using biotin-labelled probes, *EMBO J.*, 4, 2493, 1985.

56. **Nigon, V.,** Effects de la polyploidie chez un nematode libre, *C. R. Séances Acad. Sci. Ser. D.,* 228, 1161, 1949.

57. **Nigon, V.,** Polyploidie experimentale chez un nematode libre, *Rhabditis elegans* Maupas, *Bull. Biol. Fr. Belg.,* 85, 187, 1951.

58. **Madl, J. E., and Herman, R. K.,** Polyploids and sex determination in *Caenorhabditis elegans, Genetics,* 93, 393, 1979.

59. **Herman, R. K.,** Genetics, in *The Nematode Caenorhabditis elegans,* Monograph 17, Wood, W. B., Ed., Cold Spring Harbor Laboratory, Cold Spring Harbor, NY, 1989, chap. 2.

60. **Triantophyllou, A. C.,** Cytogenetics of root-knot nematodes, in *Root-Knot Nematodes (Meloidogyne Species) Systematics, Biology and Control,* Lamberti, F., and Taylor, C. E., Eds., Academic Press, London, 1979, 85.

61. **Goldstein, P., and Triantophyllou, A. C.,** The ultrastructure of sperm development in the plant-parasitic nematode, *Meloidogyne hapla, J. Ultrastruc. Res.,* 71, 143, 1980.

62. **Edgley, M. L., and Riddle, D. L.,** *Caenorhabditis elegans Genetic Map,* Caenorhabditis Genetics Center, Columbia, MO, USA, 1988.

63. **Lewis, J. A., Wu, C. H., Berg, H., Levine, J. H.,** The genetics of levamizole resistance in the nematode *Caenorhabditis elegans, Genetics,* 95, 905, 1980.

64. **Sanford, T., Golomb, M., and Riddle, D. L.,** RNA polymerase II from wild type and alpha-amanitin resistant strains of *Caenorhabditis elegans, J. Biol. Chem.,* 258, 1280, 1983.

65. **Ward, S.,** Chemotaxis by the nematode *Caenorhabditis elegans*: identification and analysis of the response by use of mutants, *Proc. Natl. Acad. Sci. (USA),* 70, 881, 1973.

66. **Hedgecock, R. M., and Russell, R. L.,** Normal and mutant thermotaxis in the nematode *Caenorhabditis elegans, Proc. Natl. Acad. Sci.,* 72, 4061, 1973.

67. **Culotti, J. G., and Russel, R. A.,** Osmotic avoidance defective mutants of the nematode *Caenorhabditis elegans, Genetics,* 86, 243, 1975.

68. **Chalfie, M., and White, J.,** The nervous system, in *The Nematode Caenorhabditis elegans,* Monograph 17, Wood, W. B., Ed., Cold Spring Harbor Laboratory, Cold Spring Harbor, NY, 1989, chap. 11.

69. **Heine, U., and Blumenthal, T.,** Characterization of regions of the *Caenorhabditis elegans* X chromosome containing vitellogenin genes, *J. Mol. Biol.,* 188, 301, 1986.

70. **Files, J. G., Carr, S., and Hirsh, D.,** Actin gene-family of *Caenorhabditis elegans, J. Mol. Biol.,* 164, 355, 1983.

71. **Anderson, P., and Brenner, S.,** A selection for myosin heavy chain mutants in the nematode *Caenorhabditis elegans, Proc. Natl. Acad. Sci.,* 81, 4470, 1984.

72. **Rose, A. M., Baillie, D. L., Candido, E. P. M., Beckenbach, K. A., and Nelson, D.,** The linkage mapping of cloned restriction fragment length differences in *Caenorhabditis elegans, Mol. Gen. Genet.,* 188, 286, 1982.

73. **Files, J. G., and Hirsh, D.,** Ribosomal DNA of *Caenorhabditis elegans, J. Mol. Biol.,* 149, 223, 1988.

74. **Cox, G. N., and Hirsh, D.,** Stage-specific patterns of gene-expression during development of *Caenorhabditis elegans, Mol. Cell. Biol.,* 5, 363, 1985.

75. **Waterston, R. H.,** A second informational suppressor, *sup-7* X, in *Caenorhabditis elegans, Genetics,* 97, 307, 1981.

76. **Fire, A., Moerman, D., White, J., Harrison, S., Albertson, D. G., and Waterston, R. W.,** personal communication, 1989.

77. **Hodgkin, J., Papp, A., Pulch, R., Ambros, V., Anderson, P.,** A new informational suppressor in the nematode *Caenorhabditis elegans, Genetics,* 123, 301, 1989.

78. **Fodor, A., and Deak, P.,** The isolation and genetic analysis of a *Caenorhabditis elegans* translocation (szT1) strain bearing an X-chromosome balancer, *J. Genet.,* 64, 143, 1986.

79. **Rosenbluth, R. E., Cuddeford, C., and Baillie, D. L.,** Mutagenesis in *Caenorhabditis elegans, Mutat. Res.,* 110, 39, 1983.

80. **Rosenbluth, R. E., and Baillie, D. L.,** The genetic analysis of reciprocal translocation eT1(III;V) in *Caenorhabditis elegans, Genetics,* 99, 415, 1981.

81. **Ferguson, E. L., and Horvitz, H. R.,** Identification and characterization of 22 genes that affect the vulval cell lineage of the nematode *Caenorhabditis elegans, Genetics,* 110, 17, 1985.

82. **Waterston, R. H.,** Muscle, in *The Nematode Caenorhabditis elegans,* Monograph 17, Wood, W. B., Ed., Cold Spring Harbor Laboratory, Cold Spring Harbor, NY, 1989, chap. 10.

83. **Emmons, S. W.,** The genome, in *The Nematode Caenorhabditis elegans,* Monograph 17, Wood, W. B., Ed., Cold Spring Harbor Laboratory, Cold Spring Harbor, NY, 1989, chap. 3.

84. **Levitt, A., and Emmons, S. W.,** The Tc2 transposon in *Caenorhabditis elegans, Proc. Natl. Acad. Sci.,* 1990, in press.

85. **Fodor, A., and Riddle, D. L.,** unpublished data, 1982.

86. **Fodor, A., Riddle, D. L., Udvardy, A., Sofi, J.,** unpublished data, 1989.

87. **Farkas, T., and Fodor, A.,** unpublished data, 1989.

88. **Coulson, A., Sulston, J. E., Brenner, S., and Karn, J.,** Towards a physical map of the genome of the nematode *Caenorhabditis elegans, Proc. Natl. Acad. Sci.,* 83, 7821, 1986.

89. **Fodor, A.,** unpublished data, 1988.

90. **Cohill, T., Marshall, T., Schubert, W., and Nelson, G.,** Ultraviolet mutagenesis of radiation-sensitive (rad) mutants of the nematode *Caenorhabditis elegans, Mut. Res.,* 209, 99, 1988.

91. **Akhurst, R. J., and Bedding, R. A.,** A simple cross-breeding technique to facilitate species determination in the genus *Steinernema, Nematologica,* 24, 328, 1983.

14. Physiology and Biochemistry of *Xenorhabdus*

Kenneth H. Nealson, Thomas M. Schmidt, and Bruce Bleakley

I. INTRODUCTION

The entomopathogenic nematode/bacterial system has attracted much recent interest with regard to its use in the control of insect pests.[1-3] As with any complex system, the ability to put it to effective use will be enhanced by an understanding of both the nematode and bacterial components as well as how they interact. This is no small task, however, as *Xenorhabdus* is a diverse and unusual bacterial group,[4-6] whose various species produce many unusual products, including bioluminescence (for *X. luminescens*),[7-10] antibiotics,[11-15] pigments,[15,16] extracellular enzymes,[17] and intracellular protein crystals.[18-21] To further complicate matters, all *Xenorhabdus* species are apparently capable of forming spontaneous colony form variants (phases) which are commonly altered in one or more of the products discussed above.[22-26]

As information has accumulated, it has become increasingly apparent that the ecophysiology of *Xenorhabdus* is very complex. Unfortunately, only a few laboratories have been involved with basic research of *Xenorhabdus*, and the questions generated have frequently outnumbered the answers obtained. Here, we summarize the present status of knowledge on the physiology and biochemistry of *Xenorhabdus*, focusing on growth and secondary metabolite production of this unique bacterial group.

II. ISOLATION, GROWTH, AND MAINTENANCE OF *XENORHABDUS*

Xenorhabdus isolates can be obtained in the field using methods in which highly susceptible "trap" insects are exposed to the soil.[27] Insects dying from entomopathogenic nematode infections are then used as sources for the isolation of *Xenorhabdus*, either from the hemocoel of the infected insect, or from infective-juvenile nematodes.[16,27,28] For *X. luminescens*, the bacteria are easily recognized by their ability to emit light;[7-10] for the nonluminous species and strains, other properties may be used for preliminary identification. These include: (1) colony color on NBTA agar plates (the colonies should be blue due to the uptake of bromothymol blue, with an area of clearing around them),[6] (2) the production of antibacterial activity (nearly all isolates produce antibiotics),[11-15] (3) the production of pigments ranging from buff to brown to red,[5,15,16] and (4) the presence of intracellular protein crystals, which can be visualized as refractile bodies upon examination by phase microscopy.[18-21]

Of the five known *Xenorhabdus* species, growth rates on complex media have been reported for only *X. luminescens* and *X. nematophilus*. In these very rich undefined media containing a source of amino acids or protein, vitamins (yeast extract), and carbon, at temperatures ranging from 24-30°C, the cells grow with generation times of 1.0-2.0 hr (Table 1).[7,10,23,28,29]

Physiological studies in defined media have been even less common, although Grimont et al.[5] reported that all strains of *Xenorhabdus* tested grew on a basal salts medium supplemented with nicotinic acid, para-aminobenzoic acid, serine, tyrosine, proline, and any of several carbon sources. Bleakley and Nealson[23] developed a proline minimal medium based upon the components commonly found in insect hemolymph; for *X. luminescens* strain Hm, they found amino acid requirements similar to those reported by Grimont et al.[5] for growth in their defined media, and reported the growth and production of secondary metabolites in several defined media. In the defined media, the production of secondary metabolites was deficient in the phase two variants although they grew substantially faster than phase one; generation times varied from 4.6-9.2 hr for phase one, and 3.0-4.9 hr for phase two.[23] For any given defined medium, phase two always grew faster than phase one (Table 1). Table 2 shows data for growth and secondary metabolite production in several defined media. Optimum production of secondary metabolites was favored by conditions of high proline, low NaCl, and low phosphate.

TABLE 1. Growth Rates of *Xenorhabdus* Species under Various Culture Conditions

Growth media	Bacterial species[a]	Temp (°C)	Generation time (hr)	Ref.
Complex Media				
Lipid Broth	Xn-1	25	0.8-1.2	28
Grace's Medium	Xn-1	25	0.9-1.3	28
Luria Broth	Xl-1	24	1.5	7
Luria Broth	Xl-1	30	1.5	10
Medium X	Xl-1	30	0.9	29
Insect Hemolymph				
	Xn-1	25	2.5-3.0	29
	Xl-1	24	2.5-3.0	7
Defined Media[b]				
Proline/Minimal	Xl-1	30	4.6-9.2	23
Proline/Minimal	Xl-2	30	3.0-4.9	23

[a]   Xn = *X. nematophilus*; Xl = *X. luminescens*; 1 = phase one; 2 = phase two.
[b]   Several different minimal media were reported. For a given medium, the phase two variant always grew faster and reached a higher yield of cells than did the phase one.

TABLE 2. Growth and Activity of *Xenorhabdus luminescens* (Hm) in Defined Media

Components added[a]
(millimolar)

NaCl	PO$_4^{2-}$	Proline	Generation time (hr)	Final OD$_{550}$	Final pH	Pigment	Anti-biotic	Lumines-cence[b]
4.3	3.5	8.7	5.5	3.7	9.5	++	+	2400
4.3	3.5	0.87	9.2	1.4	9.0	+	+	1360
4.3	35.0	8.7	4.6	2.9	9.3	+	+	3200
4.3	35.0	0.87	5.3	1.6	9.0	+	+	1300
4.3	35.0	8.7*	6.0	0.2	5.3	-	-	124
43.0	3.5	8.7	4.3	2.8	9.3	+	+	800
43.0	3.5	0.87	4.8	1.5	9.4	-	-	400
43.0	35.0	8.7	4.0	2.8	9.2	-	-	400
43.0	35.0	0.87	4.3	1.6	9.1	-	-	160
43.0	35.0	8.7**	8.3	0.7	4.9	-	-	63

[a]   All media contain basal salts: 1.7 mM MgSO$_4$; 94 mM NH$_4$Cl; 0.18 mM CaCl$_2$·2H$_2$O; 6.3 μM FeCl$_3$ · 6H$_2$O; 0.15 mM MnCl$_2$ · 4H$_2$O; and 0.12 mM Na$_2$MoO$_4$ 2H$_2$O. PO$_4^{2-}$ added as 0.35 M potassium phosphate buffer, pH = 7.0. Carbon for growth supplied as malic acid (30 mM), or D-glucose at either 11 mM (*) or 22 mM (**).

[b]   Luminescence is scored as relative light units as previously described.[23]

Growth has also been documented in the insect hemolymph.[7,29] The growth rates here can be estimated to be 2.5-3.0 hr for the doubling time, slower than in the complex media but faster than that achieved in the defined minimal media of Bleakley and Nealson.[23]

In several cases the efficiency of plating of *Xenorhabdus* isolates is low,[29] but the reason for this has not been elucidated. A variety of factors appear to be involved,[24] including osmotic sensitivity, nutrient requirements, oxygen sensitivity, and visible and UV light sensitivity.[30,30a] These may vary with different strains of bacteria.

III. PROPERTIES AND ROLES OF SECONDARY METABOLITES PRODUCED BY *XENORHABDUS*

We will focus here on several different secondary metabolites or activities, including bioluminescence, antibiotics, pigments, extracellular enzymes, and intracellular crystalline inclusions. These properties or activities are produced during growth of phase one *Xenorhabdus* spp., but are often not produced by phase two.[8,22-26] With the exception of bioluminescence, for which there is an easy and sensitive assay, it is difficult to quantitate most of the other activities;

much of the information is thus qualitative. In general, these metabolites or activities appear during mid- to late-logarithmic stage of growth.

There has been much speculation as to the function of the various metabolites in the infective process and/or nematode life cycle. While most of the hypotheses, some of which are discussed below, are quite reasonable, none are compelling considering the data available to support them.

A. Bioluminescence

The bioluminescence of *X. luminescens* is catalyzed by an enzyme similar in substrate requirements[7-9] and subunit size[10] to other bacterial luciferases. When *X. luminescens* luciferase was purified by methods used for other bacterial luciferases, and its alpha and beta subunits separated by urea chromatography, the subunits formed active enzyme hybrids with subunits from the marine luminous bacterium *Vibrio harveyi*,[10] suggesting that the luciferases from these ecologically widely separated groups are functionally and structurally closely related. Further evidence for this relationship is now available from hybridization analysis of the cloned luminescence (*lux*) genes.[31]

While the structural genes for luminescence from *X. luminescens* appear to be similar to those of its marine relatives, the regulation may be different. Marine luminous bacteria are subject to regulation by a variety of factors, including autoinduction, catabolite repression, iron concentration, osmolarity, and oxygen concentration (Table 3).[32] With the possible exception of autoinduction,[8-10] no indication of regulation by any of these factors has been seen for *X. luminescens*.[33] Only a few strains have been studied so far, and even among these there is considerable variation (e.g., Hk was constitutive for luminescence at a high level, while others exhibited various levels of inhibition of synthesis of the *lux* system during early stages of growth).[34] Any generalizations regarding the physiology or regulation of luminescence for the group should be put aside until more results are available.

Light emission of most strains is suboptimal during growth,[9,10] and addition of an exogenous long-chain aldehyde, one of the substrates involved in the luminescent reaction, stimulates light emission *in vivo* (Figure 1). The explanation for this apparent inability to synthesize sufficient quantities of aldehyde is unknown, but it was not observed when the cloned *X. luminescens lux* genes were expressed in *E. coli*.[31]

Phase two *X. luminescens* exhibit greatly reduced light emission (Figure 1),[8,23] although the reasons for this decrease are unknown. In some strains the amount of luciferase synthesized is not noticeably different between phases, and the difference in luminous activity is due to differences in cellular expression of luciferase.[34] In others, there are major differences in the amounts of luciferase present,[8,34] suggesting that the difference in observed light emission resides in either the synthesis or degradation of luciferase. In Figure 1, the luminescence *in vivo* with and without added aldehyde and the extractable luciferase activities are shown for strain Hm. For phase two of this strain both

TABLE 3. Factors that Regulate Bacterial Bioluminescence

Regulatory Factor	Vibrio fischeri	Vibrio harveyi	Photo-bacterium leiognathi	Photo-bacterium phosphoreum	Xenorhabdus luminescens
Autoinduction	+	+	+a	+a	?b,c
Oxygen					
high oxygen	-	+	+	-	-c
low oxygen	+	-	-	+	-c
Osmolarity					
high salt	-	-	-	+	-c
low salt	-	+	+	-	-c
Catabolite repression	-	+	NT	NT	-c
Iron repression	-	+	+	+	-c

a Some strains are constitutive.[37]

b Autoinduction pattern seen in complex medium.[8,10]

c No autoinduction or other regulation seen in minimal medium.[33]

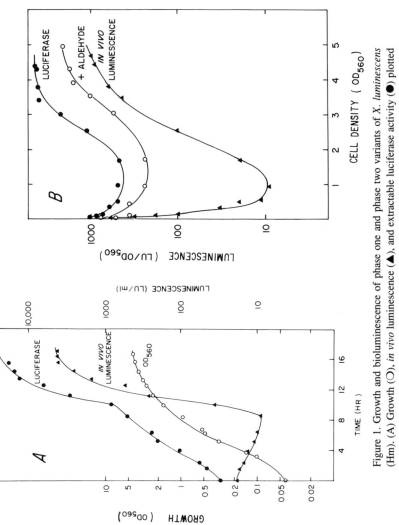

Figure 1. Growth and bioluminescence of phase one and phase two variants of *X. luminescens* (Hm). (A) Growth (O), *in vivo* luminescence (▲), and extractable luciferase activity (●) plotted as a function of time for the phase one variant. (B) Extractable luciferase per OD (●), *in vivo* luminescence per OD (▲), *in vivo* luminescence per OD with added decanal (O), for the phase one variant. (C) Identical to A for the phase two variant. (D) Identical to B for the phase two variant.

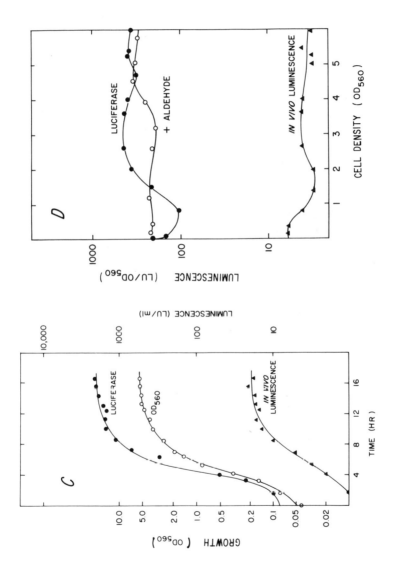

the amount of luciferase and its expression *in vivo* are diminished in comparison with phase one.

Finally, of the five characterized *Xenorhabdus* species, *X. luminescens* is the only bioluminescent group, so luminescence is clearly not a requirement for the success of this genus. Furthermore, Akhurst and Boemare[35] have recently reported a strain of *X. luminescens* that is nonluminous. Since cloned *lux* genes are now available,[31] it should be possible to identify by hybridization those strains that have *lux* genes, whether functional or not. Such analyses may lead to new appreciation for the distribution and role of bioluminescence in this genus.

Speculation on the role of bioluminescence is that it is involved with the attraction of other organisms to the infected insect cadaver — either organisms that might aid in distribution of the nematodes, or other potential host organisms. This hypothesis has not been rigorously tested, but this could presumably be done using dark mutants of *X. luminescens* in competition with wild type strains.

B. Antibiotic Production

With few exceptions,[12] phase one *Xenorhabdus* produce antibiotics capable of inhibiting the growth of many different bacteria and fungi.[11-15] Little is known of the production rates of these antibiotics, or their role in infection or insect toxicity, but several structures have been purified and chemically characterized (Figure 2).[11,13-15] The hydroxystilbene structure has been identified from two different organisms,[13,15] while the others are known from only single organisms. The structures of other inhibitors have not been determined, but since there is a wide range in the activity spectrum of these antibiotic activities,[11,12] it may well be that more antibiotic types are yet to be found.

Antibiotic production appears to be an essential feature of the *Xenorhabdus* infective process. Almost all strains produce antibiotic activity of one type or another, and the phase two variants, which usually lack this activity are susceptible to overgrowth by other bacteria. This leads to putrefaction of the insect cadaver and low nematode yields. This hypothesis is reasonable, and as with bioluminescence, it should be testable using point mutants blocked in antibiotic production.

C. Pigment Production

Xenorhabdus species display an array of pigmentation ranging from buff to brown to red.[5,6] For one strain of *X. luminescens* a red pigment was isolated and identified as an anthraquinone derivative (Figure 3).[15] This compound is in the general class known as polyketides, chemicals that are common in *Streptomyces* species but relatively rare in eubacteria. Grimont et al.[5] have used pigment presence and color as a taxonomic character. However, the color of the pigment shown in Figure 3 is sensitive to pH, so that at pH values of 9.0 or above it is

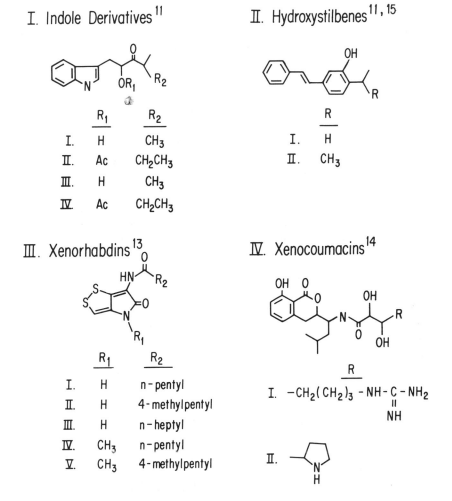

I. Indole Derivatives[11]

	R_1	R_2
I.	H	CH_3
II.	Ac	CH_2CH_3
III.	H	CH_3
IV.	Ac	CH_2CH_3

II. Hydroxystilbenes[11,15]

	R
I.	H
II.	CH_3

III. Xenorhabdins[13]

	R_1	R_2
I.	H	n–pentyl
II.	H	4–methylpentyl
III.	H	n–heptyl
IV.	CH_3	n–pentyl
V.	CH_3	4–methylpentyl

IV. Xenocoumacins[14]

I. $-CH_2(CH_2)_3-NH-C-NH_2$, with $\parallel$ NH

II.

Figure 2. Structure of antibiotics produced by four different strains of *Xenorhabdus* spp.

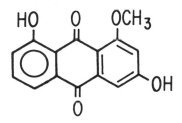

Figure 3. Structure of anthraquinone pigment produced by *Xenorhabdus luminescens* (Hk).

red, but at pH values of 8.0 or below it is yellow to colorless.[15] Thus, if pigmentation is to be used as a taxonomic tool, the colonies should be adjusted to the same pH. We have found that spraying colonies with a dilute solution of NaOH enhances the red color of the pigment from strain Hm.[15]

The genes for pigment production have been isolated from *X. luminescens* (Hb), and cloned into and expressed in *E. coli*.[31] Using these cloned genes it should be possible to examine the distribution of similar genes in other *Xenorhabdus* species, and to compare the system with similar biosynthetic pathways in other bacterial groups.

Pigment production is characteristic of many, but not all, species of *Xenorhabdus*, so, like bioluminescence, it is clearly not required for the infective process. Since the infected insects are usually easily identified by changes in color due to the bacterial pigment, the pigment may play a role similar to that proposed for bioluminescence — attraction of organisms that aid nematode dispersal or are potential hosts. As with bioluminescence, however, it is not obvious how a visual signal would be used for soil insects.

D. Extracellular Enzymes
1. Protease

Xenorhabdus species are notoriously proteolytic, a property to be expected from organisms growing in a protein-rich environment such as insect hemolymph. For *X. luminescens* (Hm), a protease was purified and partially characterized,[17] and shown to be an alkaline metalloprotease of molecular weight 61,000, which was not produced by the phase two variant.[23] Only a single protease activity was identified for strain Hm. In contrast, Boemare and Akhurst reported multiple protease activities, some of which were absent in phase two, and others which were present in different *Xenorhabdus* species and strains.[24] It may be valuable for understanding the role of the protease in bacterial virulence to identify the specific proteases in these pairs of strains, and document their association with phase one and phase two variants. Genetic approaches may prove particularly valuable for understanding the distribution and role of proteases in *Xenorhabdus*.[31]

Proteases could play a role both in inactivation of the insects' defense systems and in the digestion of the insect, and both roles have been suggested.[17] The situation with proteases is complex, however, with some strains producing several different types of activities. As with other functions, point mutants altered in single protease activities may be useful for defining the components required or important for successful infection.

2. Lipase

Lipase activity is commonly determined by the observation of hydrolysis of Tween 80 on agar plates, and while phase one *Xenorhabdus* isolates tend to be very lipolytic there are no reports of purification or characterization of any lipases. For most species, phase one show strong lipase activity and phase two show weak or absent activity.[24] However, the situation is reversed for *X. nematophilus* with phase one lacking lipase activity and phase two acquiring it.[6,24] The role of lipase in the infective process is unknown, but one may speculate that these enzymes are involved with digestion of insect lipids to the good of the bacteria, and thus to the symbiosis.

E. Crystalline Inclusions

Another feature of phase one *Xenorhabdus* is the presence of refractile intracellular inclusions identified as crystalline proteins.[18-21] As with many of the other cellular products described here, there is apparently great diversity in the types of crystals. For *X. nematophilus*,[19] it was shown that two different crystal types were composed of separate protein subunits of 26 and 22kd. Antibodies to the larger of these were used to screen other strains of *Xenorhabdus*, and it was present in some but not all species.[20] Heterogeneity was also noted with regard to the ultrastructure of the crystals.[18-20] Crystal formation generally occurred in phase one but not in phase two. However, in one strain of *X. nematophilus* neither phase produced crystals, and both phases produced crystals for a strain of *X. luminescens*.[19-20]

Based on negative experiments with injected crystalline proteins, most workers have dismissed the role of these proteins in the infective process.[19-21] At present, there is no evidence for any direct role in bacterial parasitism for the protein crystals, although it remains enigmatic why they are absent in phase two variants of most species. Boemare et al.[18] suggested that the crystals were organelles involved with light emission and that they be called lumisomes. However, on the basis of biochemical[10] and genetic data,[31] and the presence of abundant crystals in nonluminous species, it seems unlikely that this intriguing hypothesis is correct.

IV. GROWTH AND ACTIVITY OF *XENORHABDUS* IN THE INSECT HOST

Growth and/or activity of bacteria in infected insects have been measured in two studies.[7,29] In the first, cells of *X. luminescens* were injected into wax moth (*Galleria mellonella*) larvae, and bacterial growth and luminescence monitored over time. At 24°C, the generation time of *X. luminescens* in the insect was estimated to be 2.5-3.0 hr;[7] the same strain of bacterium grew with a generation time of 1.5 hr in rich medium at the same temperature.[7] In the second study, the bacterial strains grew with generation times of 0.9-1.5 hr in rich medium, and approximately 2.5-3.0 hr in the hemocoel at 25°C.[29] Observation of infected insects and the bacteria from them clearly indicated that the bacteria were active in the production of secondary metabolites during their growth phase. The infected insects were bioluminescent, pigmented, and contained bacteria with intracellular protein crystals.[29] Since no putrefaction of the insect occurred, it can also be assumed that antibiotics were produced.

The mechanisms involved in insect toxicity are unclear and may be complex. In nonimmune insects, *Xenorhabdus* can be highly pathogenic, with LD_{50} values as low as a few cells per insect.[2,7] Dunphy and Webster[36] reported that lipopolysaccharides with strong hemocyte toxicity were produced by *Xenorhabdus*. The production of these components has not been reported for phase two. The complexity of the situation is indicated by Götz et al.,[29] who found that the bacteria were aided in their ability to kill the insects by components produced by the nematodes, with the nematodes inactivating parts of the host immune system.

V. CONCLUSIONS

A few years ago, when only the barest of knowledge was available, it seemed as if *Xenorhabdus* was a rather well-defined group of symbiotic/pathogenic bacteria whose lifestyle was circumscribed around the life cycle of nematodes and the infection of insects. The products produced by the phase one variants of the bacteria were well defined, and the properties of the phase two variants were reasonably predictable. In short, although there was a small database, it was internally consistent. As more information has appeared the simple explanations have faded, and the situation has become much more subtle and variable than was first envisioned.

As more detailed biochemical and physiological data appear it should be possible to address important questions on what features of the bacteria: (1) allow them to grow in the insect, (2) are advantageous for nematode replication, (3) lead to insect death, (4) are important in the specificity of the bacterial/nematode association, and (5) are involved with phase conversion?

The approaches that will be needed to answer such questions will include detailed physiological, biochemical, and genetic work (especially with well-defined point mutants), as well as collaborative studies between bacteriologists, nematologists, and entomologists, which examine the bacteria/nematode/insect system as a whole. Substantial advances have been made, and it is anticipated that with the tools and knowledge presently available these advances will continue.

REFERENCES

1. **Poinar, G. O., Jr.,** *Nematodes for Biological Control of Insects*, CRC Press, Boca Raton, FL, 1979.
2. **Poinar, G. O., Jr.,** *The Natural History of Nematodes*, Prentice-Hall, Englewood Cliffs, NJ, 1983, chap. 10.
3. **Gaugler, R.,** Ecological considerations in the biological control of soil-inhabiting insects with entomopathogenic nematodes, *Agric. Ecosyst. Environ.*, 24, 351, 1988.
4. **Akhurst, R. J.,** Taxonomic study of *Xenorhabdus*, a genus of bacteria symbiotically associated with insect pathogenic nematodes, *Int. J. Syst. Bacteriol.*, 33, 38, 1983.

5. Grimont, P. A. D., Steigerwalt, A. G., Boemare, N., Hickman-Brenner, F. W., Deval, C., Grimont, F., and Brenner, D. J., Deoxyribonucleic acid relatedness and phenotypic study of the genus *Xenorhabdus*, *Int. J. Syst. Bacteriol.*, 34, 378, 1984.

6. Akhurst, R. J., and Boemare, N. E., A numerical taxonomic study of the genus *Xenorhabdus* and proposed elevation of the subspecies of *X. nematophilus* to species, *J. Gen. Microbiol.*, 134, 1853, 1988.

7. Poinar, G. O., Jr., Thomas, G., Haygood, M., and Nealson, K. H., Growth and luminescence of the symbiotic bacteria associated with the terrestrial nematode, *Heterorhabditis bacteriophora*, *Soil Biol. Biochem.*, 12, 5, 1980.

8. Nealson, K., Schmidt, T. M., and Bleakley, B., Luminescent bacteria: symbionts of nematodes and pathogens of insects, in *Cell to Cell Signals in Plant, Animal and Microbial Symbiosis*, NATO ASI Series, Vol. H17, S. Scannerini, Ed., Springer-Verlag, Berlin, 1988, 101.

9. Colepicolo, P., Cho, K., Poinar, G. O., Jr., and Hastings, J. W., Growth and luminescence of the bacterium *Xenorhabdus luminescens* from a human wound, *Appl. Environ. Microbiol.*, 55, 2601, 1989.

10. Schmidt, T. M., Kopecky, K., and Nealson, K. H., Bioluminescence of the insect pathogen *Xenorhabdus luminescens*, *Appl. Environ. Microbiol.*, 55, 2607, 1989.

11. Paul, V. J., Frautschy, S., Fenical, W., and Nealson, K. H., Antibiotics in microbial ecology: isolation and structure assignment of several new antibacterial compounds from the insect-symbiotic bacteria *Xenorhabdus* spp., *J. Chem. Ecol.*, 7, 589, 1981.

12. Akhurst, R. J., Antibiotic activity of *Xenorhabdus* spp., bacteria symbiotically associated with insect pathogenic nematodes of the families Heterorhabditidae and Steinernematidae, *J. Gen. Microbiol.*, 128, 3061, 1982.

13. Rhodes, S. H., Lyons, G. R., Gregson, R. P., Akhurst, R. J., and Lacey, M. J., Xenorhabdin antibiotics, Aust. Patent PCT/AU83/00156, 1983.

14. Gregson, R. P., and McInerney, B. V., Xenocoumacins, Aust. Patent PCT/AU85/00215, 1985.

15. Richardson, W. H., Schmidt, T. M., and Nealson, K. H., Identification of an anthraquinone pigment and a hydroxystilbene antibiotic from *Xenorhabdus luminescens*, *Appl. Environ. Microbiol.*, 54, 1602, 1988.

16. Khan, A., and Brooks, W. M., A chromogenic bioluminescent bacterium associated with the entomophilic nematode *Chromonema heliothidis*, *J. Invertebr. Pathol.*, 29, 253, 1977.

17. Schmidt, T. M., Bleakley, B., and Nealson, K. H., Characterization of an extracellular protease from the insect pathogen *Xenorhabdus luminescens*, *Appl. Environ. Microbiol.*, 54, 2793, 1988.

18. Boemare, N. E., Louis, C., and Kuhl, G., Etude ultrastructurale des cristaux chez *Xenorhabdus* spp. bacteries inpeodees aux nematodes entomophages Steinernematidae et Heterorhabditidae, *C. R. Soc. Biol.*, 177, 107, 1983.

19. Couche, G. A., Lehrbach, P. R., Forage, R. G., Cooney, G. C., Smith, D. R., and Gregson, R. P., Occurrence of intracellular inclusions and plasmids in *Xenorhabdus* spp., *J. Gen. Microbiol.*, 133, 967, 1987.

20. Couche, G. A., and Gregson, R. P., Protein inclusions produced by the entomopathogenic bacterium *Xenorhabdus nematophilus* subsp. *nematophilus*, *J. Bacteriol.*, 169, 5279, 1987.

21. Bowen, D. J., and Ensign, J. C., Intracellular protein crystal of the insect pathogen *Xenorhabdus luminescens*, *Proc. Ann. Mtg. Am. Soc. Microbiol.*, I-66, 183, 1987.

22. Akhurst, R. J., Morphological and functional dimorphism in *Xenorhabdus* spp., bacteria symbiotically associated with the insect pathogenic nematodes *Neoaplectana* and *Heterorhabditis*, *J. Gen. Microbiol.*, 121, 303, 1980.

23. Bleakley, B., and Nealson, K. H., Characterization of primary and secondary forms of *Xenorhabdus luminescens* strain Hm, *FEMS Microbiol. Ecol.*, 53, 241, 1988.

24. Boemare, N. E., and Akhurst, R. J., Biochemical and physiological characterization of colony form variants in *Xenorhabdus* spp., *J. Gen. Microbiol.*, 134, 751, 1988.

25. **Hurlbert, R. E., Xu, J., and Small, C. L.,** Colonial and cellular polymorphism in *Xenorhabdus luminescens, Appl. Environ. Microbiol.*, 55, 1136, 1989.
26. **Akhurst, R. A.,** Biology and taxonomy of *Xenorhabdus*, in *Entomopathogenic Nematodes in Biological Control*, Gaugler, R., and Kaya, H. K., Eds., CRC Press, Boca Raton, FL, 1990, chap. 4.
27. **Akhurst, R. J., and Brooks, W. M.,** The distribution of entomophilic nematodes (Heterorhabditidae and Steinernematidae) in North Carolina, *J. Invertebr. Pathol.*, 44, 140, 1984.
28. **Dunphy, G. B., Rutherford, T. A., and Webster, J. M.,** Growth and virulence of *Steinernema glaseri* influenced by different subspecies of *Xenorhabdus nematophilus, J. Nematol.*, 17, 476, 1985.
29. **Götz, P., Boman, A., and Boman, H. G.,** Interactions between insect immunity and an insect-pathogenic nematode with symbiotic bacteria, *Proc. R. Soc. London B.*, 212, 333, 1981.
30. **Nealson, K. H.,** unpublished observations, 1989.
30a. **Xu, J., and Hurbert, R. E.,** Toxicity of irradiated media for *Xenorhabdus* spp., *Appl. Environ. Microbiol.*, 56, 815, 1990.
31. **Frackman, S., and Nealson, K. H.,** The molecular genetics of *Xenorhabdus*, in *Entomopathogenic Nematodes in Biological Control*, Gaugler, R., and Kaya, H. K., Eds., CRC Press, Boca Raton, FL, 1990, chap. 15
32. **Hastings, J. W., Makemson, J., and Dunlap, P.,** How are growth and luminescence regulated independently in light organ symbionts, *Symbiosis*, 4, 3, 1987.
33. **Levisohn, R., and Nealson, K. H.,** Regulation of bioluminescence in *Xenorhabdus luminescens, Proc. Ann. Mtg. Am. Soc. Microbiol.*, I-117, 200, 1988.
34. **Schmidt, T. M., and Nealson, K. H.,** unpublished data, 1986.
35. **Akhurst, R. J., and Boemare, N. E.,** A non-luminescent strain of *Xenorhabdus luminescens, J. Gen. Microbiol.*, 132, 1917, 1986.
36. **Dunphy, G. B., and Webster, J. M.,** Lipopolysaccharides of *Xenorhabdus nematophilus* and their haemocyte toxicity in non-immune *Galleria mellonella* larvae, *J. Gen. Microbiol.*, 134, 1017, 1988.
37. **Rosson, R. A., and Nealson, K. H.,** Autoinduction of bacterial bioluminescence in a carbon limited chemostat, *Arch. Microbiol.*, 129, 299, 1981.

15.   The Molecular Genetics of *Xenorhabdus*

Susan Frackman and Kenneth H. Nealson

I. INTRODUCTION

The techniques of molecular biology and genetics are increasingly being applied to study the biology of various organisms, including bacteria in the genus *Xenorhabdus*. In this chapter, we review the development and utilization of molecular genetics in *Xenorhabdus* research.

II. TRANSFORMATION SYSTEMS IN *XENORHABDUS*

The introduction and maintenance of a plasmid are important components of any investigation into the molecular biology of that organism. The ability to clone genes, manipulate them *in vitro*, and study the effects of those manipulations *in vivo* has provided rapid advances in the fields of microbial physiology, genetics, and biochemistry. For many organisms the limiting factor in this scheme is the development of a transformation system, but fortunately a number of *Xenorhabdus* species have been transformed with plasmids commonly used in molecular biology.[1,2]

Xu et al.[1] have developed a system for the transformation of *X. nematophilus* strain 19061 with the broad host range plasmid pHK17. This transformation procedure yields $1\text{-}10 \times 10^5$ transformants per microgram of pHK17 DNA. Xu et al.[1] optimized this procedure with respect to growth media, growth phase of cells, washing conditions, cell to DNA ratio, addition of various cations, pH of the transformation buffer, and duration and temperature of the incubations. They found that when the transforming plasmid had been grown in *Escherichia coli* the efficiency of transformation was only 0.4% of that obtained when the transforming plasmid had been grown in *X. nematophilus*. This result strongly suggests that *X. nematophilus* contains a restriction modification system. The stability of the plasmid, pHK17, in *X. nematophilus* (19061) was determined; after 30 generations of growth without selection for the plasmid only 2% of the cells retained the plasmid, suggesting that pHK17 might be useful as a pseudosuicide plasmid for transposon mutagenesis. The transformation of several other *X. nematophilus* strains and one *X. luminescens* strain with a variety of plasmids was also studied (Table 1).

We have developed a system for the transformation of *X. luminescens* (Hm)[2] by modifying the $CaCl_2/RbCl$ procedure commonly used in the transformation of *E. coli*.[3] The modifications are (1) the addition of 0.5% NaCl to the transformation buffers to prevent cell lysis of *X. luminescens*, and (2) a change in the temperature and duration of the heat shock from 43-44°C for 30 sec to

TABLE 1. Transformation of *Xenorhabdus* spp. with Plasmids[a]

Xenorhabdus spp. (strain)	Phase	Plasmid	Source[b]	No. of transformants/ μg of DNA
X. nematophilus (19061)	1	pBR325	*E. coli*	2.7×10^3
X. nematophilus (19061)	2	pHK17	*X. nematophilus*	4.6×10^5
X. nematophilus (IM)	1	pHK17	*X. nematophilus*	0
X. poinarii	1	pHK17	*X. nematophilus*	2.0×10^3
X. luminescens (RH)	1	pHK17	*X. nematophilus*	2.0×10^3
X. luminescens (RH)[c,d]	1	pBR322	*E. coli*	1.0
X. luminescens (RH)[c,d]	1	pIMH43	*E. coli*	1.0
X. luminescens (RH)[c]	1	pIMH43	*X. nematophilus*	5.8×10^1
X. luminescens (RH)	1	pUC18	*E. coli*	5.7×10^2
X. luminescens (RH)	1	pP1	*E. coli*	4.3×10^3

[a]   From Xu et al.[1]

[b]   Bacterial hosts for the transforming plasmids were *Xenorhabdus nematophilus* (19061/1) and *E. coli* (HB101).

[c]   Cells were made competent by treatment with a $MgCl_2$-$CaCl_2$ solution.

[d]   Greater than 1 μg of DNA was used, and only one transformed colony was obtained.

37°C for 2 min to increase the transformation efficiency. Using this procedure, the phase one and two variants of *X. luminescens* (Hm) have been transformed with a number of plasmids commonly used in recombinant DNA research, including pBR322, pACYC184 and pSF60 (a derivative of pUC18 carrying a gene for kanamycin resistance). These plasmids are maintained in *X. luminescens* extrachromosomally and in high copy number. The development of plasmid based transformation systems for *Xenorhabdus* spp. will enable investigators to utilize the tools of molecular biology to further our understanding of the diverse elements of *Xenorhabdus* biology.

III. BIOLUMINESCENCE

X. luminescens is different from the other *Xenorhabdus* species in its ability to produce visible luminescence. This luminescence is due to the presence of an enzyme system with physical and biochemical properties similar to those of other bioluminescent bacteria.[4-8] The enzyme luciferase, an alpha-beta heterodimeric mixed function oxidase, oxidizes a reduced flavin and a long-chain aldehyde in the generation of light. Purified luciferase from *X. luminescens* is structurally similar to the luciferases from other luminous bacteria[7,8] and there is DNA homology between the gene for the alpha subunit of luciferase (*lux*A) of *X. luminescens* and *lux*A genes from other bacteria.[2]

Despite the biochemical similarities between the luminous system of *X. luminescens* and other luminous bacteria, bioluminescence in *X. luminescens* may not respond to environmental and physiological changes in the same way.[8] Most other bioluminescent bacteria are found in marine environments either in symbiotic association with marine organisms or living saprophytically;[9] *X. luminescens* has been found only in terrestrial environments, usually in association with nematodes. It seems likely that *X. luminescens* has evolved mechanisms to regulate bioluminescence in response to its environment that meet the special needs of a terrestrial nematode symbiont and are, therefore, different from those used by the marine bacteria.[8]

The luminous system of *X. luminescens* appears to respond to at least two different regulatory systems. In rich medium (L-broth), light production is minimal until the cells reach late logarithmic or stationary phase growth.[2,7] Figure 1a shows a typical growth curve for *X. luminescens* (Hm); Figure 1b shows the generation of light as a function of growth. Light production during growth is expressed in arbitrary light units (LU) that have been normalized for optical density providing a measure of the relative amount of light produced per cell during growth. Figure 1 clearly illustrates the late onset of luminescence in *X. luminescens*; light begins to increase after the culture has reached an optical density of >2.0 which is late logarithmic phase, with maximal light production after the culture is in stationary phase (O.D. $_{560}$ >4.0).

Several investigators have shown that the luminous system of *X. luminescens* is also controlled by the phase one to phase two variation that occurs in all *Xenorhabdus* species.[10-14] Phase one *Xenorhabdus* is generally isolated from infective nematodes, but upon prolonged culture in various media, spontaneous variants called phase two appear. In contrast to phase one, the secondary variant is dim (0.1-1% of the luminescence of phase one) and produces no measurable antibiotic activity, pigment, intracellular inclusions, extracellular protease, or extracellular lipase.[10-14] The two phases also exhibit differences in colony morphology and staining properties.[14] A number of bacterial characteristics showing differential expression in phase one and phase two are normally expressed late in the growth cycle, including bioluminescence, pigment production, and antibiotic production. The production of extracellular protease and lipase may also occur late in the growth phase since the assays for these proteins are generally not positive until after prolonged incubation. Understanding the mechanism of phase variation in *Xenorhabdus* may increase our knowledge of the regulation of late gene expression in bacteria. Bioluminescence, because it is easily and sensitively measured, is a powerful tool for such studies.

We note here that the terminology of "phase variation" is generally applied to a reversible alteration of bacterial characterisitics; since reversibility has not yet been demonstrated in the case of the variants of X. luminescens, the terminology adopted for the chapters in this volume may be subject to revision in the future.

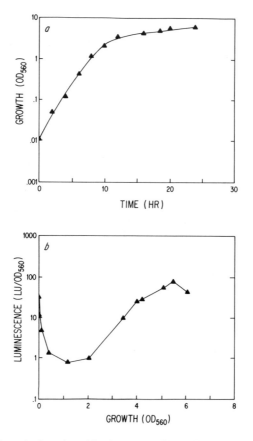

Figure 1. Growth and luminescence of *Xenorhabdus lumines-cens* (Hm) in L broth. (a) Growth is plotted as cell density (O.D.$_{560}$) as a function of time. (b) Luminescence is plotted as light units normalized to cell density (LU/O.D.$_{560}$) as a function of cell density (O.D.$_{560}$).

IV. CLONING THE *LUX* GENES OF *XENORHABDUS LUMINESCENS*

In an effort to understand the mechanism of phase one to phase two conversion and the genetic regulation of the luminous system in *X. lumines-cens*, we have cloned the genes necessary for bioluminescence (*lux* genes) from *X. luminescens* (Hm).[2] A genomic library of *X. luminescens* DNA in plasmid pUC18 was constructed and screened as outlined in Figure 2. *X. luminescens* DNA was partially digested with *Sau*3A and enriched for fragments of 9-20 kb by agarose gel electrophoresis. These fragments were ligated into the *Bam*HI site of pUC18 and transformed in *E. coli*. The resulting ampicillin resistant transformants were screened for light production. One luminous colony was identified which carried a plasmid (pCGLS1) with an approximately 11 kb insert of *X. luminescens* DNA. A map of the restriction endonuclease sites of this insert is presented in Figure 3. It is known from the cloning of other luminous systems that in order to produce light without the addition of exogenous aldehyde in *E. coli* at least five genes must be expressed.[15] These are the genes (*lux*C, *lux*D, and *lux*E) for the three polypeptides of the fatty acid reductase complex required for the synthesis of the long-chain aldehyde substrate, and the genes (*lux*A and *lux*B) for the alpha and beta subunits of the bacterial luciferase. It is therefore apparent that pCGLS1 carries at least the five genes which are analogous to *lux*A-*lux*E in other systems.

V. EXPRESSION OF THE *LUX* GENES OF *XENORHABDUS LUMINESCENS* IN *ESCHERICHIA COLI*

The production of light from *E. coli* carrying pCGLS1 was measured during growth in L-broth (Figure 4).[2] The level of light per cell remains low until mid- to late-logarithmic growth phase, and the maximal level of luminescence per cell occurs when the cells are in stationary phase. While the pattern of luminescence development from *E. coli* carrying pCGLS1 during growth is qualitatively similar to that of *X. luminescens* (Hm) (Figure 1), the amount of light produced is 10-15 times greater in *E. coli* carrying pCGLS1, probably because pCGLS1 is present in many copies per cell. The *E. coli* strain used for the experiments was DHα5F′I^Q and the cultures were grown at 28-30°C. At these temperatures the copy number of pUC18 is <25 copies/cell, which is much reduced from the approximately 128 copies/cell present when it is grown at 37°C.[16] This strain carries the *lac*I^Q repressor gene on an F′ episome producing tenfold more repressor than is found in most host strains. The combination of the relatively low copy number at 28°C and the *lac*I^Q gene should maintain the *lac* promoter present on pUC18 in a repressed state unless an inducer is present. Furthermore, the F′ episome carries a gene imparting kanamycin resistance to this strain so that the episome can be maintained in media containing kanamycin. Therefore, the expression of the *lux* genes in these experiments is not a result of the activity of the plasmid *lac* promoter.

Another plasmid, pCGLS1R, contains the same *X. luminescens* DNA insert as pCGLS1 but in the opposite orientation with respect to the vector. The emission of light during growth of *E. coli* strain DHα5F′I^Q carrying pCGLS1R (data not shown) is virtually identical to that produced by the same strain carrying pCGLS1 shown in Figure 4.[2] These results suggest that the *lux* genes are being expressed from a promoter(s) within the *X. luminescens* DNA insert rather than promoters in the vector DNA.

VI. DELETION ANALYSIS OF THE *LUX* GENES OF *XENORHABDUS LUMINESCENS*

The organization of the *lux* genes of *X. luminescens* was studied by con-

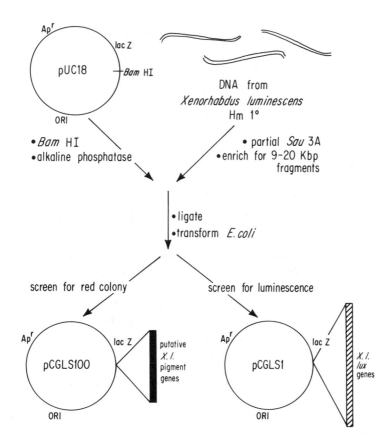

Figure 2. Diagram of the construction of a genomic library of *Xenorhabdus luminescens* (Hm) DNA in pUC18 and the screening for plasmids containing the bioluminescence genes (pCGLS1) and the pigment genes (pCGLS100).

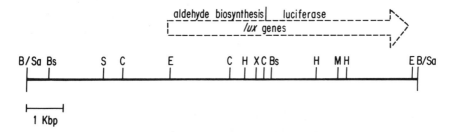

Figure 3. Map of the *lux* genes of *Xenorhabdus luminescens*. The solid line represents the insert of *Xenorhabdus luminescens* DNA in pCGLS1. The location of the *lux*A-E is given above the line and the direction of transcription is indicated by the arrow. Restriction sites are abbreviated as follows: *Bs, BstEII; C, ClaI; E, EcoRI; H, HindIII; M, MluI; S, ScaI; X, XhoI*; B/Sa indicates the ligation of *Sau3A* digested DNA to *BamHI* digested DNA.

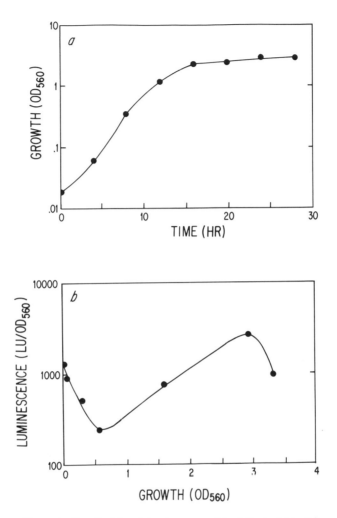

Figure 4. Growth (a) and luminescence (b) of *Escherichia coli* (DHa5F'I°) carrying pCGLS1 in L broth. The data are expressed as described in Figure 1.

structing plasmids with various fragments of *lux* DNA. The *lux* DNA fragments have been cloned into either pUC18 or pUC19, and the ability of these plasmids to produce aldehyde-independent or -dependent light in *E. coli* has been determined (Figure 5).[2] Those plasmids that are capable of producing aldehyde independent luminescence in *E. coli* contain *lux*A *lux*E; those that produce aldehyde-dependent luminescence contain the genes for luciferase (*lux*A and *lux*B) but have deletions that include at least part of one of the three aldehyde biosynthesis genes; those not producing light have deletions which include at least part of either of the genes for the luciferase subunits (*lux*A and *lux*B). The plasmids which are capable of producing aldehyde-independent lu-

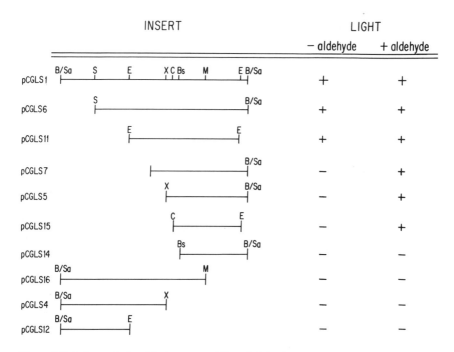

Figure 5. Deletion analysis of the *lux* genes of *Xenorhabdus luminescens*. Solid lines indicate the region of *X. luminescens* DNA which has been inserted into either pUC18 or pUC19. Restriction enzymes are abbreviated as described in Figure 3. The ability (+) or inability (-) to produce light in *E. coli* either in the absence (-aldehyde) or after the addition (+aldehyde) of exogenously added decanal is indicated for each plasmid.

minescence in *E. coli* are pCGLS1, pCGLS6, and pCGLS11. The smallest DNA region carrying all five of the genes (*lux*A-*lux*E) required for luminescnece in *E. coli* is the 6.9 kb insert of pCGLS11. The amount of light from *E. coli* carrying a plasmid with this insert is different depending upon the orientation of the insert with respect to the plasmid sequences (data not shown), suggesting that the expression of these genes is under the control of plasmid promoters. It is therefore possible that part or all of the natural *X. luminescens* promoter for the *lux* genes has been deleted in pCGLS11. When the plasmids pCGLS7, pCGLS5, and pCGLS15 are introduced into *E. coli*, there is light emission only when exogenous aldehyde is added. These plasmids carry 5.6, 4.6, and 4.3 kb, respectively, of *lux* DNA from the right end of the pCGLS1 DNA insert. This suggests that one or more of the genes for the fatty acid reductase complex reside in the 2.6 kb between the left *Eco*R1 site and the *Cla*I site. Because bacteria containing any of the plasmids which are deleted for the region to the right of the *Cla*I site (pCGLS14, pCGLS16, pCGLS4, and pCGLS12) are unable to produce luminescence, and bacteria carrying any of the plasmids which contain the DNA between the *Cla*I site and the right *Eco*R1 site (pCGLS1, pCGLS6, pCGLS7, pCGLS5, pCGLS11, pCGLS15) produce luminescence, the genes for the two subunits of luciferase, *lux*A and B, must be located on the ~4.25 kb region between the *Cla*I site and the right *Eco*RI site. These studies on the order and position of the *lux* genes indicate that the genes for aldehyde biosynthesis are located to the left of the genes for the two luciferase subunits as shown in Figure 3. This analysis of the organization of *X. luminescens lux* genes is not yet complete; studies are in progress to further define the order and location of these genes.

The 2.6 kb region shown by these studies to contain genes for the polypeptides of the fatty acid reductase complex may not be the location of all three of these genes (*lux*C, *lux*D and *lux*E), since in other bacterial bioluminescent systems these genes require ~3.5 kb. The 4.25 kb region in which the genes for the alpha and beta subunits of luciferase (*lux*A and *lux*B) are located is sufficient to code for more than those two polypeptides. The *lux*A and *lux*B genes should be present on about 2 kb of DNA since the alpha and beta subunits of luciferase from *X. luminescens* have molecular weights of about 36 and 40 kd.[7,8] It is therefore possible that one of the genes required for aldehyde biosynthesis is located to the right of *lux*A and B genes. This is an interesting possibility because it suggests that the organization of the *X. luminescens lux* genes may be similar to other bacterial *lux* genes. The organization of *lux*A-*lux*E is the same for at least three marine bioluminescent bacteria. The order of the *lux* genes from *Vibrio fischeri*, *Vibrio harveyi*, and *Photobacterium phosphoreum* is, beginning promoter proximal, *lux*C, *lux*D, *lux*A, *lux*B, *lux*E.[15,17,18] In contrast to the similarity in the organization of *lux*A-*lux*E in various organisms, the organization of the genes involved in regulation of bacterial luminescence appears to not be conserved. The two systems in which regulatory genes were defined, *V. fischeri* and *V. harveyi*, are different.[17-21] The regulatory genes of *X. luminescens* have not been identified.

VII. DIRECTION OF TRANSCRIPTION OF THE *LUX* GENES OF *XENORHABDUS LUMINESCENS*

To determine the direction of transcription of the *lux* genes of *X. luminescens*, plasmids have been constructed which carry the *E. coli lac* promoter either to the left or to the right of the *lux* genes of *X. luminescens*.[2] The *lac* promoter is oriented so that it directs transcription through the inserted *lux* genes. The activity of the *lac* promoter is regulated by the presence or absence of the inducer isopropylthio-β-galactoside (IPTG). When the *lac* promoter is oriented in the same direction as the natural *lux* promoter, the expectation is that there will be an increase in light emission in the presence of IPTG, whereas there will not be an increase in light emission in the presence of IPTG if the *lac* promoter is oriented in the opposite direction. The results of these experiments are shown in Table 2,[2] with lines 3-6 clearly showing that the genes required for bioluminescence in *E. coli* are transcribed from left to right. The plasmids pCGLS11 and pCGLS5 (Figure 5) have the *lac* promoter to the left of the *lux* genes. Strains carrying these two plasmids produce, respectively,

TABLE 2. The Direction of Transcription of the *lux* Genes

Species (Strain)[a]	Luminescence		
	LU/O.D.$_{560}$ (-IPTG)[b] (10^3)	LU/O.D.$_{560}$ (+IPTG[c])[b] (10^3)	+IPTG/-IPTG
1. *E. coli* (pCGLS1)[df]	3.80	58.00	15.00
2. *E. coli* (pCGLS1R)[ef]	4.80	35.00	7.30
3. *E. coli* (pCGLS11)[ef]	34.00	230.00	6.70
4. *E. coli* (pCGLS11R)[ef]	5.50	4.70	0.85
5. *E. coli* (pCGLS5)[efg]	10.00	120.00	12.00
6. *E. coli* (pCGLS5R)[efg]	0.27	0.19	0.70
7. *X. luminescens*	0.15	0.13	0.90

[a]   *Escherichia coli* (DHα5F'·I^Q) and *Xenorhabdus luminescens* (Hm) were used.
[b]   Light was measured when cultures were O.D.$_{560}$ ~2.0; each number represents the average of three independent samples.
[c]   IPTG was added to media at a concentration of 4×10^{-4} *M*.
[d]   The *lux* DNA is in plasmid pUC18.
[e]   The *lux* DNA is in plasmid pUC19.
[f]   The orientation of the *lux* insert with respect to the *lac* promoter are indicated by the presence or absence of an R in the name. Those without an R have the *lac* promoter on the left and those with an R have the *lac* promoter on the right. Plasmids with the same number have the same *lux* DNA insert.
[g]   Light measurements were taken after the addition of 0.1% decanal (in DMSO) to a final concentration of 0.002%.

6.7- and 12.0-fold more light when the *lac* promoter is active (Table 2, lines 3 and 5). The plasmids pCGLS11R and pCGLS5R have the same *lux* DNA inserts but the *lac* promoter is to the right of the *lux* genes. Strains carrying these plasmids do not show any increase in luminescence when the *lac* promoter is induced (Table 2, lines 4 and 6). In agreement with the above results, luminescence is increased 15-fold when the *lac* promoter is induced on the plasmid, pCGLS1 (containing the entire 11 kb insert), in which the *lac* promoter is to the left of *lux* genes (Table 2, line 1). Unexpectedly, luminescence is also increased 7.3-fold when the *lac* promoter is induced in the plasmid, pCGLS1R, in which the *lac* promoter is to the right of the *lux* genes. There is no increase in luminescence from *X. luminescens* in the presence of IPTG (Table 2, line 7), indicating that the inducer has no effect on luminescence in the absence of the *E. coli lac* promoter. This result may indicate that there is transcription in both directions from the ll kb *lux* DNA insert. Rightward transcription (Figure 3) produces the two luciferase subunits (the products of *lux*A and *lux*B) and the three proteins involved in aldehyde biosynthesis (the products of *lux*C, *lux*D, and *lux*E) and leftward transcription may produce a product which positively affects the expression of *lux*A-*lux*E genes. The presence of a leftward transcript is very speculative and requires further investigation.

VIII. EXPRESSION OF THE CLONED *LUX* GENES IN *XENORHABDUS LUMINESCENS*

The cloned *lux* genes from *X. luminescens* have been introduced into *X. luminescens* (Hm) phase one and phase two variants.[2] The plasmid, pCGLS1, has been modified slightly for these experiments by adding a gene for kanamycin resistance as shown in Figure 6. This gene was necessary for the selection of transformants because *X. luminescens* (Hm) is naturally ampicillin resistant. Phase one and phase two of *X. luminescens* were transformed with this plasmid, pCGLS2, and light emission was measured during growth (Figure 7).[2] The phase one and phase two carrying pCGLS2 produce the same amount of light, and the production of light as a function of growth in these strains is similar to *E. coli* carrying the cloned *lux* genes on pCGLS1. The maximum amount of light produced by either phase one or two *X. luminescens* carrying pCGLS2 is approximately 10-fold greater than the bioluminescence seen in the phase one variant of *X. luminescens*. Light emission from either phase one or phase two *X. luminescens* carrying pCGLS2 also increases earlier in growth than in phase one *X. luminescens*. The increase in amount and earlier induction of bioluminescence is probably because the plasmid, pCGLS2, is maintained in high copy in *X. luminescens*. Interestingly, light emission from the the phase two variant of *X. luminescens* carrying pCGLS2 increases slightly later in growth than in phase one carrying pCGLS2. This lag in induction may reflect a difference in the regulation of gene expression in phase one and two *X. luminescens*. This possibility is currently under investigation.

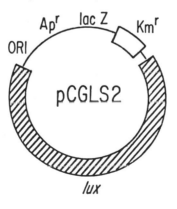

Figure 6. Diagram of the plasmid pCGLS2. pCGLS2 was constructed by the insertion of a gene for kanamycin resistance (Km^r) into the PstI site of pCGLS1. The PstI site of pCGLS1 is part of the multiple cloning site of pUC18.

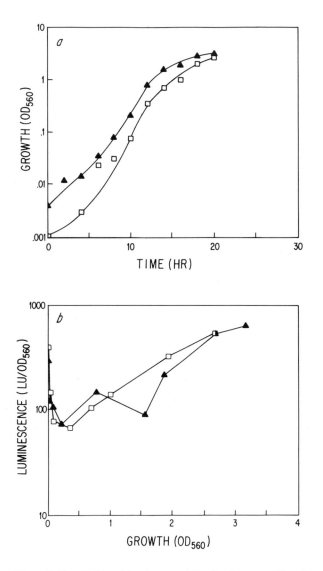

Figure 7. Growth (a) and luminescence (b) of a phase one (□) and phase two (▲) variants of *Xenorhabdus luminescens* (Hm) carrying pCGLS2 in L broth. The data are expressed as described in Figure 1.

IX. CLONING THE PIGMENT GENES OF *XENORHABDUS LUMINESCENS*

The phase one variant of *X. luminescens* produces a red pigment late in growth.[22-25] This pigment was isolated from *X. luminescens* (Hk) and identified as an anthraquinone.[25] Although the pigment has no antibiotic activity, it is related to the polyketide group of antibiotics produced by *Streptomyces* spp. The genes for the synthesis of one of these antibiotics, actinorhodin, have been cloned from *Streptomyces coelicolor*. A plasmid with a 32 kb insert of *S. coelicolor* DNA containing at least eight genes is capable of producing actinorhodin when transformed into *S. parvulus*, a strain which does not normally produce the antibiotic.[26]

We have isolated a plasmid from the *X. luminescens* genomic library which produces a red pigment in *E. coli* (Figure 2).[27] This plasmid carries 8.1 kb of *X. luminescens* DNA.[27] It appears likely that all of the genes required for the biosynthesis of the *X. luminescens* pigment are not present on this plasmid because: (1) the color of the *E. coli* carrying pCGLS100 is a darker red than that of *X. luminescens* (Hm);[27] (2) the pigment of *E. coli* carrying pCGLS100 does not exhibit the same solubility characteristics as the pigment of *X. luminescens*;[27] and (3) by analogy to the antibiotic genes of *S. coelicolor* it seems unlikely that 8.1 kb is sufficient coding capacity for all of the gene products necessary for the biosynthesis of the anthraquinone pigment. The pigment produced by *E. coli* carrying pCGLS100 may be an intermediate in the pathway to the final anthraquinone structure found in *X. luminescens*.

X. PLASMIDS AND BACTERIOPHAGES

The occurrence of plasmids in a variety of *Xenorhabdus* strains has been studied by several investigators (Table 3).[1,2,28,29] Couche et al.[28] isolated plasmids from ten different strains of *X. nematophilus* in an effort to determine whether plasmids played any role in the production of intracellular inclusions. Seven of the strains were found to contain plasmids, ranging in size from 3.6-12 kb. In the four strains in which the plasmid patterns were studied for both the phase one and phase two variants, identical plasmid patterns were seen in both phases (Table 3). Xu et al.[1] attempted unsuccessfully to isolate plasmids which could serve as cloning vectors from several *X. nematophilus* strains and one *X. luminescens* strain. Poinar et al.[29] found a plasmid of 50-56 kb in *X. luminescens*. We have isolated a plasmid (pHM1) from *X. luminescens* (Hm) which is 7.1 kb in length.[2] This plasmid has been isolated from phase one *X. luminescens* (Hm), but all attempts to isolate it from a phase two have been unsuccessful. Studies in which pHM1 was used as a probe in colony hybridization of the phase two variant of *X. luminescens* indicate that this phase contains DNA homologous to pHM1. We are presently extending these studies to determine whether phase two DNA which is homologous to pHM1 is present on a plasmid or on the chromosome.

Bacteriophages have been useful in mobilizing bacterial mutations from one host to another, as well as in the construction of genetic maps. Furthermore, an enormous amount of information has been obtained about bacterial structures and functions by analyzing the interactions between bacteriophages and their

TABLE 3. The Occurrence of Plasmids in Strains of *Xenorhabdus*

Species (strain)	Phase	Plasmid size (kb)[a]	Ref.
X. nematophilus (All)	1	12, 3.6	28
	2	12, 3.6	28
X. nematophilus (BK)	1	12, 3.6	28
	2	12, 3.6	28
X. nematophilus (A24)	1	12, 3.6	28
	2	12, 3.6	28
X. nematophilus (XnT)	1	10	28
	2	10	28
X. nematophilus (Mex)	1	7	28
X. nematophilus (AN6)	1	-	28
X. nematophilus (T319)	1	7, 3.6	28
X. nematophilus (SK2)	1	5, 3.6	28
X. nematophilus (Dan5)	1	-	28
	2	-	28
X. nematophilus (Q385)	1	-	28
	2	-	28
X. nematophilus (19061)	1	-	1
	2	-	1
X. nematophilus (IM)	1	-	1
X. poinarii (XU)	1	-	1
X. luminescens (RH)	1	-	1
X. luminescens	1	50-56	29
X. luminescens (Hm)	1	7.1	2
	2	-	2

[a]   - indicates that no plasmids were detected.

hosts. A bacteriophage, XLP, infecting phase one *X. luminescens* has recently been isolated,[29,30] and may serve as an important tool in studying the biology of this bacterium.

XI. CONCLUSIONS

We have reviewed the various aspects of *Xenorhabdus* biology which have been studied using the techniques of genetics and molecular biology. While much of the research described here is in its initial stages, this work has already produced transformation systems and identified plasmids and a bacteriophage. Genes have been cloned from *Xenorhabdus* and expressed in both *E. coli* and *X. luminescens*. As these powerful techniques are improved and adapted for *Xenorhabdus* they can be used in combination with the methods of microbial physiology, biochemistry, and ecology to answer many basic questions about *Xenorhabdus* biology, including (1) what role does *Xenorhabdus* play in the life cycle of the nematode, (2) what role does *Xenorhabdus* play in insect pathogenesis, (3) how is the production of secondary metabolites genetically regulated, and (4) what is the mechanism of phase conversion.

REFERENCES

1. **Xu, J., Lohrke, S., Hurlbert, I. M., and Hurlbert, R. E.,** Transformation of *Xenorhabdus nematophilus*, *Appl. Environ. Microbiol.*, 55, 806, 1989.
2. **Frackman, S., and Nealson, K. H.,** unpublished data, 1989.
3. **Kushner, S. R.,** Improved method for transformation of *E. coli* with *col*E1 derived plasmids, in *Genetic Engineering*, Boyer, H. B., Ed., Elsevier, Amsterdam, 1978, 17.
4. **Poinar, G. O., Jr., Thomas, G., Haygood, M., and Nealson, K. H.,** Growth and luminescence of the symbiotic bacteria associated with the terrestrial nematode, *Heterorhabditis bacteriophora*, *Soil Biol. Biochem.*, 12, 5, 1980.
5. **Nealson, K. H., Schmidt, T. M., and Bleakley, B.,** Luminescent bacteria: symbionts of nematodes and parasites of insects, in *Cell to Cell Signals in Plant, Animal and Microbial Symbiosis*, NATO ASI Series, Vol. H17, Scannerini, S., Ed., Springer-Verlag, Berlin, 1988, 101.
6. **Colepicolo, P., Cho, K., Poinar, G. O., Jr., and Hastings, J. W.,** Growth and luminescence of the bacterium *Xenorhabdus luminescens* from a human wound, *Appl. Environ. Microbiol.*, 55, 2601, 1989.
7. **Schmidt, T. M., Kopecky, K., and Nealson, K. H.,** Bioluminescence of the insect pathogen *Xenorhabdus luminescens*, *Appl. Environ. Microbiol.*, 55, 2607, 1989.
8. **Nealson, K. H., Schmidt, T. M., and Bleakley, B.,** Physiology and biochemistry of *Xenorhabdus*, in *Entomopathogenic Nematodes in Biological Control*, Gaugler, R., and Kaya, H. K., Eds., CRC Press, Boca Raton, FL, 1990, chap. 14.
9. **Nealson, K. H., and Hastings, J. W.,** Bacterial bioluminescence: its control and ecological significance, *Microbiol. Rev.*, 43, 496, 1979.
10. **Akhurst, R. J., and Boemare, N. E.,** Biology and taxonomy of *Xenorhabdus*, in *Entomopathogenic Nematodes in Biological Control*, Gaugler, R., and Kaya, H. K., Eds., CRC Press, Boca Raton, FL, 1990, chap. 4.
11. **Akhurst, R. J.,** Morphological and functional dimorphism in *Xenorhabdus* spp. bacteria symbiotically associated with the insect pathogenic nematodes *Neoaplectana* and *Heterorhabditis*, *J. Gen. Microbiol.*, 121, 241, 1988.
12. **Bleakley, B., and Nealson, K. H.,** Characterization of primary and secondary forms of *Xenorhabdus luminescens* strain Hm, *FEMS Microbiol. Ecol.*, 53, 241, 1988.
13. **Hurlbert, R. E., Xu, J., and Small, C. L.,** Colonial and cellular polymorphism in *Xenorhabdus luminescens*, *Appl. Environ. Microbiol.*, 55, 1136, 1989.
14. **Boemare, N. E., and Akhurst, R. J.,** Biochemical and physiological characterization of colony form variants in *Xenorhabdus* spp. (Enterobacteriaceae), *J. Gen. Microbiol.*, 134, 151, 1988.
15. **Meighen, E. A.,** Enzymes and genes from the *lux* operons of bioluminescent bacteria, *Annu. Rev. Microbiol.*, 42, 151, 1988.
16. **Miki, T., Yasukochi, T., Nagani, H., Furuno, M., Orita, T., Yamada, H., Imoto, T., and Horiuchi, T.,** Construction of a plasmid vector for regulatable high level expression of eukaryotic genes in *Escherichia coli*: an application to overproduction of chicken lysozyme, *Prot. Engin.*, 1, 327, 1987.

17. **Engebrecht, J., Nealson, K. H., and Silverman, M.,** Bacterial bioluminescence: isolation and genetic analysis of functions from *Vibrio fischeri, Cell,* 32, 773, 1983.
18. **Engebrecht, J., and Silverman, M.,** Identification of genes and gene products necessary for bacterial bioluminescence, *Proc. Natl. Acad. Sci. USA,* 81, 4154, 1984.
19. **Engebrecht, J., and Silverman, M.,** Nucleotide sequence of the regulatory locus controlling expression for the bacterial genes of bioluminescence, *Nucleic Acids Res.,* 15, 10455, 1987.
20. **Devine, J. H., Countryman, C., and Baldwin, T.,** Nucleotide sequence of the *luxR* and *luxI* genes and structure of the primary regulatory region of *lux* regulon of *Vibrio fischeri* AKC 7744, *Biochemistry,* 27, 837, 1988.
21. **Martin, M., Showalter, R., and Silverman, M.,** Identification of a locus controlling expression of luminescence genes in *Vibrio harveyi, J. Bacteriol.,* 171, 2361, 1989.
22. **Akhurst, R. J.,** Taxonomic study of *Xenorhabdus,* a genus of bacteria symbiotically associated with insect pathogenic nematodes, *Int. J. Syst. Bacteriol.,* 33, 38, 1983.
23. **Grimont, P. A. D., Steigenwalt, A. G., Boemare, N., Hickman-Brenner, W., Deval, C., Grimont, F., and Brenner, D. J.,** Deoxyribonucleic acid relatedness and phenotypic study of the genus *Xenorhabdus, Int. J. Syst. Bacteriol.,* 34, 378, 1984.
24. **Khan, A., and Brooks, W. M.,** A chromogenic bioluminescent bacterium associated with the entomophilic nematode *Chromonema heliothidis, J. Invertebr. Pathol.,* 29, 253, 1977.
25. **Richardson, W. H., Schmidt, T. M., and Nealson, K. H.,** Identification of an anthraquinone pigment and a hydroxystibene antibiotic from *Xenorhabdus luminescens, Appl. Environ. Microbiol.,* 54, 1602, 1988.
26. **Malpartida, F., and Hopwood, D. A.,** Molecular cloning of the whole biosynthetic pathway of a *Streptomyces* antibiotic and its expression in a heterologous host, *Nature,* 309, 462, 1984.
27. **Frackman, S., Ragudo, P., and Nealson, K. H.,** unpublished data, 1989.
28. **Couche, G. A., Lehrbach, P. R., Forage, R. G., Coonry, G. C., Smith, D. R., and Gregson, R. P.,** Occurrence of intracellular inclusions and plasmids in *Xenorhabdus* spp., *J. Gen. Microbiol.,* 133, 967, 1987.
29. **Poinar, G. O., Jr., Hess, R. T., Lanier, W., Kinney, S., and White, J. H.,** Preliminary observations of bacteriophage infecting *Xenorhabdus luminescens* (Enterobacteriaceae), *Experientia,* 45, 191, 1989.
30. **Poinar, G. O., Jr.,** Taxonomy and biology of Steinernematidae and Heterorhabditidae, in *Entomopathogenic Nematodes in Biological Control,* Gaugler, R., and Kaya, H. K., Eds., CRC Press, Boca Raton, FL, 1990, chap. 2.

16.   Insect Immunity

Gary B. Dunphy and Graham S. Thurston

I. INTRODUCTION

Steinernema carpocapsae and *Heterorhabditis bacteriophora* and their respective bacterial symbionts, *Xenorhabdus nematophilus* and *X. luminescens*, although effective against a diversity of insect species, show considerable variation in strain efficacy. This variation has been attributed to environmental adaptations, behavioral differences, and variation in the number of bacteria within the infective juveniles and the proportion of infective juveniles retaining bacteria.[1-5] In addition, the association of the nematode-bacterium complex with the homeostatic systems in the host's hemocoel that respond to foreign matter (=nonself-response systems) will contribute to variation in the efficacy and, ultimately, the success of the nematode-bacterium complex. To elucidate the intricate and well-adapted physiologies of the components of these nematode-bacterium complexes with the host's hemolymph, it is essential that this review be divided into two parts: (1) an overview of the antibacterial and antiparasite systems of the hemolymph of nonimmune and immune insects, and (2) the interaction of the nematode-bacterium complexes with the nonself defenses of nonimmune and immune insects.

II. HEMOLYMPH SYSTEMS RESPONDING TO NONSELF MATTER

Insects exist in a myriad of environments where the potential for infection by microorganisms and parasites is great. As part of a survival strategy, insects have evolved numerous and effective defense mechanisms to resist infection. The defenses include structural and passive barriers (e.g., cuticle, gut physicochemical properties, and peritrophic membrane),[6,7] constitutive cellular and humoral factors in the hemolymph, and induced antibacterial proteins.[8-11] The definition of immunity proposed by Boman and Hultmark,[10] (i.e., "resistance to or protection against a specified disease; the power to resist disease") while applicable to the described homeostatic strategies, does not recognize differences in the relative contribution of the constitutive humoral and cellular factors and antibacterial proteins. There are pronounced differences in the antiforeign matter responses between insects which have and have not been induced to produce antibacterial proteins.[6,10-12] In keeping with the traditional definition, it is proposed that insects lacking induced antibacterial proteins be regarded as nonimmune, and those with such proteins produced in response to prior exposure to nonself material be considered as immune.

In nonimmune insects the nonself-response components of the hemolymph consist of humoral factors (e.g., lysozyme, lectins, and the prophenoloxidase cascade [in most dipteran larvae, *Manduca sexta*, and *Bombyx mori*]) and cellular factors (e.g., hemocytes, pericardial cells, prophenoloxidase cascade, and fixed phagocytic organs).[6,9-11,13] As research continues it is becoming apparent that humoral and cellular responses cannot be viewed as separate.

A. Hemocyte Nonself Activity in Nonimmune Insects

The granular cells and plasmatocytes are the major effector cells of Lepidoptera, Coleoptera, and Diptera which participate in phagocytosis, nodule formation, and encapsulation.[8,14,15] These reactions have been studied in insects reared primarily in the laboratory, although Ratcliffe has documented these responses in feral insects.[8]

1. Phagocytosis

Plasmatocytes are the predominant phagocytic cells engulfing bacteria. In *Galleria mellonella*, plasmatocytes respond to *Escherichia coli* K12 by pseudopodial engulfment and to larger particles by invagination of the cell membrane.[16] Initiation of phagocytosis occurs in response to bacterial cell wall/envelope components, elements of the prophenoloxidase cascade, and the production of phagocyte-stimulating factors.[6,17] Phagocytic hemocytes of *Blaberus craniifer* require energy produced by glycolysis.[18] However, attachment of foreign particles to insect hemocytes is energy-independent, and in *G. mellonella*, may be mediated by opsonic proteins of the prophenoloxidase cascade.[8,17,19] In Lepidoptera, hemocyte nonself responses are also mediated by the microfilaments.[20,21]

Details of the bactericidal mechanisms are lacking. However, Anderson et al.[22] detected selective bactericidal activity by *B. craniifer* hemocytes that was not related to the family classification or Gram stain reaction of the bacteria. A putative antibacterial enzyme is the inducible and constitutive enzyme lysozyme with activity against the sacculus of Gram-positive and Gram-negative bacteria with defective cell envelopes.[6,23-25] It has also been detected in hemocytes of *Spodoptera frugiperda* and *Locusta migratoria*.[6,24] Walters and Ratcliffe[26] attributed the bactericidal activity of *G. mellonella* hemocytes to β-glucuronidase and β-glucosaminidase. Intracellular digestion may involve several factors.[8]

2. Biphasic Hemocyte Responses

Depending on the insect species, type of nonself test particles, virulence of the bacteria, and bacterial concentration, phagocytosis may be augmented by nodule formation.[6,8] Nodule formation represents random contact of granular cells with particles, the triggering of hemocyte degranulation, the production of sticky granular cells, and localized clotting (Figure 1).[6,10] Concomitant with this, in *G. mellonella*, is the release of a plasmatocyte-depletion factor, the removal of plasmatocytes from the hemolymph, and the activation of the prophenoloxidase cascade system.[6,10,11] The bacterial-hemocyte coagulum is then walled off by plasmatocytes. In *Heliothis virescens* nodule production is energy-dependent.[25]

With objects too large to be phagocytosed (e.g., parasites), the hemocytes of Lepidoptera, Coleoptera, and Dictyoptera initiate what is essentially nodule formation but on a larger scale, resulting in a multilayer capsule of plasmatocytes surrounding a necrotic mass of granular cells and the foreign object.[27] Most dipterans exhibit humoral rather than cellular encapsulation; however, cellular capsules have been detected in *Culex territans* and *Anopheles quadrimaculatus* parasitized by *Romanomermis culicivorax* and *Brugia pahangi*, respectively.[9] In the latter insect, melanization preceded plasmatocyte attachment.

Granular cell degeneration in both nodules and cellular capsules is usually accompanied by extensive melanization. Toxic products associated with melanization, suffocation, and the restriction of nutrient uptake and/or waste removal may kill biotic agents. Phenoloxidase associated with melanization may be directly released onto the alien from hemocytes or indirectly released from oenocytoids into plasma and then contact nonself matter.[13]

Plasmatocytes may be chemotactically attracted to granular cell-bacterial aggregates.[27,28] Hemocyte invasion of thin-walled giant cells of the gut epithelium of the cockroach *Blatella germanica* containing the nematode *Physaloptera maxillaris*, as opposed to the thick-walled giant cells of *Acheta pennsylvanicus*, may also be explained by chemotaxis.[29] Nappi and Stoffolano[30] suggested that changes in hemocyte profiles of *Musca domestica* parasitized by *Heterotylenchus autumnalis* represented chemotaxis, since changes in hemocyte types and levels occurred prior to nematode encapsulation. In view of increased hemocyte activity of *Periplaneta americana* and *Schistocerca gregaria* to prophenoloxidase-activated hemocyte lysate,[31,32] the results with *M. domestica* may represent hemocyte mobility induced by nematode metabolites initiating activation of the prophenoloxidase cascade.

B. Constitutive Nonself Humoral Factors in Nonimmune Insects

1. Lectins and Lysozyme

Lectins have been found in the hemolymph plasma and fat body and in and on hemocytes of many insects.[6,11,33-36] They are proteins with a highly specific multivalent capacity to bind to sugars. Although the role of constitutive lectins is not unequivocal, it may have a role in nonself recognition.[8,35] Ratcliffe and Rowley,[37] using lectins of *Clitumnus extradentatus* and *P. americana*, reported a reduction in the number of erythrocytes phagocytosed by the plasmatocytes on hemocyte monolayers. It was proposed that the role of lectins may be the agglutination of bacteria, enhancing the nodule formation capacity of the hemocytes.

Low levels of lysozyme have been found in nonimmune *Hyalophora*

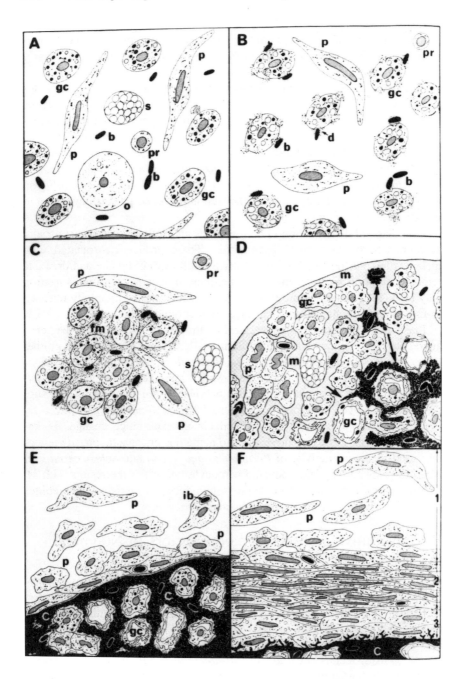

cecropia,[10] *G. mellonella*, and *Manduca sexta*.[6] Dunn[6] proposed that lysozyme may assist in the induction of antibacterial proteins by digesting peptidoglycan fragments into moieties that indirectly or directly stimulate the fat body.

2. Humoral Encapsulation

The production of a melanin layer without direct hemocyte participation as a response to nonself agents has been detected in five families of larval dipterans that contain relatively few hemocytes.[9,38] Humoral encapsulation is a biphasic process. Electron microscopy has established phase one to consist of the deposition of a homogeneous matrix (sticky proteins and melanization intermediates?) around *Hydromermis contorta* in *Culex pipiens* within 1 hr of the parasite entering the insect's hemocoel.[9,38] This is followed by the solidification of the fibrillar matrix and enhanced melanization of phase two. The resulting capsule is a protein-polyphenol complex that physically isolates the alien. Humoral encapsulation was accompanied by hemocyte lysis in adult *Aedes trivittatus* in response to parasitism by *Dirofilaria immitis*, and in the Culicidae the melanotic sheaths surrounding the nematode are eventually covered by the fragmented hemocytes.[39-42] It has been proposed that the nonself response of adult female *Aedes* spp. against microfilariae is a hemocyte-dependent phenomenon with hemocyte lysis preceding or concomitant with melanization.[43]

Melanization is triggered by nematodes,[9] β-1,3-glucans,[19,28] bacterial capsules,[38] peptidoglycan fragments and anionic polydextran,[9,13,40,44] and lipopolysaccharides.[45] In *Bombyx mori* and *Chironomus* spp., melanization is initiated by the conversion of prophenoloxidase to phenoloxidase by means of calcium-dependent serine protease(s).[9,13]

The degree of humoral encapsulation varies with the strain and species of insect as well as insect age and sex.[9] Sutherland et al.[40] reported that the ability of *Aedes aegypti* to respond to microfilariae of *Brugia pahangi* diminished with age. Christensen[41] speculated that reduced humoral encapsulation may represent a reduction in hemocytes which contain prophenoloxidase or its cascade components, a reduction in phenoloxidase activity, changes in the availability of phenoloxidase substrate, or an alteration in the nonself recognition system.

Figure 1 (Opposite page). Stages of nodule formation in larval *Galleria mellonella*.[96] A. Types of freely circulating hemocytes encountering bacteria (b): granular cell (gc), plasmatocyte (p), spherule cell (s), prohemocyte (pr), and oenocytoid (o). B. Granular cell discharge (d) entrapping bacteria. C. Formation of an extensive coagulum (fm) entrapping bacteria and granular cells. D. Melanization (arrow) around bacteria contained within a compact matrix (m). E. Second phase in nodule formation with the attachment of plasmatocytes. Some contain intracellular bacteria (ib). F. Nodule containing loosely attached plasmatocytes (1), a flattened plasmatocyte layer (2), and a melanotic layer of necrotic granular cells (3).

C. Recognition of Nonself

The means by which the hemolymph nonself-response systems recognize foreign material is presently under debate. Lackie proposed a two-tier system with the recognition of abiotic particles based on physicochemical properties (e.g., electrostatic charge and hydrophobicity) and the recognition of biotic elements by specific recognition receptors with possible contribution by lectins.[46,47] Leonard et al.[19] reported increased attachment of *Bacillus cereus* to granular cells and plasmatocytes and increased phagocytic activity of the plasmatocytes in response to the activated prophenoloxidase cascade of *G. mellonella*, *Blaberus craniifer*, and *Leucophaea maderae*. A partial correlation has been obtained between inhibition of cellular encapsulation and phenoloxidase activity in Lepidoptera parasitized by the virus of ichneumonid parasitoids.[13] Collectively, the data imply that the prophenoloxidase system plays a role in nonself recognition.[11,17,44] Whether phenoloxidase itself acts as an opsonin is not clear. As reviewed by Dunn,[6] inhibition of phenoloxidase activity impaired encapsulation in *Drosophila euronotus*, but according to Ratner and Vinson[25] this was not the case in *Heliothis virescens*. Dularay and Lackie,[47] using negatively charged Sephadex beads, reported that although phenoloxidase and four other hemolymph proteins attached to the beads, they were not encapsulated in *S. gregaria*. It was suggested that the putative opsonin did not attach or that none of the components is opsonic in this system. Lipopolysaccharide also enhanced hemocyte activity and nonself responses,[6,8,10,47-50] but, with the exception of a transient increase in *M. sexta* and strong stimulation in *L. migratoria*,[45] did not activate phenoloxidase. Dunphy and Chadwick[48] found a partial correlation between carbohydrates that modify bacterial attachment to *G. mellonella* hemocytes and phenoloxidase activity. Brookman et al.[51] noted that different lipopolysaccharide mutants of *E. coli* induced nodule formation without a direct correlation with their potential to activate the prophenoloxidase system in *S. gregaria* and *L. migratoria*. These findings suggest that phenoloxidase activity is not always an adequate indicator of opsonin-mediated attachment. Other components of the prophenoloxidase cascade, the prophenoloxidase-cleaved peptide, or discharged hemocyte granules may have opsonic activity.

Nonself hemocyte responses may be triggered by direct binding of biotic agents to hemocytes via hemocyte surface lectins and/or by indirect binding mediated by humoral lectins.[6,11,34] Carbohydrates may be part of the recognition process influencing the adhesion of bacteria to hemocytes,[48] nodule formation,[35,51,52] encapsulation,[35] and phenoloxidase activity.[51]

D. Humoral and Cellular Factors in Immune Insects

The most salient attribute of the hemolymph of immune insects is a pronounced antibacterial activity. The mechanisms producing these activities are rapidly being clarified.

Lectins, although traditionally regarded as nonimmune humoral factors, have been induced in *Sarcophaga peregrina* in response to injury and pupation, and in *M. sexta* in response to a bacterial vaccine.[53-55] The 190 kd galactose-binding lectin of *S. peregrina* was synthesized in the fat body, secreted into the hemolymph, and bound to the hemocytes. Hemocytes from nonimmune *S. peregrina* had less affinity for the lectin.[55] Spence and co-workers detected an inducible 70 kd glycoprotein, M13, which, although present in low levels in nonimmune larvae, was produced in response to bacterial challenge via the hemocoel and gut.[7,56-59] M13 is a glucose-specific lectin with the ability to dedifferentiate hemocytes of both immune and nonimmune insects into a filamentous coagulum. Although an immune surveillance role was suggested for the *S. peregrina* lectin,[59] the contribution of M13 is unknown.

Lysozyme in nonimmune insects occurs in low levels. However, in immune larvae lysozyme levels may be elevated by 2- to 50-fold, depending on the immunizing agent, its method of preparation, and the insect species.[6,10,58,60,61] Low molecular weight peptidoglycan fragments effectively stimulated lysozyme induction in *M. sexta*.[6] Other inducing agents include dead Gram-positive and Gram-negative bacteria and endotoxin.[58] It is possible that lysozyme may augment the activity of other induced antibacterial proteins (e.g., cecropins and attacins).[10]

Antibacterial proteins have been induced *de novo* in eight Lepidoptera, three Diptera, and one Coleoptera.[10,53-55,60-62] The proteins are synthesized in the fat body with hemocyte mediation and possibly in the pericardial cells.[6,10,53-58] In the case of *Hyalophora cecropia* with cecropins and attacins, as well as insects with cecropin-like and attacin-like antibacterial proteins and sarcotoxins I A-C and II A-C, *de novo* immune specific mRNA is synthesized within 2 to 5 hr postvaccination. The specific lag period varies with the insect species.[10]

Both the cecropin family (3.5-4 kd) and the antibacterial proteins of the diptericin family (3-10 kd) are basic proteins produced in the fat body of *H. cecropia* and *Phorimia terranovae*, respectively.[10,60,63,64] The major cecropins vary with the insect species. Different species of cecropins vary in their activity toward Gram-positive and Gram-negative bacteria. The diptericins have a broad activity spectrum, but whether this applies to the individual five protein species and six sarcotoxins is unknown. Amino acid analysis of three major diptericins suggests they are different from cecropins, attacins, and sarcotoxins.[54,63,64]

The cecropins are synthesized in a preform, contain a signal peptide, and become activated by sequential removal of proline-containing dipeptides. Cecropins, with their basic N-terminus and hydrophobic C-terminus, form amphipathic α-helices, which Boman and Hultmark[10] suggested may confer membrane-disrupting activity against prokaryotes. Sarcotoxin I A is somewhat homologous to the cecropins in amino acid sequences.[54] Its mode of action may be similar, although activity spectra differ.

The attacins represent six species of antibacterial proteins (20-23 kd) which are divided, according to their amino acid composition, into acidic and basic attacins. The attacins also may be synthesized in a preproform.[6,10] Several attacins are active on the outer membrane of Gram-negative bacteria and may interfere with cell division.[10]

III. *STEINERNEMA* AND *HETERORHABDITIS* AND THEIR BACTERIAL MUTUALISTS IN HEMOLYMPH

Once within the hemocoel of an insect host the nematode/bacterium complex encounters numerous nonself-response systems. To be effective as biological insecticides, these complexes must either (1) tolerate the host's defense systems, (2) evade recognition as nonself, or (3) suppress the host's nonself-response systems.[65,66] Despite the contribution that modeling of the interaction of entomopathogenic nematodes with the nonself-response systems of insect hemolymph could make to explain efficacy in biological control programs, such research has been limited.[67-70]

A. Nematodes Interacting with Hemolymph Nonself-Response Systems

Steinernema and *Heterorhabditis* are capable of killing many insect species, including some dipterous larvae able to mount nonself responses, by either overloading those defenses or triggering nonself reactions toxic to the host insect.[71,72] Bronskill[73] described humoral encapsulation of *S. carpocapsae* (DD-136) in the hemocoel of *Aedes aegypti*, *A. stimulans*, and *A. trichurus* within 5 hr of penetration. The nematodes generally remained in the hemocoel and did not influence insect metamorphosis. Humoral encapsulation has been reported for steinernematids in *Culex pipiens* larvae.[74] Nematode encapsulation did not always prevent insect death (e.g., larvae of *A. aegypti* in which all of the nematodes were encapsulated often died)[71] possibly because the capsules may have been incomplete, and therefore the nematodes were able to release *Xenorhabdus*, or release of the bacterium occurred before encapsulation was complete. Parasite load is important in host survival. Instances of low parasite load in *C. pipiens* have been correlated with highly developed humoral capsules and larval survival; however, with two or more *S. carpocapsae* humoral encapsulation was not sufficiently developed to preclude release of *X. nematophilus*.[71]

The salient point of most steinernematid-dipteran studies is the absence of initial intervention by host hemocytes.[68,73,74] *S. carpocapsae*, isolated from the codling moth, *Cydia pomonella*, initiated humoral encapsulation in the form of a homogeneous matrix within 25 min of entering the hemocoel of *C. pipiens*.[74] By 1-2 hr the capsule restricted nematode movement as the pigment granules coalesced. Pigmentation was pronounced by 2 hr and by 5-10 hr a rigid dark sheath encased the nematode. Interestingly, the nematodes were viable if freed from the capsule. The resulting sheath was composed of two layers, an innermost highly melanized layer closely abutting the nematode, and an outer less melanized layer. Occasionally, an outermost third layer containing cellular and tracheolar debris was observed, but not during the early stages of melanosis. The only other host-parasitic interaction beyond humoral encapsulation that has been described in Diptera is a reduction in total hemocyte counts and phenoloxidase-positive hemocytes in *A. aegypti* challenged with an undefined strain of *S. carpocapsae*.[75] Although a decline in hemocyte number occurred concomitant with melanization of the parasite, ligation established the two events to be independent. This is contrary to events in *Musca domestica*, in which the hemocyte levels increased in response to *Heterotylenchus autumnalis*.[76] It was suggested that hemocytopenia in the former study was due to host wound repair in response to lesions induced by the nematodes.

In the lepidopteran *G. mellonella* and the coleopteran *Leptinotarsa decemlineata*, there was no evidence of encapsulation of the Mexican and DD-136 strains of *S. carpocapsae* or *H. bacteriophora* in the first 4 hr postinjection with either axenic or monoxenic third-stage juveniles.[65,69,70,77,78] However, in the western corn rootworm, *Diabrotica virgifera virgifera*, a cellular encapsulation response was detected against four strains of *S. carpocapsae* but was unrelated to host susceptibility.[79] *Hylobius abietis* larvae surviving infection by *S. carpocapsae* revealed encapsulated nematodes.[80] None of the nematodes were subjected to cellular encapsulation in *G. mellonella*. The absence of response to the nematodes was not attributed to suppression of the nonself-response systems, because the nematodes did not reduce the adhesion of *Bacillus subtilis* to granular cells and plasmatocytes (Table 1), lysozyme activity (Table 2),[81] or phenoloxidase activity (Table 3) *in vitro*.[65,77,78] Studies *in vivo* established that the nematodes did not alter total or differential hemocyte counts during the initial 4-6 hr postinjection, nor impair the removal of *B. subtilis*. The absence of the suppression of nonself-response systems is advantageous for the nema-

TABLE 1. Effect of Third-Stage Juvenile *Steinernema carpocapsae* (DD-136) on Adhesion of *Bacillus subtilis* to Hemocytes of Nonimmune Larval *Galleria mellonella*[65]

Hemocyte type	Time with nematode (min)	Bacteria/ hemocyte	% hemocytes
Plasmatocytes	5	2.1	62
	20	2.4	74
Plasmatocyte control	5	2.7	78
(no nematodes)	20	2.6	69
Granular cells	5	2.4	73
	20	2.2	64
Granular cell control	5	2.3	77
(no nematodes)	20	2.6	75

TABLE 2. Effect of Third-Stage Juvenile *Steinernema carpocapsae* (DD-136) on Lysozyme Activity in Larval Serum of Nonimmune *Galleria mellonella*[65]

		Optical density of bacteria at 450 nm	
Treatment	Incubation time (min)	*Micrococcus luteus*	*E. coli*
Phosphate buffer (pH 6.5) (=PB)	5	0.90	0.91
	60	0.89	0.92
Larval serum + PB	5	0.68	0.53
	60	0.66	0.53
Larval serum + PB + *S. carpocapsae*	5	0.66	0.54
	60	0.64	0.53

TABLE 3. Effect of Third-Stage Juvenile *Steinernema carpocapsae* (DD-136) on Phenoloxidase Activity in Larval Serum of Nonimmune *Galleria mellonella* Followed by Activation by Laminarin for One Hour[81]

Treatment	Time with nematodes (min)	Phenoloxidase activity (units/ml)
Serum only	5	26
	60	29
Serum + nematodes	5	27
	60	28

todes in that the host is able to (1) repair tissue damaged by the invading nematode, and (2) respond to bacteria entering the hemocoel via a wound. This allows *S. carpocapsae* and *H. bacteriophora* to void their respective bacteria, at a later time, into a noncompetitive environment with conditions conducive to bacterial and, subsequently, nematode growth and reproduction.

Although the nematodes do not initially impair their hosts' nonself-response systems, Dunphy and Webster[77] proposed that the nematodes are not recognized as nonself, as opposed to the hemocytes being unable to adhere to the nematode cuticle after recognizing the nematode as foreign. This is because (1) activated hemocytes should have accelerated the removal of *B. subtilis* from the hemolymph, which did not occur, (2) there was no increase in the number of hemocytes, which would have indicated hemocyte mobilization as detected by Nappi and Stoffolano[76] for *Musca* parasitized by *H. autumnalis*, and (3) there was no change in differential hemocyte counts, as in *Musca* with *H. autumnalis* prior to discernable parasite-induced humoral encapsulation.

How *S. carpocapsae* and *H. bacteriophora* evade hemocyte recognition is unknown. The mechanisms proposed by Vinson[82] for endophagous parasitoids, (1) the acquisition of a coating of host materials, (2) possession of nonreactive surfaces, (3) possession of heterophilic antigen(s), or (4) innate molecular mimicry independent of a coating of host materials, may be applicable to entomopathogenic nematodes. The acquisition of host material in insect midgut resulting in blocking of nonself recognition in the insect hemocoel occurs with microfilariae in mosquitoes.[39-41] Similar changes in physicochemical properties and surface antigenicity of *Hydromermis roseus* and *Filipjevimermis leipsandra* may explain the absence of response of *Chironomus riparius* and *Diabrotica undecimpunctata*, respectively. This, however, does not appear to be the case in *S. carpocapsae* and *H. bacteriophora*. Dunphy and Webster[65,78] have shown that the absence of *G. mellonella* response to these nematodes occurs whenever the nematodes are injected and regardless of their previous culture history.

The epicuticle of infective juveniles of *S. carpocapsae* strain DD-136 prevents hemocytic encapsulation of the nematodes in *G. mellonella* larvae.[83] Removal of the sugars α-mannose, β-N-acetyl-D-galactosamine and β-N-acetyl-D-glucosamine from the epicuticle did not cause hemocyte attachment, indicating that these sugars are not important in preventing adhesion. In general, proteolytic enzymes did not alter the infective juvenile epicuticle sufficiently to elicit hemocyte-mediated encapsulation, thus showing that glycoproteins do not mask nonself recognition. Lipase enhanced nematode encapsulation and so established the lipoidal nature of the epicuticle. Lipase may have induced hemocyte attachment by (1) removing a nonreactive surface, (2) removing glycolipids recognized as self (i.e., with sugars different from the types used in the study), and/or (3) exposing molecules recognized by the hemocytes as nonself.

S. carpocapsae and *H. bacteriophora* may impair hemocyte activity later in their development as evident by a suppression of an increase in total hemocyte counts and changes in differential hemocyte counts within 4-6 hr postparasitism.[65,78] This occurred shortly before or when the bacteria are believed to be released. It is not known if hemocyte suppression is induced by a proteinaceous toxin similar to that of Burman[84] or Boemare et al.[85]

B. Virulence Mechanisms of *Xenorhabdus* in Insect Hemolymph

The time of release of *Xenorhabdus* species from steinernematids and heterorhabditids, and the cues triggering release *in vivo* are unknown. However, Dunphy et al.[3] detected the release of *Xenorhabdus poinarii* from *Stein-*

ernema glaseri in Grace's insect tissue culture medium within 2 hr of incubation followed by a 2-hr adaptation period before the bacteria initiated multiplication. Matha and Mrácek[67] reported elevated hemocyte counts in *G. mellonella* within 4 hr of an undefined *Steinernema* sp. (isolated from the sawfly *Cephalcia abietis*) penetrating the insect gut, and Dunphy and Webster[78] reported hemocytopenia in *G. mellonella* within 5 hr of injecting surface-sterilized monoxenic *H. bacteriophora*. Since both are responses to *Xenorhabdus* species, it is proposed that bacteria are released from the nematode vesicle within 5 hr postparasitism.

The fate of *Xenorhabdus* injected into the hemocoels of insects varied with the physiological status of the insect, the insect species, and bacterial species. In immune *Hyalophora cecropia*, the cecropins lysed the Mexican strain of *X. nematophilus* unless the nematode was present to inactivate them.[68] Dunn[86] reported similar results with *M. sexta* infected by the DD-136 strain of *X. nematophilus*. Bacterial survival was dependent upon the nematode releasing a serine protease.

In nonimmune insects, the lethal dose of bacteria required to kill 50% of the insects varied from one to ten, depending on bacterial and insect species.[5,65,66] Such low doses might be attributed to (1) toxins produced by the bacteria, (2) lack of effective antibacterial activity in the insects, or (3) the bacteria tolerate the nonself responses.

Toxin elaboration may not be a virulence component in the traditional sense because neither strains of *X. nematophilus* nor *X. bovienii* produced metabolites toxic to *G. mellonella* in routine bacteriological media or insect tissue culture media.[84,85,87] However, this could be an artifact of culture since all species of *Xenorhabdus* produce hemocytotoxins *in vivo* (see below). An alkaline metalloprotease isolated from *X. luminescens* (Hm) may have a role in bacterial virulence mechanisms in view of the hemocytotoxic activity of proteases of *Serratia marcescens*.[88]

The high levels of bacterial virulence could not be attributed to the absence of antibacterial responses because nodule formation occurred to varying extents at different rates for all test bacteria. *X. nematophilus* (DD-136) was not removed from the hemolymph during the initial 1 hr exposure in the hemocoel of *G. mellonella* (Figure 2).[87] Rapid, albeit partial, removal of *X. bovienii, X. nematophilus* (Mexican), and *X. luminescens* was effected within minutes postinjection.[77,78] Shortly thereafter, all *Xenorhabdus* species reentered the hemolymph. The absence of removal of *X. nematophilus* (DD-136) was correlated to low numbers of bacteria attached to the plasmatocytes and granular cells *in vitro*, and low levels of plasmatocytes and granular cells with bacteria. Hemocytopenia did not occur.[87] The reverse situation occurred for *X. bovienii* and *X. luminescens* (Table 4).[81] In the gypsy moth, *Lymantria dispar, X. nematophilus* was rapidly cleared from the hemolymph (Figure 2).

The absence of immediate nodule formation about the DD-136 strain of *X. nematophilus* was not attributed to suppression of hemocyte activity, because

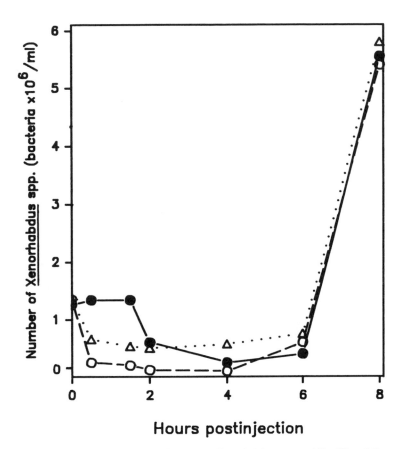

Figure 2. Changes in the level of phase one *Xenorhabdus nematophilus* (●) and *X. bovienii* (Δ) in nonimmune larval *Galleria mellonella*, and *X. nematophilus* (○) in larval *Lymantria dispar*.

spent culture filtrates did not reduce the ability of the hemocytes to respond to *B. cereus*.[87] Suppression of hemocyte activity is relatively rare, being restricted to a few hymenopterous parasitoids and their associated polydna viruses,[86-90] and an acanthocephalan.[92]

The interaction of traditional humoral factors with *X. nematophilus* (DD-136) is not as clear cut. Lysozyme was not inhibited during the first 30 min of infection, based upon the lytic activity of infected serum toward *Micrococcus luteus* and *E. coli* (RR1).[78,87] However, the level of phenoloxidase activity was reduced by more than 80%.[81] It is not known if reduced phenoloxidase activity represents blocked release of antiphenoloxidase inhibiting factors from the hemocytes (a possible step in self-nonself recognition in *Locusta migratoria*).[44] This, however, is considered unlikely since the bacteria cause hemocyte degranulation and eventual lysis.[87] Suppression of phenoloxidase has rarely been reported and is predominantly limited to polydna viruses injected into insects

TABLE 4. Interaction of Selected *Xenorhabdus* Species with the Nonself Response Systems of the Hemolymph of Nonimmune Larval *Galleria mellonella*

Xenorhabdus sp.	#*Xenorhabdus*/hemocyte		Percentage of hemocytes with *Xenorhabdus*		Bacterial removal rate 20 min postinjection ($\times 10^6$/ml/min)	Phenoloxidase activity[81] (units/mg protein/ml serum)
	Plasmatocyte	Granulocyte	Plasmatocyte	Granulocyte		
X. nematophilus[87]	4.8	4.3	9	4	0.1	17
X. bovienii[81]	7.2	15.7	20	37	30.2	—[a]
X. luminescens[78]	6.7	28.2	21	48	58.8	12
Control (no bacteria)	—	—	—	—	—	85

[a] not determined.

from the calyx fluid of *Hyposoter exiguae*,[91] lipopolysaccharides from *E. coli* in *Schistocerca gregaria*,[31,32] and the acanthocephalan *Moniliformis moniliformis* in *Periplaneta americana*.[92] Inhibition of phenoloxidase may be conducive to nematode survival since melanization of steinernematids usually results in nematode death[79,81,93] and the prophenoloxidase cascade would be expected to be released with hemocyte damage.[87] The requirement of 1 hr incubation of the DD-136 bacterial isolate *in vivo* or *in vitro* in larval serum prior to the initiation of early nodule formation in *G. mellonella* implies humoral factors other than phenoloxidase altered the bacterial envelope, facilitating hemocyte contact.[87] However, these factors and the enzymes associated with the hemocytes appear to have been tolerated since, in the overall, the bacteria proliferated and killed the host.

In all cases the bacteria reemerged into the hemolymph (Figure 2) and for *X. nematophilus* (DD-136 and Mexican) and *X. luminescens*, this was independent of ongoing bacterial metabolism.[77,87] The level of bacteria re-emerging into the hemolymph exceeded the inoculum level, suggesting bacterial multiplication. This was confirmed for *X. nematophilus* (DD-136 strain) by detecting (1) an increase in muramic acid (a component unique to bacterial cell walls) in infected insects, and (2) observing bacterial multiplication during their association with the hemocytes.[87] Similar multiplication was detected for *X. luminescens* on *G. mellonella* hemocytes.[78]

Re-emergence of *Bacillus* spp. and lipopolysaccharide mutants of *Pseudomonas aeruginosa* in *G. mellonella* revealed that the hemocytes of *G. mellonella* did not successfully contain virulent pathogens, in general resulting in secondary bacteremia.[26,50] Similar results have been reported for *Pseudaletia unipunctata* infected by *B. thuringiensis*.[87] Dunn[6] documented the inability of *Manduca sexta* hemocytes to contain an infection of *P. aeruginosa* (11-1).

The reemergence of all *Xenorhabdus* species and strains in *G. mellonella* was associated with a rapid increase in the total hemocyte counts. Although Seryczynska and Kamionek,[69] Kamionek,[94] and Seryczynska et al.[70] reported similar results, they interpreted the increase to be stimulation of the insect's defenses; Dunphy and Webster[87] established that the hemocytes were too severely damaged to respond to foreign matter and thus the increase did not represent hemocyte stimulation.

Collectively, the nonself-response systems did not restrict the pathology. That is not to say that some bactericidal activity did not occur. Bacteriolytic activity was demonstrated using dead *X. nematophilus* (DD-136) which did not reemerge to the inoculum level even though dead bacteria were more effective at damaging the hemocytes.[81] Bacteriolytic activity may be a strain variable since the Mexican strain of antibiotic-killed *X. nematophilus* did not exhibit lysis.[77]

The most definitive evidence of hemocytotoxins produced by *Xenorhabdus* isolates in *G. mellonella* and *Leptinotarsa decemlineata* is for lipopolysaccharides or endotoxin.[49,70] Lipopolysaccharides are any molecular species consisting of a main polysaccharide chain of varying length linked to a core composed of heptose and 3-deoxymannooctulosonic acid and anchored to the outer membrane of Gram-negative bacteria by lipid A.[95]

Lipopolysaccharide was determined to be the hemocytotoxin based on (1) the presence of a phenol-water extract from *Xenorhabdus*-infected *G. mellonella* serum (from which the bacteria had been removed) which elevated hemocyte counts, (2) the similarity in SDS-polyacrylamide gel electrophoretic profile between lipopolysaccharide from species of *Xenorhabdus* and the hemolymph extract, (3) the ability of lipopolysaccharide from *Xenorhabdus* spp. to elevate the hemocyte counts, and (4) the appearance of a factor capable of inducing gelation of *Limulus* amebocyte lysate (an indicator of lipopolysaccharide in biological fluids).[49] Using the amebocyte lysate assay, Dunphy and Webster[49,78] detected a correlation in the appearance of lipopolysaccharide from *X. nematophilus* and *X. luminescens in vivo*, bacterial reemergence into the hemolymph, and elevation in hemocyte levels (Figure 3), concluding that endotoxin is a significant part of the virulence mechanisms of the bacteria.

The toxic moiety of the endotoxin was identified as lipid A for all lipopolysaccharide species from *Xenorhabdus*.[49] The electrophoretic profiles of the endotoxin from the DD-136 and Breton strains of *X. nematophilus* differed, even though both endotoxin species were equally toxic, implying that the oligosaccharide side chain (O-side chains) and core were not toxic to insect hemocytes. Injections of the O-side chains alone induced nodule formation. Lipid A injections elevated hemocyte counts. This effect was eliminated by co-injecting lipid A and polymyxin B (a nonapeptide antibiotic that binds to lipid A) establishing that lipid A may bind to the hemocytes. This was further confirmed by inhibition studies of ^{32}P-labelled lipid A by polymyxin B.[49] Binding of both the ^{32}P-labelled lipopolysaccharide and lipid A from *X. nematophilus* (DD-136) and *X. luminescens* to isolated granular cells was blocked by low concentrations of N-acetylated and nonacetylated glucosamine suggesting that lipid A bound to glucosaminyl receptors on the hemocytes.[49,87] Total fatty acid extracts from the glucosaminyl glucosamine dissacharide backbone of lipid A were as toxic to the hemocytes as a corresponding amount of lipid A and endotoxin. The major fatty acids detected in Enterobacteriaceae lipid A vary in their hemocyte toxicity in *G. mellonella* with 3-hydroxy tetradecanoic acid and *n*-tetradecanoic acid being the most toxic.[87]

IV. CONCLUSIONS

Insects are capable of recognizing a diversity of foreign objects in their hemolymph and initiating a myriad of nonself responses to contain the aliens, including phagocytosis, nodule formation, cellular and humoral encapsulation, and the induction of antibacterial proteins. The types of responses made to the *Steinernema/Xenorhabdus* and *Heterorhabditis/Xenorhabdus* complexes and their effectiveness vary with the insect species, physiological status, and the strain of nematode/bacterium complex. In nonimmune Lepidoptera, the com-

317

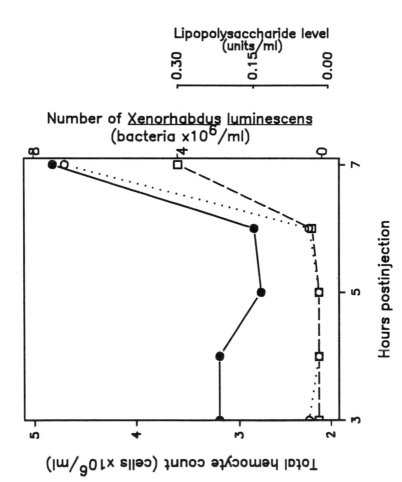

Figure 3. Changes in the levels of *Xenorhabdus luminescens* (●), lipopolysaccharide activity (□), and total hemocyte counts (○) in nonimmune larval *Galleria mellonella*.

plex successfully kills the host because the nematode evades recognition as nonself, allowing time to release its bacterium. In turn, the bacteria tolerate the insect's cellular defenses and lysozyme, and inhibit phenoloxidase activity. The bacteria eventually destroy the host's hemocytes by releasing the hemocytotoxin, lipopolysaccharide, from the bacterial outer membrane into the hemolymph. The toxin binds to the hemocytes, in part, by the lipid A moiety; lipid A contains the toxic fatty acids that damage the hemocytes. In the initial stages of parasitism the nematodes do not impair the host's defenses, allowing the insect to repair nematode-induced damage and/or contain contaminating bacteria inadvertently introduced by the nematode. The host's hemolymph is thus maintained in a bacterial-free state, allowing *Xenorhabdus* to grow and establish conditions favoring nematode development. In immune insects the antibacterial proteins, cecropins, are capable of lysing *Xenorhabdus*. This is prevented by protein secretions from the nematodes destroying the cecropins.

Future research to produce effective models of host-nematode/bacterium interaction to allow more realistic projections of pest control will require (1) a better understanding of insect nonself recognition in a greater range of insect species, (2) greater elucidation of constitutive humoral factors (e.g., carbohydrases and proteases) and their contribution to antibacterial and antinematode activity, and (3) mechanisms of tolerance by *Xenorhabdus* within the hemocytes. From the point of view of the ability of the steinernematids and heterorhabditids to survive in both nonimmune and immune insects, further research into the cuticular lipids and enzymes released from the nematode, and their association with insects with different nonself response mechanisms is a major concern.

REFERENCES

1. **Bedding, R. A., Molyneux, A. S., and Akhurst, R. J.,** *Heterorhabditis* spp., *Neoaplectana* spp. and *Steinernema kraussei*: interspecific and intraspecific differences in infectivity for insects, *Exp. Parasitol.,* 55, 249, 1983.
2. **Molyneux, A. S., Bedding, R. A., and Akhurst, R. J.,** Susceptibility of larvae of the sheep blowfly *Lucilia cuprina* to various *Heterorhabditis* spp., *Neoaplectana* spp., and an undescribed steinernematid (Nematoda), *J. Invertebr. Pathol.,* 42, 1, 1983.
3. **Dunphy, G. B., Rutherford, T. A., and Webster, J. M.,** Growth and virulence of *Steinernema glaseri* influenced by different subspecies of *Xenorhabdus nematophilus, J. Nematol.,* 17, 476, 1985.
4. **Akhurst, R. J.,** *Neoaplectana* species: specificity of association with bacteria of the genus *Xenorhabdus, Exp. Parasitol.,* 55, 258, 1983.
5. **Griffin, C. T., Simons, W. R., and Smits, P. H.,** Activity and infectivity of four isolates of *Heterorhabditis* spp., *J. Invertebr. Pathol.,* 53, 107, 1989.
6. **Dunn, P. E.,** Biochemical aspects of insect immunology, *Annu. Rev. Entomol.,* 31, 321, 1986.
7. **Rupp, R. A., and Spence, K. D.,** Protein alterations in *Manduca sexta* midgut and haemolymph following treatment with a sublethal dose of *Bacillus thuringiensis* crystal endotoxin, *Insect Biochem.,* 15, 147, 1985.

8. **Ratcliffe, N. A.**, Insect cellular immunity and the recognition of foreignness, in *Immune Mechanisms in Invertebrate Vectors*, Lackie, A. M. Ed., Clarendon Press, Oxford, 1986, 21.

9. **Götz, P.**, Mechanisms of encapsulation in dipteran hosts, in *Immune Mechanisms in Invertebrate Vectors*, Lackie, A. M., Ed., Clarendon Press, Oxford, 1986, 1.

10. **Boman, H. G., and Hultmark, D.**, Cell-free immunity in insects, *Annu. Rev. Microbiol.*, 41, 103, 1987.

11. **Rowley, A. F., Ratcliffe, N. A., Leonard, C. M ., Richards, E. H., and Renwrantz, L.**, Humoral recognition factors in insects, with particular reference to agglutinins and the prophenoloxidase system, in *Hemocytic and Humoral Immunity in Arthropods*, Gupta, A. P., Ed., John Wiley & Sons, New York, 1986, 381.

12. **Morton, D. B., Dunphy, G. B., and Chadwick, J. S.**, Reactions of hemocytes of immune and non-immune *Galleria mellonella* larvae to *Proteus mirabilis*, *Dev. Comp. Immunol.*, 11, 47, 1987.

13. **Yoshida, H., and Ashida, M.**, Microbial activation of two serine enzymes and prophenoloxidase in the plasma fraction of hemolymph of the silkworm, *Bombyx mori*, *Insect Biochem.*, 16, 539, 1986.

14. **Ratcliffe, N. A.**, Invertebrate immunity - a primer for the non-specialist, *Immunol. Lett.*, 10, 253, 1985.

15. **Kaaya, G. P., Ratcliffe, N. A., and Alemu, P.**, Cellular and humoral defenses of *Glossina*: reactions against bacteria, trypanosomes and experimental implants, *J. Med. Entomol.*, 23, 30, 1986.

16. **Ratcliffe, N. A., and Rowley, A. F.**, In vitro phagocytosis of bacteria by blood cells, *Nature*, 252, 391, 1974.

17. **Brookman, J. L., Ratcliffe, N. A., and Rowley, A. F.**, Optimization of a monolayer phagocytosis assay and its application for studying the role of the prophenoloxidase system in the wax moth, *Galleria mellonella*, *J. Insect Physiol.*, 34, 337, 1988.

18. **Anderson, R. S., Holmes, B., and Good, R. A.**, Comparative biochemistry of phagocytizing insect hemocytes, *Comp. Biochem. Physiol.*, 46B, 595, 1973.

19. **Leonard, C. M., Ratcliffe, N. A., and Rowley, A. F.**, The role of prophenoloxidase activation in non-self recognition and phagocytosis by insect blood cells, *J. Insect Physiol.*, 31, 789, 1985.

20. **Wago, H.**, Involvement of microfilaments in filopodial function of phagocytic granular cells of the silkworm, *Bombyx mori*, *Dev. Comp. Immunol.*, 6, 655, 1982.

21. **Davies, D. H., and Preston, T. M.**, Effect of disruption of plasmatocyte microfilaments on encapsulation *in vitro*, *Dev. Comp. Immunol.*, 11, 353, 1987.

22. **Anderson, R. S., Holmes, B., and Good, R. A.**, In vitro bactericidal capacity of *Blaberus craniifer* hemocytes, *J. Invertebr. Pathol.*, 22, 127, 1973.

23. **Anderson, R. S., and Cook, M. L.**, Induction of lysozyme-like activity in the hemolymph and hemocytes of an insect, *Spodoptera eridania*, *J. Invertebr. Pathol.*, 33, 197, 1979.

24. **Zachary, D., and Hoffman, J.A.**, Lysozyme is stored in the granules of certain hemocyte types in *Locusta.*, *J. Insect Physiol.*, 30, 405, 1984.

25. **Ratner, S., and Vinson, S. B.**, Phagocytosis and encapsulation: cellular immune responses in arthropods, *Amer. Zool.*, 23, 185, 1983.

26. **Walters, J. B., and Ratcliffe, N. A.**, Studies on the *in vivo* cellular reactions of insects: fate of pathogenic and non-pathogenic bacteria in *Galleria mellonella* nodules, *J. Insect Physiol.*, 29, 417, 1983.

27. **Ratcliffe, N. A., and Rowley, A. F.**, Role of hemocytes in defense against biological agents, in *Insect Hemocytes: Development, Forms, Functions and Techniques*, Gupta, A. P., Ed., Cambridge University Press, Cambridge, 1979, 331.

28. **Gunnarsson, S. G. S.**, Infection of *Schistocerca gregaria* by the fungus *Metarhizium anisopliae*: cellular reactions in the integument studied by scanning electron microscopy and light microscopy, *J. Invertebr. Pathol.*, 52, 9, 1988.

29. **Cawthorn, R. J., and Anderson, R. C.**, Cellular reactions of field crickets (*Acheta pennsylvanicus* Burmeister) and German cockroaches (*Blatella germanica* L.) to *Physaloptera maxillaris* Molin (Nematoda: Physalopteroidea), *Can. J. Zool.*, 55, 368, 1977.

30. **Nappi, A. J., and Stoffolano, Jr., J. G.**, Distribution of haemocytes in larvae of *Musca domestica* and *Musca autumnalis* and possible chemotaxis during parasitization, *J. Insect Physiol.*, 18, 169, 1972.

31. **Tackle, G. B., and Lackie, A. M.**, Chemokinetic behavior of insect haemocytes *in vitro*, *J. Cell Sci.*, 85, 85, 1986.

32. **Huxham, I. M., and Lackie, A. M.**, Behavior *in vitro* of separated fractions of haemocytes of the locust *Schistocerca gregaria*, *Cell Tissue Res.*, 251, 677, 1988.

33. **Götz, P., and Boman, H. G.**, Insect immunity in *Comprehensive Insect Physiology, Biochemistry and Pharmacology*, Vol. 3, Kerkut, G. A., and Gilbert, L. I., Eds., Pergamon Press, Toronto, 1985, 453.

34. **Pendland, J. C., Heath, M. A., and Boucias, D. G.**, Function of a galactose-binding lectin of *Spodoptera exigua* larval hemolymph: opsonization of blastospores from entomogenous hyphomycetes, *J. Insect Physiol.*, 34, 533, 1988.

35. **Lackie, A. M., and Vasta, G. R.**, The role of galactosyl-binding lectin in the cellular immune response of the cockroach *Periplaneta americana* (Dictyoptera), *Immunology*, 64, 353, 1988.

36. **Bradley, R. S., Stuart, G. S., Stiles, B., and Hapner, K. D.**, Grasshopper haemagglutinin: immunochemical localization in haemocytes and investigation of opsonic properties, *J. Insect Physiol.*, 35, 353, 1989.

37. **Ratcliffe, N. A., and Rowley, A. F.**, Opsonic activity in insect haemolymph, in *Comparative Pathobiology (Cells and Serum Factors)*, Vol. 6, Cheng, T. C., Ed., Plenum Press, New York, 1984, 187.

38. **Götz, P., and Vey, A.**, Humoral encapsulation in insects, in *Hemocytic and Humoral Immunity in Arthropods*, Gupta, A. P., Ed., John Wiley & Sons, New York, 1986, 407.

39. **Christensen, B. M., Sutherland, D. R., and Gleason, L. N.**, Defense reactions of mosquitoes to filarial worms: comparative studies on the response of three different mosquitoes to inoculated *Brugia pahangi* and *Dirofilaria immitis* microfilariae, *J. Invertebr. Pathol.*, 44, 267, 1984.

40. **Sutherland, D. R., Christensen, B. M., and Forton, K. F.**, Defense reactions of mosquitoes to filarial worms: role of the microfilarial sheath in the response of mosquitoes to inoculated *Brugia pahangi* microfilariae, *J. Invertebr. Pathol.*, 44, 275, 1984.

41. **Christensen, B. M.**, Immune mechanisms and mosquito-filarial worm relationships, in *Immune Mechanisms in Invertebrate Vectors*, Lackie, A. M., Ed., Clarendon Press, Oxford, 1986, 145.

42. **Christensen, B. M., Huff, B. M., Miranpari, G. S., Harris, K. L., and Christensen, L. A.**, Hemocyte population changes during the immune response of *Aedes aegypti* to inoculated microfilariae of *Dirofilaria immitis*, *J. Parasitol.*, 75, 119, 1989.

43. **Nappi, A. J., Christensen, B. M., and Tracy, J. W.**, Quantitative analysis of hemolymph monophenol oxidase activity in immune reactive *Aedes aegypti*, *Insect Biochem.*, 17, 685, 1987.

44. **Brookman, J. L., Ratcliffe, N. A., and Rowley, A. F.**, Studies on the activation of the prophenoloxidase system of insects by bacterial cell wall components, *Insect Biochem.*, 19, 47, 1989.

45. **Bréhelin, M., Drif, L. B., and Boemare, N.**, Insect haemolymph: cooperation between humoral and cellular factors in *Locusta migratoria*, *Insect Biochem.*, 19, 301, 1989.

46. **Lackie, A. M.**, Effect of substratum wettability and charge on adhesion *in vitro* and encapsulation *in vivo* by insect haemocytes, *J. Cell Sci.*, 63, 181, 1983.

47. **Dularay, B., and Lackie, A. M.**, Haemocytic encapsulation and the prophenoloxidase-activation pathway in the locust *Schistocerca gregaria* Forsk., *Insect Biochem.*, 15, 827, 1985.

48. **Dunphy, G. B., and Chadwick, J. S.**, Effects of selected carbohydrates and the contribution of the prophenoloxidase cascade system to the adhesion of strains of *Pseudomonas aeruginosa* and *Proteus mirabilis* to hemocytes of nonimmune larval *Galleria mellonella*, *Can. J. Microbiol.*, 35, 524, 1989.

49. **Dunphy, G. B., and Webster, J. M.**, Lipopolysaccharides of *Xenorhabdus nematophilus* (Enterobacteriaceae) and their haemocyte toxicity in nonimmune *Galleria mellonella* (Insecta: Lepidoptera) larvae, *J. Gen. Microbiol.*, 134, 1017, 1988.

50. **Dunphy, G. B., Morton, D. B., Kropinski, A., and Chadwick, J. S.**, Pathogenicity of lipopolysaccharide mutants of *Pseudomonas aeruginosa* for larvae of *Galleria mellonella*: bacterial properties associated with virulence, *J. Invertebr. Pathol.*, 47, 48, 1986.

51. **Brookman, J. L., Rowley, A. F., and Ratcliffe, N. A.**, Studies on nodule formation in locusts following injection of microbial products, *J. Invertebr. Pathol.*, 53, 315, 1989.

52. **Rizki, T. M., and Rizki, R. M.**, Surface changes on hemocytes during encapsulation in *Drosophila melanogaster* Meigen, in *Hemocytic and Humoral Immunity in Arthropods*, Gupta, A. P., Ed., John Wiley & Sons, New York, 1986, 157.

53. **Okada, M., and Natori, S.**, Purification and characterization of an antibacterial protein from haemolymph of *Sarcophaga peregrina* (fleshfly) larvae, *Biochem. J.*, 211, 727, 1983.

54. **Shiraishi, A., and Natori, S.**, Humoral factor activating the *Sarcophaga* lectin gene in cultured fat body, *Insect Biochem.*, 19, 261, 1981.

55. **Komano, H., Nozawa, R., Mizumo, D., and Natori, S.**, Measurement of *Sarcophaga peregrina* lectin under various physiological conditions by radioimmunoassay, *J. Biol. Chem.*, 258, 2143, 1983.

56. **Minnick, M. F., Rupp, R. A., and Spence, K. D.**, A bacterial-induced lectin which triggers hemocyte coagulation in *Manduca sexta*, *Biochem. Biophys. Res. Commun.*, 137, 729, 1986.

57. **Hurlbert, R. E., Karlinsey, J. E., and Spence, K. D.**, Differential synthesis of bacteria-induced proteins of *Manduca sexta* larvae and pupae, *J. Insect Physiol.*, 31, 205, 1985.

58. **Minnick, M. F., and Spence, K. D.**, Tissue site and modification of a bacteria-induced coagulation protein from *Manduca sexta*, *Insect Biochem.*, 18, 637, 1988.

59. **Spies, A. G., Karlinsey, J. E., and Spence, K. D.**, Antibacterial hemolymph proteins of *Manduca sexta*, *Comp. Biochem. Physiol.*, 83B, 125, 1986.

60. **Postlethwait, J. H., Saul, S. H., and Postlethwait, J. A.**, The antibacterial immune responses of the medfly *Ceratitis capitata*, *J. Insect Physiol.*, 34, 91, 1988.

61. **Trenczek, T., and Faye, I.**, Synthesis of immune proteins in primary cultures of fat body from *Hyalophora cecropia*, *Insect Biochem.*, 18, 299, 1988.

62. **George, J. F., Karp, R. D., Rellahan, B. L., and Lessard, J. L.**, Alteration of the protein composition in the haemolymph of American cockroaches immunized with soluble proteins, *Immunology*, 62, 505, 1987.

63. **Dimarcq, J. L., Keppi, E., Lambert, J., Zachary, D., and Hoffman, D.**, Diptericin A, a novel antibacterial peptide induced by immunization or injury in larvae of the dipteran insect *Phormia terranovae* (Diptera), *Dev. Comp. Immun.*, 10, 626, 1986.

64. **Keppi, E., Zachary, D., Robertson, M., Hoffmanan, D., and Hoffman, J. A.**, Induced antibacterial proteins in the haemolymph of *Phormia terranovae* (Diptera), purification and possible origin of one protein, *Insect Biochem.*, 16, 395, 1986.

65. **Dunphy, G. B., and Webster, J. M.**, Influence of *Steinernema feltiae* (Filipjev) Wouts, Mracek, Gerdin and Bedding DD-136 strain on the humoral and haemocytic responses of *Galleria mellonella* (L.) larvae to selected bacteria, *Parasitology*, 91, 369, 1985.

66. **Nappi, A. J., and Christensen, B. M.**, Insect immunity and mechanisms of resistance by nematodes, in *Vistas on Nematology*, Veech, J. A., and Dickson, D. W., Eds., Society of Nematologists, Hyattsville, 1987, 285.

67. **Matha, V., and Mrácek, Z.**, Changes in haemocyte counts in *Galleria mellonella* (L.) (Lepidoptera: Galleriidae) larvae infected with *Steinerema* sp. (Nematoda: Steinernematidae), *Nematologica*, 30, 86, 1984.

68. **Götz, P., Boman, A., and Boman, H. G.**, Interactions between insect immunity and an insect-pathogenic nematode with symbiotic bacteria, *Proc. Roy. Soc. Lond.*, 212B, 333, 1981.

69. **Seryczynska, H., and Kamionek, M.**, Defensive reactions of *Leptinotarsa decemlineata* Say in relation to *Neoaplectana carpocapsae* Weiser (Nematoda: Steinernematidae) and *Pristionchus uniformis* Fedorko et Stanuszek (Nematoda: Diplogasteridae), *Bull. Acad. Pol. Sci.*, 22, 95, 1974.

70. **Seryczynska, H., Kamionek, M., and Sander, H.**, Defensive reactions of caterpillars of *Galleria mellonella* L. in relation to bacteria *Achromobacter nematophilus* Poinar et Thomas (Eubacteriales: Achromobacteriaceae) and bacteria-free nematodes *Neoaplectana carpocapsae* Weiser (Nematoda: Steinernematidae), *Bull. Acad. Pol. Sci.*, 22, 193, 1974.

71. **Welch, H. E., and Bronskill, J. F.**, Parasitism of mosquito larvae by the nematode, DD-136 (Nematoda: Neoaplectanidae), *Can. J. Zool.*, 40, 1263, 1962.

72. **Beresky, M. A., and Hall, D. W.**, The influence of phenylthiourea on encapsulation, melanization, and survival in larvae of the mosquito *Aedes aegypti* parasitized by the nematode *Neoplectana carpocapsae*, *J. Invertebr. Pathol.*, 29, 74, 1977.

73. **Bronskill, J. F.**, Encapsulation of rhabditoid nematodes in mosquitoes, *Can. J. Zool.*, 40, 1269, 1962.

74. **Poinar, G. O., Jr., and Leutenegger, R.**, Ultrastructural investigation of the melanization process in *Culex pipiens* (Culicidae) in response to a nematode, *J. Ultrastruct. Res.*, 36, 149, 1971.

75. **Andreadis, T. G., and Hall, D. W.**, *Neoaplectana carpocapsae*: encapsulation in *Aedes aegypti* and changes in host hemocytes and hemolymph proteins, *Exp. Parasitol.*, 39, 252, 1976.

76. **Nappi, A. J., and Stoffolano, J. G., Jr.**, Haemocytic changes associated with the immune reaction of nematode infected larvae of *Orthellia caesarion*, *Parasitology*, 62, 295, 1972.

77. **Dunphy, G. B., and Webster, J. M.**, Influence of the Mexican strain of *Steinernema feltiae* and its associated bacterium *Xenorhabdus nematophilus* on *Galleria mellonella*, *J. Parasitol.*, 72, 130, 1986.

78. **Dunphy, G. B., and Webster, J. M.**, Virulence mechanisms of *Heterorhabditis heliothidis* and its bacterial associate, *Xenorhabdus luminescens*, in nonimmune larvae of the greater wax moth, *Galleria mellonella*, *Int. J. Parasitol.*, 18, 729, 1988.

79. **Jackson, J. J., and Brooks, M. A.**, Susceptibility and immune response of western corn rootworm larvae (Coleoptera: Chrysomelidae) to the entomogenous nematode, *Steinernema feltiae* (Rhabditida: Steinernematidae), *J. Econ. Entomol.*, 82, 1073, 1989.

80. **Pye, A. E, and Burman, M.**, *Neoaplectana carpocapsae*: infection and reproduction in large pine weevil larvae, *Hylobius abietis*, *Exp. Parasitol.*, 46, 1, 1978.

81. **Dunphy, G. B., and Webster, J. M.**, unpublished data, 1986.

82. **Vinson, S. B.**, Insect host responses against parasitoids and the parasitoid's resistance: with emphasis on the Lepidoptera-Hymenoptera association, in *Comparative Pathobiology (Invertebrate Immune Responses)*, Vol. 3, Bullasir, L. A., and Cheng, T. C., Eds., Plenum Press, New York, 1977, 103.

83. **Dunphy, G. B., and Webster, J. M.**, Partially characterized components of the epicuticle of dauer juvenile *Steinernema feltiae* and their influence on hemocyte activity in *Galleria mellonella*, *J. Parasitol.*, 73, 584, 1987.

84. **Burman, M.**, *Neoaplectana carpocapsae*: toxin production by axenic insect parasitic nematodes, *Nematologica*, 28, 62, 1982.

85. **Boemare, N., Laumond, C., and Luciani, J.**, Mise en évidence d'une toxicogenèse provoquée par le nématode axénique entomophage *Neoaplectana carpocapsae* Weiser chez l'insecte axénique *Galleria mellonella* L., *C. R. Acad. Sci., Paris*, 295, 543, 1982.

86. **Dunn, P. E.**, personal communication, 1989.

87. **Dunphy, G. B., and Webster, J. M.**, Interaction of *Xenorhabdus nematophilus* subsp. *nematophilus* with the haemolymph of *Galleria mellonella, J. Insect Physiol.*, 30, 883, 1984.

88. **Schmidt, T. M., Bleakley, B., and Nealson, K. H.**, Characterization of an extracellular protease from the insect pathogen *Xenorhabdus luminescens, Appl. Environ. Microbiol.*, 54, 2793, 1988.

89. **Stoltz, D. B., and Guzo, D.**, Apparent haemocytic transformations associated with parasitoid-induced inhibition of immunity in *Malacosoma disstria* larvae, *J. Insect Physiol.*, 32, 377, 1986.

90. **Rizki, R. M., and Rizki, T. M.**, Selective destruction of a host blood cell type by a parasitoid wasp, *Proc. Natl. Acad. Sci., USA*, 81, 6154, 1984.

91. **Stoltz, D. B., and Cook, D. I.**, Inhibition of host phenoloxidase activity by parasitoid Hymenoptera, *Experientia*, 39, 1022, 1983.

92. **Lackie, A. M., and Holt, R. H. F.**, Immunosuppression by larvae of *Moniliformis moniliformis* (Acanthocephala) in their cockroach host (*Periplaneta americana*), *Parasitology*, 98, 307, 1988.

93. **Zervos, S., and Webster, J. M.**, Susceptibility of the cockroach *Periplaneta americana* to *Heterorhabditis heliothidis* (Nematoda: Rhabditoidea) in the laboratory, *Can. J. Zool.*, 67, 1609, 1989.

94. **Kamionek, M.**, Effect of heat-killed cells of *Achromobacter nematophilus* Poinar et Thomas, and the fraction (endotoxin) isolated from them on *Galleria mellonella* caterpillars, *Bull. Acad. Pol. Sci.*, 23, 277, 1975.

95. **Hammond, S. M., Lambert, P. A., and Rycroft, A. N.**, *The Bacterial Cell Surface*, Kapitan Szabo, Washington, DC, 1984, 220.

96. **Ratcliffe, N. A., and Gagen, S. J.**, Studies on the *in vivo* cellular reactions of insects: an ultrastructual analysis of nodule formation in *Galleria mellonella, Tissue Cell*, 9, 73, 1977.

Conclusions

17. Perspectives on Entomopathogenic Nematology

W. M. Hominick and A. P. Reid

I. INTRODUCTION

We propose to show where entomopathogenic nematology might benefit from, or contribute to, developments in such disciplines as biological control, population and community ecology, and molecular biology. To facilitate consideration of topics ranging from molecular biology to ecology, we have divided our chapter into areas which we feel will see the most activity in the 1990s.

II. SPECIES, STRAINS, AND MOLECULAR BIOLOGY

In parasitology, the concept of a "strain" is widespread and is used to refer to a group within a species, the individuals of which have certain characteristics in common, such as preferred hosts, virulence, persistence, host-finding ability, and tolerance to environmental conditions.[1] The idea that many, perhaps most, species of parasites, including entomopathogenic nematodes, exist as a complex of strains has become widely accepted in the past decade.

Curran[2] documents the use of the two most common molecular biology techniques, enzyme electrophoresis and restriction endonuclease analysis of DNA, to confirm that species and strains of entomopathogenic nematodes are genetically distinct. The philosophy behind such techniques is that (1) they can prove useful to separate morphologically similar groups, and (2) genetic differences between strains might correlate with differences that are biologically important. The logic behind this argument is that populations within species of parasites are isolated from each other. Therefore, they should gradually diverge through genetic drift and natural selection by exposure to particular hosts and local environments. Thompson and Lymbery[1] question this premise because it is naive to assume that variation is always neatly packaged into strains. The question of frequency of parasite outbreeding remains unanswered and there may be no groups, but a continuum, with each parasite slightly different from the next. Studies on genetic variability within species of entomopathogenic nematodes will contribute to this developing debate. Nevertheless, the notion of a strain has provided a starting point for attempts to control parasitic diseases, and has contributed to the development of entomopathogenic nematodes for biological control.

An important aspect of entomopathogenic nematology concerns identification of species and strains of the nematodes and their bacterial associates. Akhurst and Boemare[3] address the taxonomic problems associated with the bacteria, and the nomenclature and classification have clearly become more complex as more strains are studied. Recently, bacteria identified as *Xenorhabdus luminescens* have been identified from human clinical cases.[4] This raises the spectre of regulatory agencies requiring stringent precautions or preventing mass culture until resolution of the source of the bacteria and their identity occurs. However, association with clinical cases does not necessarily mean that the bacteria caused the pathology. Nealson and Hastings[5] commented that luminescent wounds sometimes noticed at battlefield hospitals in the 19th century were thought to be a helpful indication for the patient, possibly because antibiotic production by *Xenorhabdus* would inhibit pathogenic organisms. Moreover, Farmer et al.[4] concluded that *X. luminescens* strains used in insect control are probably not important in public health. Apparently, the strains can be placed into five distinct DNA hybridization groups, all of which could be considered as distinct species.[4] It is essential that the taxonomy of those bacteria associated with clinical cases be confirmed and their relationship with entomopathogenic nematodes clarified.

Poinar[6] has provided a benchmark publication for the taxonomy and identification of entomopathogenic nematodes. There has been much taxonomic confusion in the past, and it is important that his proposals are followed or refuted in the literature so that a universal system evolves. Correct identity of entomopathogenic nematodes is essential so that work can be replicated and compared meaningfully. To this end, the large numbers of isolates available now and designated by code numbers must be correctly identified and new species described. For example, Akhurst's[7] database for entomopathogenic nematodes lists 61 *Heterorhabditis* spp., but only 15 are identified. Also, the origin and isolate designations of the nematodes being studied must be provided in all publications so that future workers can identify them as taxonomic changes occur.

Identification of species will continue to be based on morphological criteria, supplemented by molecular methods.[2] It is at the level of sibling species and subspecific groups that molecular methods will come into their own. Of the molecular techniques available, DNA sequence analysis appears to offer the most promise to characterize isolates below the species level. However, the major gap in our knowledge on the usefulness of DNA techniques for taxonomy is the level associated with different degrees of sequence divergence for genera, species, and intraspecific groups.[2] Although differences have been observed for isolates of *Heterorhabditis*, lack of cross-breeding data prohibits correlation of observed differences with the taxonomic level of these differences. Our own DNA work[8] has concentrated on *Steinernema* spp. because these nematodes are common in the U.K.[9] and cross-breeding is easily accomplished. About half the isolates from 80 sampling sites produce typical *S. feltiae (=bibionis)* patterns when Southern blots are hybridized with a *Caen-*

orhabditis elegans rDNA probe. The other isolates show three other distinct patterns and two types have proved to be a second species that is as yet undescribed. Nematodes are now being evaluated for ability to interbreed and for biological differences such as infectivity at low temperatures. Such studies will have to be extended to other species to ascertain the extent of natural intraspecific variation. The same applies to studies on the bacteria.[3] Eventually, assignment to a particular taxonomic grouping will require acceptance that an arbitrary degree of sequence divergence is representative of a given grouping.[2] This has already been employed at the species level for results based on starch gel electrophoresis. Thus, Poinar[6] concluded that *H. bacteriophora* and *H. heliothidis* are conspecific, partly because an electrophoresis study revealed only a 10% dissimilarity. He also used Akhurst's results showing 41% dissimilarity, together with morphological details, to describe *H. zealandica*.

In bacteria, DNA/DNA hybridization studies showing relatedness in excess of 60-70% are indicative of conspecificity, where as 20-60% DNA relatedness indicates that isolates are from closely related species.[3] Also, no significant differences between the phase one and two *Xenorhabdus* variants are detectable by DNA/DNA hybridization, demonstrating the value of the technique in that two physiologically different phases were correctly identified. For bacteria, it is recommended that a DNA homology group should not be described as a distinct species unless it can also be differentiated by some phenotypic character. This recommendation need not apply to amphimictic nematodes such as *Steinernema* spp. where cross-mating experiments are easily performed. It is likely that some sibling *Steinernema* species will be characterized by DNA homology in the absence of morphological differences. However, for the hermaphroditic *Heterorhabditis* species, the recommendation is prudent until more information is accumulated on natural variability in *Steinernema* or until cross-mating of the amphimictic generation becomes less problematical. Below the species level, decisions on the level of sequence divergence which identifies particular groups await further study. This will be fundamental to the organization of information on the increasing numbers of species, strains, and isolates accumulating in laboratories around the world. To this end, it would be extremely useful for one laboratory to act as a reference center for identification and deposition of living material, made possible by storing infective stages in liquid nitrogen.[10] Candidates would require relatively stable funding as well as expertise with bacteria, nematodes, and DNA techniques. The CSIRO laboratory in Canberra, Australia, which maintains a database of entomopathogenic nematode isolates,[7] is one of the few sites meeting those requirements.

It is now possible to clone nucleic acids from entomopathogenic nematodes and to produce species- and strain-specific probes derived from repetitive DNA. Curran[2] has identified a species-specific probe for *S. carpocapsae*, and we[8] have one specific for *S. feltiae*. No strain-specific probes have been reported for entomopathogenic nematodes but it is only a matter of time before they become available.

Intraspecific differences in DNA sequence analyses may reflect differences in biology, and these might be used to provide foundation populations for selecting worthwhile characters. Thus DNA characterization may provide an initial screen to identify useful strains. It is more efficient to bioassay populations which differ genetically than to randomly test a number of different isolates, many of which will be identical. For example, Curran and Webster[11] reported that a genotypic difference between two *Heterorhabditis* isolates, as determined by restriction fragment length differences, was associated with biological differences in their ability to control strawberry root weevils. However, they caution that genotypic "fingerprinting" of isolates cannot be used exclusively because isolates can show identical patterns, yet differ in their efficacy as control agents. Biological differences such as infectivity, host-seeking, and desiccation tolerance need not be under the control of single genes and need not be associated with the particular probe or restriction enzyme being used. It may be possible in the future to develop probes to identify useful biological characteristics.

As the powerful tools of DNA technology are applied to entomopathogenic nematodes, they will provide greater insight into the mechanisms of parasitism and speciation. Such contributions will be welcome in the wider context of biological control because current understanding of the population genetics of natural enemies is so limited that realistic predictions for applied biological control are generally not available.[12]

III. GENETIC IMPROVEMENT OF STRAINS

Genetic improvement through selective breeding, as opposed to genetic engineering, is a possible means of increasing the efficacy of entomopathogenic nematodes. This possibility has been neglected while other methods to increase efficacy such as evaporation retardants, baits, and ultraviolet protectants have been investigated with various degrees of success. Recent investigations[13] using *S. carpocapsae* showed that all strains tested showed both low tolerance to ultraviolet radiation and poor host-finding ability suggesting that more extensive sampling of natural populations would not yield better strains. However, there were significant phenotypic differences in host-finding abilities, suggesting that this trait possessed sufficient genetic variability to be improved by selective breeding. A subsequent paper[14] showed that 13 rounds of selection produced a 20- to 27-fold increase for host-finding ability. If the selection was relaxed, the population gradually reverted back toward the much lower wild-type level. Thus, the concept of genetic improvement of the nematodes appears sound. Which characters are amenable to improvement and how practical it is remain to be determined. In this light, genetic drift of strains toward less efficacy after commercial production of a number of generations or repeated culturing under laboratory conditions should also be considered. *C. elegans*[15] provides a valuable technological base for investigating genetic improvement of entomopathogenic nematodes, particularly *Heterorhabditis* spp. which, like *C. elegans*, are hermaphroditic but also produce functional males. Thus, both inbreeding and outbreeding are possible.

The technology may also soon be available to improve entomopathogenic nematodes by genetic engineering,[2] although this poses regulatory problems. Most countries are still evolving regulations for controlling the release of genetically engineered organisms and costs of demonstrating the safety of such organisms will likely be borne by the licensee. This increased cost, together with the generally specialized use of strains against specific target species and hence limited market, may preclude this approach in the near future. It will be more cost-effective to test and use the large numbers of strains and species presently available or still to be isolated. It will also be possible to insert characteristic sequences into the genome to act as a marker to identify particular strains for commercial or regulatory purposes.[2] This would still encounter regulatory restrictions and it is doubtful whether a particular strain would be worth the investment.

IV. BACTERIA/NEMATODE INTERACTIONS

Knowledge of bacteria/nematode interactions are based mainly on the *S. carpocapsae*/*X. nematophilus* model. However, these do not necessarily apply to other species and particularly not to *Heterorhabditis*/*X. luminescens*, a different family of nematode and probably a different genus of bacterium from the model.[3] Critical aspects of the interactions include:[16] (1) bacteria other than *Xenorhabdus* serve less well in supporting nematode development in monoxenic culture, (2) the phase one variant of the symbiont is preferred by the nematode, and (3) nematode strains have evolved close and specific interactions with their symbionts and grow best with the symbionts with which they have evolved. On the last point, Akhurst and Boemare[3] state that the best nutrient conditions *in vitro* for the nematode are not necessarily produced by its natural symbiont. This illustrates that the biochemistry and physiology of the bacteria are not well described, a view supported by the list of important questions posed by Nealson et al.[17] The most fundamental practical question is the basis for the existence of the phase one and two variants of the bacteria. The development of plasmid-based transformation systems for *Xenorhabdus* will help to elucidate *Xenorhabdus* biology.[18] This is essential for reliable commercial production of entomopathogenic nematodes, especially heterorhabditids.

V. POPULATION BIOLOGY

We are almost completely ignorant of the population biology of entomopathogenic nematodes, yet such information is fundamental to understanding their persistence, distribution, effect on insect populations, and to the development of predictive models for control programs. Ehler[12] places entomopathogenic nematology into the context of modern biological control theory and lists some contemporary issues in biological control that are rarely considered by nematologists. Here, we address some of these issues, but in a more specific way.

Studies on the natural occurrence of entomopathogenic nematodes show that they are ubiquitous, with 25-50% of random soil samples proving positive.[9,19] Surveys are "snapshots" in time and provide no information on persistence or recycling. Most information on persistence relates to field releases, but generally such introductions do not persist long. Persistence under more natural conditions has rarely been studied. Georgis and Hague[20] showed that *S. carpocapsae* in *Cephalcia lariciphila* (larch sawfly) prepupae in a Welsh forest varied in prevalence from 7-17% from July to April. Temperature was a limiting factor for infection and reproduction, with the nematodes persisting in both the soil and hosts during cold periods. Klein[21] mentions that *S. glaseri* could maintain itself in the field for years with a white grub population of less than 54 per m^2, and survived for 24 years when the grub population was maintained by periodic restocking. Harlan et al.[22] recovered *S. carpocapsae* (DD-136) from Louisiana grass plots 16 months after they had been applied for control of white-fringed beetle larvae. They also noted that the larval population was high in all plots, so the nematode may not have been present in sufficient numbers to control the insect population. These studies contradict short-term observations on persistence under more artificial conditions, and stress the importance of an adequate host population for maintenance of a parasite population. Recent developments in staining the infective stage[2] will provide a valuable tool in addressing ecological problems such as persistence.

Because of the lack of data on persistence and variability of natural populations, 15 sites in southern England, where *S. feltiae* predominates, were monitored for 28 months.[23] Soil samples taken on ten occasions over this period were assayed with *Galleria* traps. Two sites yielded nematodes on all sampling dates, while the other sites converted unpredictably from positive to negative or negative to positive. Whether negative bioassays reflect the absence of nematodes or an inability of nematodes present to infect a host is unknown. These alternatives are important considerations because they have important consequences for population genetics and could contribute to the founder effect.

Variations at some sites could reflect extinctions and reintroductions. Theoretical considerations[24] predict an oscillatory character for the relationship between host and pathogen populations if the pathogen is a major cause of host mortality. There is a risk that the pathogen becomes extinct during the phase of low host abundance after an epizootic. High transmission potential will accentuate the amplitude of oscillations and hence decrease the likelihood of long-term persistence. No studies on entomopathogenic nematodes provide data adequate to test these predictions; but it may be that in field trials, where a particular insect predominates, suppression with entomopathogenic nematodes reduces the host population to such a low level that the pathogen is unable to persist. This probably also occurs in natural habitats. However, some habitats provide a constant supply of insects, and the low host specificity of the nematodes would contribute to their persistence. A working hypothesis is that low insect populations, or populations made up of a single species with a long and synchronized life cycle, whose stages are not equally susceptible to the nematodes, will show severe oscillations in the nematode population, leading to periodic extinctions. Sites will show reduced oscillations if they provide a constant supply of hosts, frequently in the form of multiple species making up the insect community. Low, but not freezing, temperatures will also contribute to persistence because nematode survival, both within and outside hosts, is increased at low temperatures.

An alternative explanation for the absence of nematodes in a bioassay is that the infective stage may enter a dormant or quiescent phase and are not infectious.[25] This type of adaptation is not unusual for nematodes[26] but has not been explored for entomopathogenic species. For such lethal pathogens, an effective survival strategy might be for infectivity to be phased over time. Thus, upon emergence from a host, some individuals may be immediately infectious, while others become dormant for a time. Gaugler et al.[13] provide support for this possibility. Their laboratory assessment of *S. carpocapsae* for host-finding showed that only a small proportion of the infective juveniles were aggressive seekers of hosts; most remained inactive. However, they did not follow the less aggressive individuals over a long period of time to see if these juveniles became more active. They later showed that host finding could be improved through genetic selection and that if the selection pressure was relaxed, the nematode population slowly reverted to the less aggressive state.[14] An experiment in our laboratory [27] shows that the possibility of "phased host-seeking" is worth investigating.

Our experiment documented establishment of *Steinernema* sp. (designated Nashes strain) in *Galleria* larvae after different periods of storage in sand at 15 or 5°C. Infective juveniles were used to inoculate 25 cm³ of sand in each of 300 tubes, with 250 infective juveniles per tube. The tubes were divided into 15 groups of 20, with seven groups incubated at 5°C, seven at 15°C, and one was used as day 0 controls. On day 1 and weeks 2, 4, 6, 8, 12, and 16, one group of tubes was removed from each temperature, one *Galleria* larva was added to each of 10 tubes from the two temperatures, and the tubes were incubated at 15°C. Fresh *Galleria* larvae were added every 4 days until infections ceased, and successful establishment was assessed by counting the number of adult nematodes in each dead host. Of the remaining ten tubes from each temperature, five were subjected to "mini" Whitehead tray extractions and five to sucrose extraction to assess the numbers of surviving nematodes.

Survival of the nematodes was best at the cooler temperature. After 16 weeks at 5°C, a mean of 162 nematodes was extracted compared with only six after 16 weeks at 15°C (Figure 1A). Regardless of the storage temperature, there was a remarkable drop in nematodes establishing in hosts even after 2 weeks of storage (Figure 1B). However, nematodes stored at 5°C subsequently

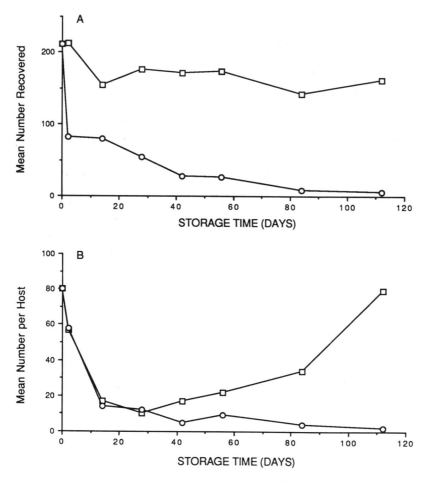

Figure 1. Data from an experiment in which tubes containing 25 cm³ sand were each inoculated with 250 infective *Steinernema* sp. (Nashes strain) juveniles on day 0 and were then stored at 5°C (□) or 15°C (○) for 1-112 days. After the indicated periods of storage, tubes were subjected to "mini" Whitehead tray extractions or bioassays with *Galleria* larvae. (A) Mean number of nematodes recovered by "mini" Whitehead trays, 5 tubes extracted on each occasion. (B) Mean number of nematodes that established in individual *Galleria* larva. One larvae was placed in each tube for 4 days at 15°C, with 10 tubes on each occasion. Data from Fan.[27]

regained the ability to infect and establish, so that after 16 weeks, the number of nematodes/host was back to that obtained on day 0. By contrast, the nematodes stored at 15°C showed a high mortality rate, and establishment continued to decline. The pattern at 5°C was not changed if nematodes were allowed additional time to infect: cumulative numbers in hosts until infections ceased showed the same pattern. These results suggest that cold temperatures induce many nematodes to temporarily lose their ability to cause host infection, although they move normally. For example, after 4 weeks at 5°C, a mean of only 17 nematodes was recorded from the insect cadaver in each tube, while 176 nematodes migrated through the Whitehead sieve. After an obligatory period of cooling, they regain their infectivity. These results show that much remains to be learned about the biology of entomopathogenic nematodes. Studies such as these should be repeated and extended to other species because they have practical implications. For example, storing nematodes at low temperatures until needed for biological control programs[28] may cause reduced virulence. The period of storage may be critical. Important aspects of the bionomics of entomopathogenic nematodes have been neglected because most work has concentrated on their application for biological control programs. Most information on basic biology relates to *S. carpocapsae* or *S. feltiae*, and it is dangerous to extrapolate the information to other species.

To understand population dynamics, it is necessary to quantify the numbers of infective juveniles in the environment. Such information is rarely available, with most workers assuming that percent mortality of hosts in bioassays is related to numbers of nematodes in the soil. Recent studies at our field station at Silwood Park[27] documented the numbers of infective juveniles of a *Steinernema* sp. from a 10 m² plot in a natural grassland habitat. Soil samples were tested with *Galleria* traps, and larvae were added continuously until infections ceased. The total number of infective juveniles accumulated in all larvae represented the infective juveniles in that sample. Samples were taken in June and October 1987 and April 1988, from adjacent 2 × 2 m plots; four samples (each 5 cm in diam. to a depth of 20 cm) were taken randomly within each plot. The nematodes were highly aggregated in distribution. For example, the four samples from one plot yielded 2495 infective juveniles in June, 57 in October, and 571 in April, compared with 112, 3, and 38, respectively, in the samples from an adjacent plot. Because of this clumping, three cores to depths of 10 cm (200 cm³ of soil) were taken randomly within some of the plots and each sample was assessed separately to determine the degree of aggregation on a smaller scale. Results showed that nematode distributions were highly aggregated and heterogeneous even within 2 × 2 m plots. For example, one set of samples in April 1988 yielded 3, 376, and 2237 nematodes, but another set yielded 1, 9, and 53. In August 1988, results from the same two plots were 3, 5, and 7 nematodes, and 0, 10, and 26 nematodes. These results are also noteworthy because they show that the numbers of infective juveniles in a natural population are minute compared with the numbers applied for control programs.

Our results raise important questions. What level of sampling will be meaningful for quantifying the number of infective juveniles in the environment? Another way to assess populations is to count the numbers of infected insects. This may work for field trials utilizing test insects which can be recovered quickly. However, in natural conditions, discovery of infected hosts is rare and certainly inadequate to assess nematode populations. Our results agree with observations of others that entomopathogenic nematodes tend to remain in the place where they were applied under laboratory or field conditions.[13,19,25] Why do the nematodes remain aggregrated when presumably it would be advantageous to disperse to find hosts? Is it more advantageous to remain clumped and wait until a host moves into the immediate vicinity? Can this natural propensity for clumping, and therefore for reduced immediate effectiveness in the field, be overcome? Gaugler[13] proposes genetic improvement of populations, while Ishibashi and Kondo[25] raise the intriguing possibility of increasing infectivity by adding chemicals such as pesticides to the application medium, thereby increasing nematode activity. Are there pheromones and/or behavioral responses which cause this effect[19] and, if so, can the pheromones be identified, synthesized, and utilized? Studies on the extent of clumping and reasons for it could also provide valuable information for increasing nematode effectiveness.

Attempts to model interactions between hosts and pathogens can be made for predictive purposes, to provide insights into the biology of the system, and to reveal fruitful areas for further study. Hochberg[29] recently proposed a theoretical model to explain the population dynamics of invertebrate host-pathogen interactions. His model, contrary to previous ones, allows for the heterogeneity observed in pathogen populations. The model considers the pathogen as two separate subpopulations, one transmissible and short-lived, the other nontransmissible and long-lived. The model shows that host populations may be regulated to low and relatively constant densities if sufficient numbers of pathogens are translocated from pathogen reservoirs to habitats where transmission can occur. Measures to increase the efficacy of pathogens should focus on the identification and manipulation of pathogen reservoirs and the processes that move pathogens between nontransmissible and transmissible subpopulations. The assumption in entomopathogenic nematode work has been that transmission is uniform, but this may be unwarranted because some individuals may be immediately infective while others are not. The existence and effect of such biologically different subpopulations deserve consideration.

Nearly all studies on entomopathogenic nematodes assess populations indirectly in terms of prevalence (percent of hosts infected). Studies on the population biology of intestinal helminths[30] have shown that prevalence is not the most useful statistic because it frequently approaches 100% for most of the age groups in a community. It is the number of parasites per host (intensity) which is crucial for the population dynamics of macroparasites such as helminths, because density-dependent effects occur at the level of parasites in individual hosts. An extreme example is that one steinernematid is sufficient to kill a host, but reproduction will not occur. At the other end of the spectrum, Friedman[16] mentions that the size of the females and the number of eggs produced per female depends on nematode density. The degree of aggregation in the host population will also affect population dynamics.[30] That is, do a few hosts have many parasites or are the parasites more regularly distributed in the host population, an assessment that can be made by calculating the variance to

mean ratio for numbers of parasites per host. The degree of aggregation in the environment, mentioned earlier, will also affect population dynamics. Infective nematodes tend to aggregate in proportion to an increase in population density and hence increase survival.[25] However, this aggregation will reduce infection of hosts and may account for poorer results than expected for the number of nematodes applied in control programs. Bedding[31] cites unpublished information that the dosage of nematodes required appears to be in part proportional to host density. These considerations require information on the number of parasites before population biology can even begin to be examined.

Work will develop over the next decade to determine whether entomopathogenic nematodes are exerting a regulatory effect on insect populations. Two theoretical papers by Anderson and May in 1978 on the potential of parasites to regulate host population abundance provided the framework for an explosion of research into the dynamics of parasite-host interactions.[32] Consequently, parasite ecology has progressed from a descriptive science to a more experimental quantitative one. It is important to realize that even when parasites exist at low prevalences they may exert a strong regulatory force on the host population.[33] Scott and Dobson[32] recently pointed out that several density-dependent and density-independent processes may interact to determine the size of a population, and the relative intensity with which each operates is likely to vary both spatially and temporally. Therefore, it is unlikely that field observations will lead to an understanding of the role of entomopathogenic nematodes in the ecosystem. Scott and Dobson[32] list prerequisites for an experiment designed to investigate whether or not a given parasite has the potential to regulate host population abundance. The empirical evidence for the role of entomopathogenic nematodes in affecting insect abundance will arise from carefully designed experiments based on long-term interactions without intervention. However, parasites which lack host specificity are less likely to act as regulators of host abundance because they depend on the presence and density of a variety of different host species.[32] That is, changes in parasite abundance need not correlate with changes in the abundance of any particular host species, so that it would be difficult to understand whether such parasites regulate the abundance of any particular host species. On the other hand, the obvious effects of parasitism by entomopathogenic nematodes are an advantage in such studies. These factors, plus the ease with which the nematodes can be reared and their short generation times, suggest that studies on their population dynamics will not only benefit from other work, but have the potential to contribute to this developing area of community ecology. The richness of questions concerning the role of parasites in regulating host abundance and structuring host communities will require large-scale collaborative efforts to determine the answers.[32]

VI. BIOASSAYS

No single species of entomopathogenic nematode is suitable for controlling

all or even most insect species. Initially, it is better to screen different species rather than different strains of the same species because species vary more than strains.[31] Also, the nematode of choice against particular insects may vary between countries.[21] Therefore, Bedding[31] advocates an effective, efficient screening procedure which should eliminate inappropriate insects and nematodes at an early stage while minimizing the possibility of erroneous rejection. It is also important that standardized, sensitive, and reliable techniques are used to determine the relative infectivities of different species, strains, and batches of nematodes for commercial quality control.[28] Once a technique is proven, a standard preparation with an assigned potency can be developed for each nematode considered for commercial use. Also, production by individual laboratories could be standardized against the designated standards. Bioassays will remain the principal method for determining virulence until research progresses to the point where analysis of a specific component responsible for virulence may be possible using biochemical or other methods.[28]

Inundative use of entomopathogenic nematodes has led to assessment of their effectiveness by methods applicable to chemical insecticides. The assumption is that all individuals are equally infective, and probit analysis is used to calculate an LD_{50} or LD_{95}. Filter paper tests are too removed from reality to be meaningful, and Bedding's[31] support for bioassays in sand and calculating LD_{50}s has been followed by many recent workers. Because of differences in the effectiveness of species and strains for controlling particular insect species, a universal bioassay that predicts levels of control under all conditions will probably never be achieved. Still, it is important to have a universal bioassay as a first screen to narrow the choice under particular environmental conditions (e.g., temperature, soil type), for quality control assessment when mass production is attempted, and for designation of standard preparations. The nearest we have to a universal bioassay is the *Galleria* test, but this insect is so susceptible that reliable dose-response tests are difficult or impossible. However, work in our laboratory[34] has shown that dose-response tests in which the numbers of nematodes that establish in *Galleria* are counted, provide a promising bioassay. Figure 2 shows the results of four separate tests at 15°C with a *Steinernema* sp. and two *Heterorhabditis* isolates using single *Galleria* larvae in tubes containing 25 cm³ of sand with 15 replicates for each dose. Note the highly significant regression coefficients, with no transformation of the data necessary. The slope of the line can be considered to represent virulence because it measures infection by all individuals capable of invading and establishing. This bioassay looks promising and it would be worthwhile to pursue similar studies to see if a dose/establishment bioassay has any value for comparative or predictive purposes.

VII. ECOLOGICAL PERSPECTIVE

It must be assumed that introduced entomopathogenic nematodes will have an environmental impact. The impact will not necessarily be adverse; environ-

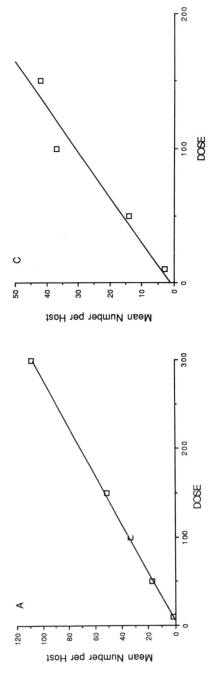

Figure 2. Mean number of nematodes (with best fit regression line) that established in individual *Galleria* larvae exposed to increasing doses of infective juveniles in 25 cm³ of sand at 15°C, 15 replicates per dose. A & B: *Steinernema* sp. (Nashes strain), (A) after 72 hr exposure, $y = -2.21 + 0.36x$, $R^2 = 0.999$; (B) after 144 hr exposure, $y = 5.94 + 0.35x$, $R^2 = 0.995$. C & D: *Heterorhabditis* sp. after 120 hr exposure (C) is for an isolate from England, $y = 0.86 + 0.30x$, $R^2 = 0.949$; (D) is for a Dutch isolate, $y = 7.82 + 0.23x$, $R^2 = 0.948$. Data from Fan and Mason.[34]

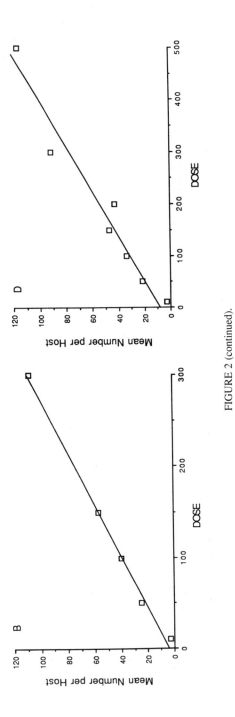

FIGURE 2 (continued).

mental impact and environmental risk are different matters. The nematodes may pose little environmental risk, but considerably more evidence is required before a generalization can be made.[12] The role of entomopathogenic nematodes in an ecosystem is unknown, yet such information is required from the regulatory point of view as well as from the ecological viewpoint of what factors impinge on community structure. Each release of nematodes can be viewed as a perturbation experiment, and if attendant changes in community structure are assessed it could enhance our understanding of the target habitat.[12]

Insect herbivory clearly affects plant species composition, vegetation cover, and structure in natural plant communities.[35] Also, these effects frequently alter the balance of species in the plant assemblage and can therefore modify both the direction and the rate of plant succession. This work, however, was based on above-ground herbivores. Only recently has the question of the influence of below-ground herbivory on the development of plant communities been addressed. It seems that below-ground herbivory does have a significant effect, as soil insecticides greatly enhance plant species richness and maintain a balance between herbs and grasses. This effect is both pleasing to the eye and conducive to the colonization, and therefore, conservation of foliage-feeding insect species dependent on specific host plants.[35] These findings may be particularly pertinent in view of proposed changes in marginal agricultural land coming out of intensive production and being put to other uses. Such a low-cost, easily implemented means of management may warrant the consideration of farmers and land owners.[35] However, this means of management depends on the use of soil chemical insecticides which are often environmentally unacceptable. Thus, a new market for entomopathogenic nematodes appears. This raises questions as to the role of entomopathogenic nematodes in community ecology. Since they are so widespread naturally in soil, do they play a significant role in reducing below-ground herbivory by insects and hence affect plant communities? It may be significant that a recent survey of the U.K. showed that the nematodes were most common in roadside verges, where there were mixed and diverse plant communities.[9] Are entomopathogenic nematodes able to direct succession by affecting insects which are soil herbivores, or does their presence rely on established patterns of insect populations?

VIII. DEVELOPMENT AND COMMERCIALIZATION

Entomopathogenic nematology has progressed from numerous field trials with limited success to the point where commercial production is not only possible, but also profitable. Indeed, commercialization has acted as a great stimulus for research. On the other hand, the need to protect proprietary information has sometimes hindered progress because information has not been exchanged freely. Regardless, the successes will continue. Specific nematodes, either species or strains, will continue to be developed and used to control specific pests under particular conditions. The necessary research for such development has been termed "near market research" by the British

Government and, as such, must rely on funding from the companies concerned. While perhaps not so blatantly expressed by other governments, this is probably the pattern that will emerge in countries where private enterprise will attempt to profit from biological control with entomopathogenic nematodes. This approach does not apply to collaborative projects between developed and developing countries. Bedding[31] has provided an excellent outline for development of an entomopathogenic nematode biological control program for developing countries which should serve as a guide for everyone.

Friedman[16] mentions that both steinernematid and heterorhabditid nematodes can be produced on a commercial scale and makes no attempt to distinguish between them. While actual methods of large-scale production used in industry are proprietary, the impression is that large-scale and consistent production of heterorhabditids lags behind that of steinernematids. Perhaps this is because liquid fermentation, which is the most cost-effective means of producing the nematodes in developed countries,[16] is less suitable for heterorhabditids. A solid phase system has produced more species more reliably than any other.[31] Production methods for heterorhabditids may have to change, and costs for them may be higher than for steinernematids. Commercial production of heterorhabditids is an area of intense study, but industrial secrecy means that advances will be communicated through patent applications rather than the scientific literature.

The ability to store nematodes in product form is critical to commercial success, but little basic work has been done on factors that limit survival.[16] Apparently, several patents have been applied for that claim various means of storing nematodes for extended periods, but the commercial development of these patents has not been seen. The use of calcined attapulgite clays[36] and superabsorbent gels[37] appear to be receiving the most attention, but again, proprietary rights inhibit knowledge of progress.

IX. GOVERNMENT REGULATIONS

The U.S. Environmental Protection Agency has determined that all species and strains of nematodes belonging to *Steinernema* and *Heterorhabditis*, with their associated bacteria, are exempted from registration requirements under the Federal Insecticide, Fungicide, Rodenticide Act (FIFRA). These nematodes are classed as macroparasites and their movement is regulated by the U.S. Department of Agriculture.[38] This does not pertain to all countries. Steinernematids and heterorhabditids are not regarded as plant or animal pathogens in the U.K. and are free of Ministry of Agriculture, Fisheries, and Food restrictions. However, the Department of the Environment has ruled that the nematodes are bound by Section 14 of the Wildlife and Countryside Act, 1981. This prohibits release of nonindigenous species or strains of animals. Permits are necessary for testing nonindigenous nematodes and the tests are bound by stringent requirements. Whether permits would ever be issued for general release and how nonindigenous strains can be identified are unknown. Even more stringent regulations will apply in every country when it becomes possible to genetically engineer organisms. The cost of complying with such regulations will probably preclude the use of genetically engineered nematodes and we will continue to rely on natural species and strains for a long time to come.

X. CONCLUSIONS

Our perspective in this chapter is that of parasitologists, while Ehler[12] presented the perspective of the science of biological control. It is clear from Ehler's contribution that we who work with entomopathogenic nematodes have a long way to go to provide a conceptual basis for our work. On the other hand, attributes of the nematodes such as short generation times, relative ease of culture, limited ability to disperse, and ability to be recovered and quantified offer exciting possibilities for experimental manipulation and development and testing of models. Thus, the fledgling science of entomopathogenic nematology has tremendous potential for contributing to the theory and practice of biological control.

REFERENCES

1. **Thompson, A., and Lymbery, A.,** Parasites keep the upper hand, *New Scientist*, 48, 8, 1988.
2. **Curran, J.,** Molecular techniques in taxonomy, in *Entomopathogenic Nematodes in Biological Control,* Gaugler, R., and Kaya, H. K., Eds., CRC Press, Boca Raton, FL, 1990, chap. 3.
3. **Akhurst, R. J., and Boemare, N. E.,** Biology and taxonomy of *Xenorhabdus,* in *Entomopathogenic Nematodes in Biological Control,* Gaugler, R., and Kaya, H. K., Eds., CRC Press, Boca Raton, FL, 1990, chap. 4.
4. **Farmer, J. J., III, Jorgensen, J. H., Grimont, P. A. D., Akhurst, R. J., Poinar, G. O., Jr., Ageron, E., Pierce, G. V., Smith, J. A., Carter, G. P., Wilson, K. L., and Hickman-Brenner, F. W.,** *Xenorhabdus luminescens* (DNA hybridization group 5) from human clinical specimens, *J. Clin. Microbiol.*, 27, 1594, 1989.
5. **Nealson, K. H., and Hastings, J. W.,** Bacterial bioluminescence: its control and ecological significance, *Microbiol. Rev.*, 43, 496, 1979.
6. **Poinar, G. O., Jr,** Taxonomy and biology of Steinernematidae and Heterorhabditidae, in *Entomopathogenic Nematodes in Biological Control,* Gaugler, R., and Kaya, H. K., Eds., CRC Press, Boca Raton, FL, 1990, chap. 2.
7. **Akhurst, R. J.,** unpublished data, 1989.
8. **Reid, A. P., and Hominick, W. M.,** unpublished data, 1989.
9. **Hominick, W. M., and Briscoe, B. R.,** Occurrence of entomopathogenic nematodes (Rhabditida: Steinernematidae and Heterorhabditidae) in British soils, *Parasitology*, 100, 295, 1990.
10. **Popiel, I., and Vasquez, E.,** Cryopreservation of *Steinernema feltiae* and *Heterorhabditis* sp. infective juveniles, *J. Nematol.*, 21, 580, 1989.
11. **Curran, J., and Webster, J. M.,** Genotypic analysis of *Heterorhabditis* isolates from North Carolina, USA, *J. Nematol.*, 21, 140, 1989.

12. **Ehler, L. E,** Some contemporary issues in biological control of insects and their relevance to the use of entomopathogenic nematodes, in *Entomopathogenic Nematodes in Biological Control,* Gaugler, R., and Kaya, H. K., Eds., CRC Press, Boca Raton, FL, 1990, chap. 1.

13. **Gaugler, R., McGuire, T. R., and Campbell, J. F.,** Genetic variability among strains of the entomopathogenic nematode *Steinernema feltiae, J. Nematol.,* 21, 247, 1989.

14. **Gaugler, R., Campbell, J. F. and McGuire, T. R.,** Selection for host-finding in *Steinernema feltiae, J. Invertebr. Pathol.,* 54, 363, 1989.

15. **Wood, W. B.,** Ed., The Nematode *Caenorhabditis elegans,* Cold Spring Harbor Monograph Series 17, Cold Spring Harbor Laboratory, Cold Spring Harbor, NY, 1988.

16. **Friedman, M. J.,** Commercial production and development, in *Entomopathogenic Nematodes in Biological Control,* Gaugler, R., and Kaya, H. K., Eds., CRC Press, Boca Raton, FL, 1990, chap. 8.

17. **Nealson, K. H., Schmidt, T. M., and Bleakley, B.,** Biochemistry and physiology of *Xenorhabdus,* in *Entomopathogenic Nematodes in Biological Control,* Gaugler, R., and Kaya, H. K., Eds., CRC Press, Boca Raton, FL, 1990, chap. 14.

18. **Frackman, S., and Nealson, K. H.,** The molecular genetics of *Xenorhabdus,* in *Entomopathogenic Nematodes in Biological Control,* Gaugler, R., and Kaya, H. K., Eds., CRC Press, Boca Raton, FL, 1990, chap. 15.

19. **Kaya, H. K.,** Soil ecology, in *Entomopathogenic Nematodes in Biological Control,* Gaugler, R., and Kaya, H. K., Eds., CRC Press, Boca Raton, FL, 1990, chap. 5.

20. **Georgis, R., and Hague, N. G. M.,** A neoaplectanid nematode in the larch sawfly *Cephalcia lariciphila* (Hymenoptera: Pamphiliidae), *Ann. Appl. Biol.,* 99, 171, 1981.

21. **Klein, M. G.,** Efficacy against soil-inhabiting insect pests, in *Entomopathogenic Nematodes in Biological Control,* Gaugler, R., and Kaya, H. K., Eds., CRC Press, Boca Raton, FL, 1990, chap. 10.

22. **Harlan, D. P., Dutky, S. R., Padgett, G. R., Mitchell, J. A., Shaw, Z. A., and Bartlett, F. J.,** Parasitism of *Neoaplectana dutkyi* in white-fringed beetle larvae, *J. Nematol.,* 3, 280, 1971.

23. **Hominick, W. M., and Briscoe, B. R.,** Survey of fifteen sites over 28 months for entomopathogenic nematodes (Rhabditida: Steinernematidae), *Parasitology,* 100, 289, 1990.

24. **Anderson, R. M., and May, R. M.,** The invasion, persistence and spread of infectious diseases within animal and plant communities, *Phil. Trans. R. Soc. Lond.,* B314, 533, 1986.

25. **Ishibashi, N., and Kondo, E.,** Behavior of infective juveniles, in *Entomopathogenic Nematodes in Biological Control,* Gaugler, R., and Kaya, H. K., Eds., CRC Press, Boca Raton, FL, 1990, chap. 7.

26. **Evans, A. A. F.,** Diapause in nematodes as a survival strategy, in *Vistas on Nematology,* Veech, J. A., and Dickson, D. W., Eds., Society of Nematologists, Hyattsville, 1987, chap. 26.

27. **Fan, X.,** Bionomics of British Strains of Entomopathogenic Nematodes (Steinernematidae), Ph.D. thesis, Imperial College of London University, London, 1989.

28. **Georgis, R.,** Formulation and application technology, in *Entomopathogenic Nematodes in Biological Control,* Gaugler, R., and Kaya, H. K., Eds., CRC Press, Boca Raton, FL, 1990, chap. 9.

29. **Hochberg, M. E.,** The potential role of pathogens in biological control, *Nature,* 337, 262, 1989.

30. **Bundy, D. A. P.,** Population ecology of intestinal helminth infections in human communities, *Phil. Trans. R. Soc. Lond.,* B321, 405, 1988.

31. **Bedding, R. A.,** Logistics and strategies for introducing entomopathogenic nematode technology into developing countries, in *Entomopathogenic Nematodes in Biological Control,* Gaugler, R., and Kaya, H. K., Eds., CRC Press, Boca Raton, FL, 1990, chap. 12.

32. **Scott, M. E., and Dobson, A.,** The role of parasites in regulating host abundance, *Parasitol. Today,* 5, 176, 1989.

33. **Anderson, R. M.**, Parasite pathogenicity and the depression of host population equilibrium, *Nature*, 279, 150, 1979.
34. **Fan, X., and Mason, J. M.**, unpublished data, 1989.
35. **Brown, V. K., and Gange, A. C.**, Differential effects of above- and below-ground insect herbivory during early plant succession, *Oikos*, 54, 67, 1989.
36. **Bedding, R. A.**, Storage of entomopathogenic nematodes, Int. Patent Appl. PJ 0630/88, 1988.
37. **Popiel, I., Holtenmann, K. D., Glazer, I., and Womersley, C.**, Commercial storage and shipment of entomogenous nematodes, Int. Patent WO 88/01134, 1988.
38. **Nickle, W. R., Drea, J. J., and Coulson, J. R.**, Guidelines for introducing beneficial insect-parasitic nematodes into the United States, *Ann. Appl. Nematol.*, 2, 50, 1988.

Index

349

INDEX

A

Acclimatization, 145
Acetylcholinesterase, 259
Acheta pennsylvanicus, 303
Acid phosphatase, 65
Actin, 259
Activated charcoal, 177
Activation, 146—148
Active dispersal, 142
Adaptations, 120, 129, see also specific
    types
  biochemical, 117, 130, 134
  metabolic, 130
Aedes
  aegypti, 225, 305, 308
  stimulans, 308
  trichurus, 308
  trivitatus, 305
Agar, 126
Agaricus bisporus, 131
Agarose, 126
Agglutination, 84, 303
Aggregation, 140, 145, 335, 336
Agrotis
  ipsilon, 175, 179
  segetum, 208—209
Aldehyde biosynthesis genes, 291
Aldehydes, 274, see also specific types
Alginate, 174—177
Alkaline metalloprotease, 312
Alphitobius diaperinus, 224
American cockroach, see *Periplaneta*
    americana
Amino acids, 134, 307, see also specific
    types
Amphimictic spp., 26
Amyelois transitella, 184, 221
Anaerobic conditions, 120, 125
Anasa tristis, 220
Anhydrobiosis, 117, 118, 140
  biochemical aspects of, 129—134
  defined, 117
  ecological considerations for evaluation
    of potential for, 118—119
  formulation and, 178—179
  in Heterorhabditidae, 122—129
  induction of, 120
  on model substrates, 125—129
  physical factors affecting, 119—122
  physiological aspects of, 129—134

potential for, 118—119
  in Steinernematidae, 122—129
  storage and, 241
Anopheles quadrimaculatus, 303
Anoxybiosis, 117
Antagonists, 103—105, 110
Anthraquinone, 278, 297
Antibacterial activity, 306, see also
    Antibiotics; Antimicrobials
Antibacterial proteins, 301, 305, 307
Antibiotics, 271, 273, 278, 281, 328, see
    also Antimicrobials
Antigenic differences, 84
Antimicrobials, 75, 80, 82, see also
    Antibiotics; specific types
Antiphenoloxidase inhibiting factors, 313
Ants, red fire, see *Solenopsis invicta*
Aphelenchus avenae, 120, 130
Aplectana kraussei, 24
Application technology, 179—185, see also
    specific types
Applied biological control, 2
Aquatic habitat, 224—226
Aquatic pseudocoelomates, 117
Arachnida, 51
Armyworms, see *Mythimna separata*;
    Pseudoletia unipuncta
  beet, see *Spodoptera exigua*
Artemia salina, 130
Arthrobotrys spp., 106
Arthropods, 4, 51, see also specific types
Artichoke plume moth, see *Platyptilia*
    carduidactyla
Attacins, 307
Attapulgite clays, 124
Attractants, 146
Attraction gradient, 239
Augmentative biological control, 2, 4—5
Australia, 233, 238, 241
Autoinduction, 274
Avermectin, 264
Axenic cultures, 155, 159—161

B

Baccharis pilularis, 2
Bacillus
  cereus, 306, 313
  popillae, 195
  subtilis, 309, 310
  thuringiensis, 105, 208, 220

C